AGROFORESTRY AND CLIMATE CHANGE

Issues and Challenges

AGROFORESTRY AND CLIMATE CHANGE

Issues and Challenges

Edited by
Manoj Kumar Jhariya
Dhiraj Kumar Yadav
Arnab Banerjee

Apple Academic Press Inc. | Apple Academic Press Inc.
3333 Mistwell Crescent | 1265 Goldenrod Circle NE
Oakville, ON L6L 0A2 | Palm Bay, Florida 32905
Canada USA | USA

First issued in paperback 2021

Exclusive worldwide distribution by CRC Press, a member of Taylor & Francis Group

No claim to original U.S. Government works

ISBN 13: 978-1-77463-465-3 (pbk)
ISBN 13: 978-1-77188-790-8 (hbk)

Library and Archives Canada Cataloguing in Publication

Title: Agroforestry and climate change : issues and challenges / edited by Manoj Kumar Jhariya, Dhiraj Kumar Yadav, Arnab Banerjee.

Names: Jhariya, Manoj Kumar, editor. | Yadav, Dhiraj Kumar, editor. | Banerjee, Arnab (Professor of environmental science), editor.

Description: Includes bibliographical references and index.

Identifiers: Canadiana (print) 20190127910 | Canadiana (ebook) 20190127937 | ISBN 9781771887908 (hardcover) | ISBN 9780429057274 (ebook)

Subjects: LCSH: Agroforestry. | LCSH: Agroforestry—Environmental aspects. | LCSH: Climate change mitigation.

Classification: LCC S494.5.A45 A47 2019 | DDC 634.9/9—dc23

CIP data on file with US Library of Congress

About the Editors

Manoj Kumar Jhariya, PhD

Manoj Kumar Jhariya, PhD, is an Assistant Professor in the Department of Farm Forestry at Sarguja University Ambikapur (Chhattisgarh), India, and is the author or co-author of more than 50 research papers in peer-reviewed journal, three books, seven book chapters, and several extension articles. Dr. Jhariya acquired his BSc (Agriculture), MSc (Forestry), and PhD (Forestry) degrees from Indira Gandhi Krishi Vishwavidyalaya, Raipur, Chhattisgarh, India. He won the University Gold Medal for securing a first-class first position for his PhD examination. He was awarded the Chhattisgarh Young Scientist Award in the year 2013 from the Chhattisgarh Council of Science and Technology. He was awarded an UGC-RGNF Fellowship, New Delhi, India. He is an editorial board member of several journals and is a life member in the Indian Science Congress Association, Applied and Natural Science Foundation, Society for Advancement of Human and Nature, and Medicinal and Aromatic Plants Association of India. He has supervised 31 MSc (Farm Forestry) students to his credit. He is dynamically involved in teaching (post-graduate) and research. He has proved himself as an active scientist in the area of forestry.

Dhiraj Kumar Yadav, PhD

Dhiraj Kumar Yadav, PhD, is an Assistant Professor and Head (Incharge) of the Department of Farm Forestry, Sarguja Vishwavidyalaya, Ambikapur, Chhattisgarh, India. He completed his graduation, postgraduation, and PhD in Forestry Jawaharlal Nehru Agricultural University, Indira Gandhi Agricultural University, and Kumaun University Nainital, respectively. He also earned an MBA in Human Resource Management from Sikkim Manipal University, and a diploma in environmental management. He has been awarded the Chhattisgarh Young Scientist Award in the year 2008 from the Chhattisgarh Council of Science and Technology. He won a merit scholarship during his BSc curriculum. He also worked as Project Co-ordinator and in other positions at various forestry institutes. He has published many research papers, articles, and book chapters in various national and internationally reputed publishers. He has one book in his credit. He is also a member of several academic societies of India. He has supervised 31 MSc (Farm Forestry) students to his credit. He is dynamically involved in teaching (postgraduate) and research. He has proved himself as an active researcher in the area of forestry.

Arnab Banerjee, PhD

Arnab Banerjee, PhD, is an Assistant Professor, Department of Environmental Science, Sarguja Vishwavidyalaya, Ambikapur, Chhattisgarh, India. He has completed his MSc and PhD (Environmental Science) at Burdwan University and his MPhil in Environmental Science at Kalyani University West Bengal. He won the University Gold Medal for securing first-class first position in MSc examination. He has been awarded the Young Scientist Award for best oral presentation at an international conference held at the University of Burdwan. He was a project fellow under a UGC-sponsored major research project. He has published 59 research papers in reputed national and international journals. To his credit, he has published five books and two book chapters. He is a life member of the Academy of Environmental Biology. He has supervised 16 PG students. He is dynamically involved in teaching (post-graduate) and research. He has proved himself as an active scientist in the area of environmental science.

Contents

Contributors

Muneesa Banday
Junior Research Fellow, Temperate Sericulture Research Institute, Mirgund, SKUAST-K, India

Arnab Banerjee
Assistant Professor, University Teaching Department, Department of Environmental Science, Sarguja Vishwavidyalaya, Ambikapur–497001 (C.G.), India, Tel.: +00-91-9926470656, E-mail: arnabenvsc@yahoo.co.in

M. F. Baqual
Associate Professor, Temperate Sericulture Research Institute, Mirgund, SKUAST-K, India

Bibek Birua
Department of Natural Resource Management, Faculty of Forestry, Birsa Agricultural University, Ranchi–834006, Jharkhand, India, Tel.: +00-91-9934958010, E-mail: vivek.forestry@gmail.com

Satpal Singh Bisht
Professor, Department of Zoology, Kumaun University, Nainital, Uttarakhand–263001, India, Tel.: +00-91-7579138434, E-mail: sps.bisht@gmail.com

Manoj Kumar Jhariya
Assistant Professor, Department of Farm Forestry, Sarguja University Ambikapur (Chhattisgarh), India, E-mail: manu9589@gmail.com

Debalina Kar
Guest Lecturer in Environmental Science, Michael Madhusudan Memorial College, Durgapur–713216, India, Tel.: 09732128980, E-mail: ka.debalina@gmail.com

Aseem Kerketta
Assistant Professor, University Teaching Department, Department of Biotechnology, Sarguja Vishwavidyalaya, Ambikapur–497001 (C.G.), India, Tel.: +00-91-8871651803, E-mail: aseem.2410@gmail.com

Irfan Latif Khan
Assistant Professor, Temperate Sericulture Research Institute, Mirgund, SKUAST-K, India

Saima Khursheed
PhD Scholar, Temperate Sericulture Research Institute, Mirgund, SKUAST-K, India

Mushtaq Rasool Mir
Associate Professor, Temperate Sericulture Research Institute, Mirgund, SKUAST-K, India, Tel.: +919596024573, E-mail: drrasool@rediffmail.com

Rojita Mishra
Lecturer, Department of Botany, Polasara Science College, Polasara, Ganjam, Odisha, India, Tel.: +00-91-8018914462, E-mail: rojitamishra@gmail.com

Prabhat Ranjan Oraon
Junior Scientist-cum-Asstt, Professor, Department of Silviculture and Agroforestry,
Faculty of Forestry, Birsa Agricultural University, Ranchi–834006, Jharkhand, India,
Tel.: +00-91-9431326222, E-mail: prabhat.ranjan.oraon@gmail.com

Debnath Palit
Associate Professor, Department of Botany, Durgapur Government College, Durgapur–713214,
India, Tel.: 0983215737, E-mail: debnath_palit@yahoo.com

Amrita Kumari Panda
Assistant Professor, University Teaching Department, Department of Biotechnology,
Sarguja Vishwavidyalaya, Ambikapur–497001 (C.G.), India, Tel.: +00-91-7669347152,
E-mail: itu.linu@gmail.com

Abhishek Raj
PhD Scholar, Department of Forestry, College of Agriculture, I.G.K.V., Raipur–492012 (C.G.),
India, Tel.: +00-91-8269718066, E-mail: ranger0392@gmail.com

Rameez Raja
Project Assistant, Temperate Sericulture Research Institute, Mirgund, SKUAST-K, India

Mahendra Rana
Assistant Professor, Department of Pharmacy, Kumaun University, Nainital, Uttarakhand–263001,
India, Tel.: +00-91-9639929116, E-mail: rana.mahendra214@gmail.com

S. Sarvade
Assistant Professor, College of Agriculture, Balaghat, Jawaharlal Nehru Krishi Vishwa Vidyalaya,
Jabalpur–481331 (MP), India, Tel.: +00-91-8989851720, E-mail: somanath553@gmail.com

Amit Sen
Assistant Professor, Department of Biotechnology, Mewar University, Chittorgarh, Rajasthan,
Tel.: +91-7357770161, E-mail: amitsenbhl28@gmail.com

Sant Kumar Singh
Professor & Head, Department of Natural Resource Management, Faculty of Forestry,
Birsa Agricultural University, Ranchi–834006, Jharkhand, India, Tel.: +00-91-9801334301,
E-mail: sksnrm@gmail.com

Vishnu K. Solanki
Assistant Professor, Department of Agroforestry, College of Agriculture, Ganjbasoda, JNKVV,
Jabalpur, Madhya Pradesh. Tel.: +91-9509536620, E-mail: drvishnu@hotmail.com

V. B. Upadhyay
Dean College of Agriculture, Balaghat, Jawaharlal Nehru Krishi Vishwa Vidyalaya,
Jabalpur–481331 (MP), India, Tel.: +00-91-9179085067, E-mail: deanbalaghat@gmail.com

Dhiraj Kumar Yadav
Assistant Professor and the Head (Incharge), Department of Farm Forestry, Sarguja
Vishwavidyalaya, Ambikapur, Chhattisgarh, India, E-mail: dheeraj_forestry@yahoo.com

Abbreviations

ACA	*Acacia auriculiformis*
ACU	adult cattle units
AEV	annual equivalent value
AF	agroforestry
AFS	agroforestry systems
AGF	agroforestry
AICRP	All India Coordinated Research Project
Al	aluminum
ALIN	average length of internode
ALL	*Albizia lebbeck*
AN	available nitrogen
ANOVA	analysis of variance
AP	available potassium
APP	available phosphate phosphorous
AS	available sodium
ATMA	Agriculture Technology Management Agency
AZI	*Azadirachta indica*
B	Boron
B:C	benefit–cost ratio
BAIF	Bhartiya Agro-Industry Foundation
BCR	benefit-cost ratio
BD	bulk density
C	carbon
Ca	calcium
CAF	*Cassia fistula*
CAFRI	Central Agroforestry Research Institute
CAMPA	Compensatory Afforestation Fund Management & Planning Authority
CBD	Convention on Biological Diversity
CDIAC	carbon dioxide information analysis center
CDM	Clean Development Mechanism
CFC	chlorofluorocarbons
CO_2	carbon dioxide

CON	conductivity
COP	conference of parties
CRA	climate resilient agriculture
CS	carbon sequestration
CSA	climate-smart agriculture
CSRTI	Central Sericulture Research and Training Institute
CW	crown width
DAC	Department of Agriculture & Cooperation
DARE	Department of Agriculture Research & Education
DAS	*Dalbergia sissoo*
Dbh/DBH	diameter at breast height
DCP	digestible crude protein
DM	dry matter
ECL	Eastern Coalfield Limited
EMO	*Emblica officinalis*
ETB	Ethiopian Birr
FAO	Food and Agricultural Organization
Fe	Iron
Fig.	Figure
FRA	*Forest* Resources Assessments
FRA	Forest Rights Act
FSI	Forest Survey of India
FYM	farmyard manure
GDP	gross domestic product/production
GFA	global forest assessment
GHG	greenhouse gas
GHs	glycosyl hydrolases
GoI	Government of India
HN	hybrid Napier
HOI	*Holoptelia integrifolia*
ICAR	Indian Council for Agriculture Research
ICCF	International Correspondence Chess Association
ICFRE	Indian Council for Forestry Research & Education
ICRAF	International Centre for Research in Agroforestry
IFS	integrated farming system
IHR	Indian Himalaya region
IINRG	Indian Institute of Natural Resins and Gums
IPCC	Intergovernmental Panel on Climate Change

Irri.	irrigated
IUCN	International Union for Conservation of Nature
IWMP	Integrated Watershed Management Programme
K	potassium
KMSP	Kisan Mahila Sashatikaran Pariyojana
KSSRDI	Karnataka State Sericulture Resource Development Institute
KVKs	Krishi Vigyan Kendras
MAI	mean annual increment
MEA	Ministry of External Affairs of India
Mg	magnesium
MgC	megagram carbon
MGNREGA	Mahatma Gandhi National Rural Employment Guarantee Programme
Mha	million hectares
MIDH	Mission for Integrated Development of Horticulture
MN&RE	new and renewable energy
MoEF	Ministry of Environment and Forest
MoRD	Ministry of Rural Development
MoU	Memorandum of Understanding
MPTs	multipurpose tree species
MT	metric tones
MT	million tones
N	nitrogen
NAB	number of auxiliary bud
NABARD	National Bank for Agriculture & Rural Development
NAMAs	Nationally Appropriate Mitigation Actions
NAP	National Agroforestry Policy
NAPAs	National Adaptation Plans of Action
NB	Number of Branches
NBM	National Bamboo Mission
NBSS & LUP	National Bureau of Soil Survey and Land Use Planning
NFP	National Forest Policy
NFTs	nitrogen-fixing tree species
NGIM	National Green India Mission
NGO	non-governmental organization
NMPB	National Medicinal Plants Board
NN	number of node

NPK	Nitrogen Phosphorous Potassium
NPV	net present value
NRAA	National Rainfed Area Authority
NRCAF	National Research Centre for Agroforestry (Newly known as Central Agroforestry Research Institute)
NRLM	National Rural Livelihood Mission
NTFPs	non-timber forest products
OC	organic carbon
OCPs	opencast coal pit
OM	organic matter
OM	organic material
P	phosphorous
PAF-AH	platelet-activating factor acetylhydrolase
PBP	payback periods
PC	planning commission
PD	particle density
PESA	Panchayats (Extension to Scheduled Areas) Act
Pg	petagram
POG	*Pongamia glabra*
Ppm	part per millions
QRT	quinquennial review team
R & D	Research and Development
RBD	randomized block design
RCF	Raniganj coalfield
REDD	reducing emissions from deforestation and forest degradation
RF	rainfed
RKVY	Rashtriya Krishi Vikas Yojana
RSRS	Regional Sericulture Research Institute
SAP	sustainable agriculture practice
SAT	semi-arid tropic
SD	sustainable development
SIC	soil inorganic carbon
SL	shoot length
SOC	soil organic carbon
SOM	soil organic matter
SP	soil porosity
Spp	species

SPs	Silvopasture systems
ST	soil texture
SWM	*Swietenia macrophylla*
t/hA	ton per hectare
tC ha^{-1}	tons carbon per unit hectare
UNCCD	United Nations Convention to Combat Desertification
UNCED	United Nations Conference on Environment and Development
UNFCCC	United Nations Framework Convention on Climate Change
VCK	Van Chetna Kendra
WBI	wood-based industry
WDRA	Warehouse Development and Regulation Act
WHC	water holding capacity
WHO	World health organization
WWF	World Wildlife Fund
Zn	zinc

Preface

Climate change is a major global issue nowadays. It has a multidimensional impact on the different components of the environment. Diverse forestry practices are prevailing in India from ancient times. Agroforestry is one such theme that involves conjoint practices of agriculture and forestry for the benefit of mankind. Climate change is imposing a huge pressure upon world food production, and is resulting in loss of biodiversity, environmental degradation, depletion of soil quality as well as a higher rate of deforestation. In such consequences, agroforestry can be effectively utilized for combating climate change. A sustainability approach in the form of agroforestry towards climate change can be implemented on various issues. Issues such as improper nutrient cycling, the rise in atmospheric temperature, inadequate amount of precipitation, the occurrence of flood, and drought, rapid decline in agricultural productivity along with the rapid loss of forest cover throughout the world are the key component of climate change. To address such diverse issues of climate change and how an agroforestry system can be effectively utilized to minimize these diverse impact is the central theme of the book.

This book, *Agroforestry and Climate Change: Issues and Challenges*, covers the diverse issues of climate change and the multidimensional role of agroforestry towards climate change and mitigation. The book comprises 11 chapters. Chapter 1 describes the recent understandings of the scenario of climate change, its issues, and its challenges and provides an in-depth analysis of the potential of agroforestry towards climate change mitigation and adaptation. Chapter 2 addresses vermicomposting as an emerging eco-friendly technology and its potential role towards combating climate change. Different examples are cited in the chapter for better discussion. Chapter 3 deals with the horticulture techniques adopted under agroforestry schemes that help to combat climate change. Chapter 4 emphasizes the potential role of agroforestry on C sequestration. Chapter 5 addresses the key issues of conservation of medicinal plant resources from Indian perspectives. Chapter 6 provides a detailed insight about the silvipastoral scheme under agroforestry. Chapter 7 addresses the most important issue of livelihood security and the role of agroforestry towards generating a livelihood. In Chapter 8 different interactions of trees with soil in arboretums have been discussed exclusively. Chapter 9 reflects the research findings of phytoremediation techniques for eco-restoration of mined wastelands. Chapter 10 addresses a novel issue of different forest products and their potentiality towards combating climate change. Chapter 11 reflects the research findings related to mulberry-based agroforestry addressing climate change mitigation and revenue generation perspectives.

This book will be a standard reference work for the disciplines such as forestry, agriculture, ecology, and environmental science as well as will be a way forward towards strategy formulation for combating climate change.

The editors would appreciate receiving comments from readers that may assist in the development of future editions.

—Manoj Kumar Jhariya, PhD
Dhiraj Kumar Yadav, PhD
Arnab Banerjee, PhD

Acknowledgments

Life is eventful, and sometimes it goes positive and sometimes negative. Humans are the architects of their own fortunes. The presence of the Almighty is everywhere and giving his spiritual bliss to all persons who are working hard.

This book project is the outcome of hard work, dedication, and spiritual support from the Almighty. We are highly grateful to the Almighty for his blessing that lead us to compile such a book addressing modern issues of applied science. It would not have been possible to publish such a book without the blessing of the Almighty.

With a profound and unfading sense of gratitude, we express our deep sense of gratitude to Prof. Rohini Prasad, Honourable Vice Chancellor, Sarguja University; Mr. Binod Kumar Ekka, Registrar, Sarguja University; and Prof. M. M. Ranga, Professor of Environmental Science, Sarguja University for their continuous support, enthusiasm, and encouragement every step of the way to help us accomplish this venture.

We are grateful to Mr. Ashish Kumar, President, Apple Academic Press, for allowing me to accomplish my dream of publishing this book. Further, we acknowledge Dr. Mohammed Wasim Siddiqui, Assistant Professor and Scientist, Bihar Agriculture University, India, and Prof. M. N. Hoda, Director Bharati Vidyapeeth's Institute of Computer Application and Management, New Delhi, for their continuous help and support for the execution of the project.

CHAPTER 1

Agroforestry and Climate Change: Issues, Challenges, and the Way Forward

M. K. JHARIYA,[1] A. BANERJEE,[2] D. K. YADAV,[1] and ABHISHEK RAJ[3]

[1]*Assistant Professor, University Teaching Department, Department of Farm Forestry, Sarguja Vishwavidyalaya, Ambikapur–497001 (C.G.), India, Mobile: +00-91-9407004814, +00-91-9926615061 (D. K. Yadav), E-mail: manu9589@gmail.com (M. K. Jhariya), dheeraj_forestry@yahoo.com (D. K. Yadav)*

[2]*Assistant Professor, University Teaching Department, Department of Environmental Science, Sarguja Vishwavidyalaya, Ambikapur-497001 (C.G.), India, Mobile: +00-91- 9926470656, E-mail: arnabenvsc@yahoo.co.in*

[3]*PhD Scholar, Department of Forestry, College of Agriculture, I.G.K.V., Raipur–492012 (C.G.), India, Mobile: +00-91-8269718066, E-mail: ranger0392@gmail.com*

ABSTRACT

Variability in climate happens to be the primary concern of the entire world as it is hampering the integrity of the whole earth ecosystem. Human civilization is mostly dependent upon food, fodder, and fuelwood. The development and progress of human civilization lie with the sustainability of the environment. Climate change imposes negative consequences in terms of food insecurity, less agricultural productivity, depletion of natural resources, spread of pests, and diseases upon human civilization. In the agricultural sector, the impact is much more severe due to imbalances in the natural processes due to variability in climate. The present situation of perturbation in the climatic elements agroforestry appears as one potential practice under sustainable agriculture system to boost up agricultural productivity as well as a strategy for combating climate change. Such land-use practices promote elevated carbon (C) sink. Under these circumstances, agroforestry seems to have very high potential to combat climate change, but successful implementation of agroforestry does require some issues. Among them, development of suitable marketing mechanism, production of high-quality planting material, to become self-sufficient in terms of transportation and harvesting of agroforestry produce, ownership of lands on a long-term basis, the small holding of land area are the key issues. Farmers need to be trained with extension activities, and awareness generation regarding long-term benefits of agroforestry should be inculcated in them. The major strategies throughout the country are to develop a uniform agroforestry policy having relaxations in the existing rules and regulations needs to be formulated, and the various agroforestry schemes under varied agro-ecological conditions should be implemented by framing some government bodies which also promotes research and development (R&D) activities in agroforestry. As India is a nation with diverse conditions, therefore, screening of suitable species along with specific agroforestry model should be done on the basis of local needs and desire. Agroforestry is basically climate-resilient agricultural practices that provide benefits in socio-economic and environmental dimension to rural people and therefore can be recommended as the most suitable strategy to work on under extreme climatic events.

1.1   INTRODUCTION

Agroforestry since pre-historic time is an emerging discipline which is under the spotlight nowadays due to anthropogenic alterations in the environment which is prevailing throughout the world. It is a tool to fulfill the farmers needs to upgrade his farming system through diversification and intensification of agriculture practice (Jhariya et al., 2015; Singh and Jhariya, 2016). Government intervention for promoting agroforestry is an essential need through knowledge-based technology of the modern century. Agroforestry is an applied aspect in terms of adjustment as an ecosystem component responding towards climatic change.

Under the background of alteration of land-use pattern, climatic extremities major focused should be emphasized upon the sustainability of the environment keeping in view about food security and quality. In order to achieve this, we need to consider nutritional insecurity facts along with climate change mitigation one needs to realize the physiological roles of plants in place of highly modernized species of crops.

It has been reported that agroforestry is such a scheme which is designed with conjoint application of agriculture and forestry principles which can function under varied ecological conditions (Jhariya et al., 2015). Agroforestry has a multidimensional role to play in terms of maintaining the stability of climate along with its associated factors that lead to overall ecological sustainability (Jhariya et al., 2015; Singh and Jhariya, 2016).

1.2 CLIMATE CHANGE

As a global issue climate change imposes its negative impact mostly on poor farming communities of less developed nations. The major solution for this problem includes proper adoption of sustainable land-use practices which may not be possible due to improper policy, strategy and other vital key factors for survivability. Climate change and food security are the most alarming issues of the world nowadays.

The greenhouse effect is a common phenomenon in which the rise of temperature occurs due to the higher presence of GHGs (greenhouse gases) in the air. Inherent properties regarding GHGs leads towards the reflection of heat energy back into the earth surface from it was addressed and thereby causing a rise in surface temperature. Increase in carbon dioxide (CO_2) level within the climate change is such an event which leads to a reduction in the cultivation period a lesser crop yield. Several research reports the influence of variability in climatic elements with respect to agricultural productivity have been reported (Saseendran et al., 2000; Aggarwal and Mall, 2002). The projected increase of ambient CO_2 level tends to be higher in the 21st century will cause a massive impact on the atmosphere. The contribution to ambient CO_2 level is through tropical deforestation and land-use as well as the anthropogenic contribution of methane appears to be from the agricultural agro-based countries. As per scientific documentation disease and pest infestation is correlated with temperature and humidity level of the environment (Nagarajan and Joshi, 1978).

Climate change is such an event which is regulated by various internal and external factors, and the balance is often dwindled by anthropogenic influence. Impacts include global warming, variability in precipitation pattern, the frequent occurrence of natural calamities and agricultural fallout. As a consequence agriculture throughout the world is the most sensitive system in terms of economic, social and ecological paradigm. With the growing human population urgent need of an increase in the production of food and fiber, agro-based ecosystems are under severe threat in the era of climate change (FAO, 2007). In this context, the availability of cultivable land is gradually being shrinking due to the rapid pace of growth of the human population (Ajayi et al., 2007). It has been found maximum GHGs contribution is done by the agricultural base of some developing countries through conventional practices. As a result of this agriculture and climate change has an influence on each other.

The major problem lies between food security and adaptability of small hold farmers towards climate change to identify key technologies that would combat the problem. Sustainable agriculture practice (SAP) is an important aspect in the modern context as it addresses bidirectional dimension that is to combat the challenge of poverty and degeneration of nature under the preview of adaptation towards climatic variability (Antle and Diagana, 2003). Agroforestry as one of the principles under SAP aims towards restoring soil fertility under the theme of natural resource management. It is a very interesting scenario that SAPs are not very much applicable or achieved success throughout the world due to the reluctance of farming community for ready uptake of the technology as well as lesser productivity due to some unexplorable reason (Ajayi et al., 2003).

1.3   CLIMATE CHANGE AND AGRICULTURE

Crop production happens to be a climate-dependent economic activity fulfilling the desired needs of human civilization. As a consequence it is obvious agriculture will have a significant impact under the influence of climatic variations (Vuren et al., 2009). GDP (Gross domestic products) contribution of agriculture is 2.0% for high-income countries and 2.9% for the globe mostly which is very significant for poorer countries such as Ethiopia, Peru, and other Latin American countries (Ackerman and Stanton, 2013).

The event of climate change influences agriculture in the form of productivity due to irregularity in temperature regimes, precipitation pattern and extreme climatic condition as well as natural disasters such as flood, drought, soil erosion leading to the epidemic spread of pest and diseases. The growth stages of the crop are significantly influenced by climatic variability as well as plants inherent properties of maturation (Hoffmann, 2013). Alteration in the precipitation regimes promotes water stress conditions for crop cultivation which also changes the farmers planning and prediction of economic gain (OECD, 2014). Temperature and moisture the crucial factors of the plant growth will reduce the fertilizer uptake and supply of mineral nutrients to plant which may be ultimately reflected through a decline in productivity and yield. Variation in climate influences the livestock through alteration of both livestock feed and the occurrence of extreme events (Hoffmann, 2013). As climate change is impacting on the pastureland ecosystem, therefore it would aggravate the impact on livestock population.

1.4 SCOPE OF AGROFORESTRY

Agroforestry has got a wider scope in terms of its application towards climate change mitigation. This was internationally supported in the Durban Conference of Intergovernmental Panel on Climate Change (IPCC) during their 17th meeting by of Conference of Parties (COP). Such issues gained international coverage through United Nations Convention to Combat Desertification (UNCCD) and CBD (Convention on Biological Diversity). Under the national level, National Adaptation Plans of Action (NAPAs) and Nationally Appropriate Mitigation Actions (NAMAs) have emphasized the transformation of agriculture system through the incorporation of agroforestry principles under traditional agricultural practice. Attempts have been made by India and abroad countries to explore the potentialities of the agroforestry system based upon information framework system of the country.

1.5 SCENARIO FOR AGROFORESTRY DEVELOPMENT

Agroforestry has its own constraints in terms of its wider applicability in the field of food crisis and alteration of environmental conditions. Even local condition makes it difficult for the farmers to adopt such eco-friendly technology. Agroforestry is an integrating framework that harmonizes tree plantation with agriculture practices and integrates people with forestry. It has got technical, economic and social dimension. Screening of suitable species is an important technical part for the successful implementation of agroforestry. Economically one should keep in mind that adoption of agroforestry doesn't lead towards financial loss of farmers. Subsequently, the farmers should be apparently made aware about the profitability of agroforestry systems.

R&D needs to be designed to explore the driving force behind the successful adoption of agroforestry from rural India perspectives (Raj and Jhariya, 2017). As it is a multidimensional event, so it varies area-to-area, place-to-place and microscale to the macro scale. Overall public participation in terms of involvement of local community people is a mandatory aspect for successful implementation of agroforestry systems.

1.6   OVERVIEW OF CLIMATE CHANGE ADAPTATION

The major task towards combating climate change lies within proper adaptation with it; therefore, UNEP (2001) has provided a scientific definition of practices to be employed to reduce the vulnerability of climate change. Further as per the recommendation of IPCC (4th) was to simply reduce the vulnerability of climate change by any means or gives stress upon ecosystem resilience (ECA, 2009). It is a fact that adaptation towards climate change is taking place in due time and in various dimensions such as spatial dimension and sectorial dimension. From the prehistoric times, human civilization has developed its own mechanism of adaptation to climate change. This includes diversity in the agricultural practices, opt for better irrigation, sustainable water management, appropriate climate forecasting and community participation for disaster management which encompasses the early prediction. Agroforestry stand to be a prominent example of agricultural diversion to cope up with climate change (ECA, 2009).

1.7   CHALLENGES FOR CLIMATE CHANGE ADAPTATION

Several mechanisms are available to enhance adaptability towards climate change which includes identification and supportive action towards the issue of climate change. In the global perspective collaboration between developed and developing nation is required with a positive problem-solving attitude in the issue of climate change. As per the records develop countries have shared their technology and provide financial assistance towards climate change adaptation (Shemsanga et al., 2010). This has helped various countries to come under the arena of fighting climate change under the scheme of this mutual collaboration.

As climate change is an international issue; therefore policy was framed globally to move forward towards adapting climate change. As per article (4.1b) of UNFCCC, every participating country should arrange for regional and national level programme under their jurisdiction to promote adaptation towards climate change. Article 4.5 refers towards sharing eco-friendly technologies along with financial assistance in the same agenda for the developing countries. The key objectives of such strategies include climate change adaptation and reduction in GHGs emissions. Many other countries arc ratifying themselves to join in the same events (Hepworth, 2010) which would stimulate other countries also. In the national context, climate change has given priority in the form of strategies, policy and plan in various developmental projects through identifying the influence of alteration of climate under NAPA (National Adaptation Programme of Action) and proper its execution (MWE, 2007). NAPA is working with a view to generate a mechanism for developing countries and to identify the necessity and importance for adapting to climate change (Hepworth, 2010). Designing a strategy often determines whether we are going for adaptation or for mitigation.

For every issue, there are some constraints and challenges. In climate change perspectives the dependency of human beings towards nature in terms of a resource often acts as berries in the proper adaptation scheme. Also, the inherent capability of agroforestry to adopt climate change is also a vital factor for the effective implementation of climate change adaptation (Raj and Jhariya, 2017). Gender disparities are also crucial factors in the issue which are most vulnerable towards climate change. As for instance women farmers of South Africa are under the severc threats of poverty, food insecurity and natural calamities.

Agriculture is the most significant sector where climate change is having its drastic effect. To combat such effect agricultural diversion is essential. Agroforestry is such a type of model which is very much promising in this context. But to study the impact of climate change through agroforestry models has got its own limitation. The major limitation includes lesser utilization under various environmental conditions. As a consequence, they are not the proper representatives of variability of climate change in all system. Therefore, some modifications are suggested in those models to get a valid result. The main problem associated with agroforestry models includes the following:

- Complex nature of agroforestry system makes it difficult to judge the effectivity of the models to predict climate change impact.
- The wide variability prevailing within the agroforestry system makes it difficult to apply in all environmental setup.
- There are errors and uncertainties in terms of prediction of impact of climate change, and therefore screening of models along with their valediction in various climatic conditions are required.
- More research and developmental activities still need to be conducted for proper development of area-specific agroforestry models.
- Within the agroforestry model one need to know about the various structural complexities such as plantation crop association, proper knowledge about various components of agroforestry models, time-oriented analysis for climate change prediction, knowledge regarding error sources generated due to complexity of models, accuracy and precision in data collection, sampling biases along with proper reliability of the data set.
- Further proper prediction of elevated CO_2 level, prediction about future climatic conditions and influence of biotic stress is not possible to predict properly.

1.8   AGROFORESTRY AND CLIMATE CHANGE ADAPTATION

Agroforestry is a holistic principle that integrates the plantation scheme with cultivation practice in order to march the ecological and economical dimensions of human civilization (Young, 2002). Therefore, such principle plays an active role over climate change (Verchot et al., 2007). Montagnini and Nair (2004) reported that the model of agroforestry with its adaptability varies regionally as well as globally. The farming community of Costa Rica has adopted agroforestry practices to cope up with a long span of summer in the form of silvi-pastoral scheme. Under this scheme, *Brachiaria brizantha* were planted for its multi-dimension uses (Oelbermann and Smith, 2011). Plantation of rotational wood lots, boundary plantation, plantation for generation, plantation for fertilizer use is some of the schemes under agroforestry which has been effectively used in Tanzania (NASCO, 2006). Such technologies can be applied to other places for solving problem-related to variability in climate. The principle behind such approaches includes developing perception and realities among the farming community that it would benefit the agriculture system as well as mitigate climate change. The principle of agroforestry jointly addresses adaptation and mitigation towards the issue of climatic variability (Roy et al., 2011).

Agroforestry practices maintain the ecological integrity of the ecosystem through livelihood generation which has economic and social return and on the other crop and tree plantation acts as a major carbon sink to reduce the ambient concentration of CO_2 to check the temperature of the earth surface (Mangala and Makoto, 2014). In the present context agroforestry through its agricultural diversion approach further addresses the problem of food security (Ekpo and Asuquo, 2012). In India agroforestry is a traditional practice which is very helpful to maintain the production rate in the drier period of season (Roy et al., 2011). In some other countries agroforestry also serve to maintain the productivity both under dry and wet condition. This thus helps the small hold farmers to move towards sustainability (Ekpo and Asuquo, 2012).

Plantation scheme under agroforestry are often aimed towards combating climate change, and thus selection of suitable species helps to combat drought and therefore prevents crop failure in the agriculture system. In India, stress is given upon developing fast-growing species such as Popular (*Populus deltoides*) to generate considerable economic gain in the form of timber within a very short span of time being compatible with the agricultural system (Chauhan et al., 2010). Such species have a multidimensional role to play in the form of generating diverse income sources, maintaining soil and water quality and nutrient balance within the ecosystem (Lasco and Pulhin, 2009). Thus agroforestry principles become very much effective for the small hold farmers of developing countries to cope up with the climate change scenario (Thorlakson and Neufeldt, 2012).

Climate change mitigation involves sequestration of C in soil ecosystem thereby reducing net C emission in the soil. GHGs sequestration potential within the agriculture sector happens to be more than 4000 mt yr^{-1} from the poor countries of the world. IPCC have explored and proposed an increase of C sequestration up to 150 tons C/ha covering more than 850 million ha area of the globe (Leakey, 2010). By developing climate resilient agricultural system, the developing countries can significantly contribute to C sequestration (Schoeneberger et al., 2012). It has been reported that agroforestry has the potential to fix higher C concentration in comparison to traditional agricultural practices. The range of C fixation may go more than 200 Mg C/ha (Mbow et al., 2014b). Higher potentiality of agroforestry towards C sequestration might be attributed towards higher coverage of the area. For example, the agroforestry system prevailing in Africa is the third largest contributor of C sequestration in comparison to other land uses (Mbow et al., 2014b). There is wide variability in the estimation process of C sequestration as well as geographical area influences both on C stock and C sequestration. Research reports reveal higher C fixation ability under agro-silvi-pastoral scheme through integrated land use (Mbow et al., 2014b). Agroforestry species has its own niche differentiation in order to avoid their competition with traditional agricultural species. For instance, deep-rooted species which belongs to the non-crop category are able to harness water and nutrient under a long dry spell. Therefore, it is very much easy for them to access the nutrients and water unavailable for other plant species. Simultaneously high water retention in the soil is an important property for agroforestry species to supply water under the absence of precipitation (Richard, 2013).

The problem with N fertilizer application lies with the release of a high concentration of non-CO_2 GHGs emission. It has been reported leguminous agroforestry species provides more N input into the soil and plant system with lesser emission of non-CO_2 GHGs emission. From an economic point of view implementation of agroforestry system seems to be cost effective in comparison to other available methods in terms of financial burden due to C sequestration (Verchot et al., 2005). Agroforestry also serves to reduce the resource dependency of community people from forests. Agroforestry has also got high bioenergy potential by producing fuelwood for the community people (Mbow et al., 2014a). Under tropical climate condition, climate change imposes a two-dimensional effect on agroecosystem. Firstly the deficit of precipitation and increase water stress conditions. Underwater stress conditions the different uses of water will further aggravate the problem by increasing the competition of water use for small hold farmers (Richard, 2013). Under this perspective climate, resilient agriculture practice is very much essential in the form of agroforestry which would maintain the agricultural production (Rivest et al., 2013). From a future perspective, agroforestry has the potential to reduce the pollution level in various agricultural sectors.

1.9 NEED OF AGROFORESTRY TOWARDS CLIMATE CHANGE MITIGATION

Agroforestry promotes eco-friendly cropping practices along with more plantations on farmland. Aerial coverage of different agroforestry species ranges between 5.0–45.0% under humid tropics (Zomer et al., 2009). Under the African region, one-tenth of farms have been reported to have tree cover of more than 25.0. This is suitable evidence that Africa represents agroforestry species with higher C sequestering potential with lesser GHGs emission from the agricultural sectors without hampering the agriculture outputs. Biomass accumulation happens to be more than thrice in agroforestry systems in comparison to the traditional agricultural system (Smith and Wollenberg, 2012). Zomer et al. (2009) reported the possibilities of land use in the form of agroforestry is much greater throughout worldwide. The various factors which influence the agroforestry systems include proper species selection, management of soil ecosystem, livelihood development and other factors. Options are there to go for climate change mitigation, but it should fulfill the need of local community stakeholders (Albrecht and Kandji, 2003).

1.10 ISSUES AND CHALLENGES IN AGROFORESTRY

Agroforestry is a fruitful alternative as it encompasses eco-friendly practices along with addressing towards food security, maintaining livelihood, and overall development of the farming community (Jhariya et al., 2015). In the year 2014, separate policy has been designed by India for agroforestry to promote afforestation at field level as well as provide a resource base for the supply of industrial timber.

Besides being a boon for the agricultural community of developing countries agroforestry has got its own inherent drawbacks which often come into the front as the biggest challenges in the era of climate change. This is due to that agroforestry is not an isolated event rather an integrated system which encompasses different factors such as choice of species, soil characteristics, meteorological conditions, addressing food security problems and overall meeting up demand of local community people (Mbow et al., 2014b). Selection of proper species along with proper management practices is an essential prerequisite for the successful implementation of agroforestry (Mbow et al., 2014a). During species selection, prioritization should be given on the local conditions and local need (Mbow et al., 2014a). Targeting optimum water use includes proper selection of shade trees along with root anatomy based species selection. It includes root structure, morphology, distribution pattern and mycorrhizal association of the intercropped trees can also serve the same purpose (Tscharntke et al., 2011). Conjoint application of deep and shallow rooted system under agroforestry is a judicious way for optimum use of water and nutrients. Favorable sites for plantation along with suitable management practices fulfilling the local need is the other essential prerequisite.

For successful implementation, farmers should be rest assures the long-term return of agroforestry. Land should be considered as a resource of common interest where communities based initiatives are taken. Studies revealed due to long tenure initial return of agroforestry remain low which may influence farmer's capability of optimum forest management (Mbow et al., 2014a; Ofori et al., 2014). This problem needs financial assistance to the farming community in the form of micro-credit, intermittent funding at various stages of agroforestry scheme at the grass root level. To improve upon the social, economic dimension of agroforestry one needs to bring gender synergies, development of proper marketing systems which maximizes benefits to the farming community who is involved in agroforestry systems. Subsequently, the transport setup should be modifying so that better infrastructure and better economic output can be generated. There may be a provision for financial assistance for optimum ecosystem service, and simultaneous certification system can be employed for that (Leakey, 2010; Tscharntke et al., 2011). In addition to this an eye on demand of agroforestry products keeping in pace with the gradual increase in human population in cities needs to be monitored (Mbow et al., 2014b).

Tree domestication is another crucial factor for the successful implementation of agroforestry (Ofori et al., 2014). This includes improvement in the traditional knowledge base by identifying suitable indigenous plant species used in traditional agroforestry system. Cultivation of new cultivars from traditional species is another breakthrough in the field of agroforestry. Proper access to improve quality plant material within India and also import from outside is very much important aspects for agroforestry implementation. Again the role of the farming community cannot be ignored in this context as most of the farmer belongs to small hold category. As per research data, more than 75.0% of agricultural lands are very least in amount for agroforestry purpose (Kumar, 2006). Such advanced approaches need to be incorporated in the agricultural sector for optimization on the benefit of agroforestry. Financial constraint is the biggest challenge for farming community who has very little option for investment in comparison to traditional agricultural practice such as cash crops. Such a problem can be addressed by improving the diverse application of agroforestry principles in terms of tree, crop, and livestock along with the generation of new products to upgrade financial stability.

Priority should be given to identify the process that would give maximum benefits both in social and economic dimensions in the field of agroforestry. The thinking approach of separating environment, agriculture and other allied fields as separate domain needs to be discouraged without which such integrated approaches cannot work properly. In India, ecology and economics are the two separate entities which are never merged with each other by any means. Government is also very much reluctant in these aspects. In order to maximize benefits from agroforestry, one needs to integrate these two concepts into one (Morgan et al., 2010). Very interestingly under the umbrella of FAO (Food and Agricultural Organization) GFA (Global Forest Assessment) considered agroforestry either is agriculture or forest principle. From a global perspective, more research database is needed regarding the possibilities of agroforestry (Mbow et al., 2014a).

1.10.1 POLICY IMPLICATION OF AGROFORESTRY TOWARDS CLIMATE CHANGE MITIGATION

Afforestation and reforestation schemes under plantation are already under action to restore forest cover as well as ecological health of the forest ecosystem. It is also used as a major strategy for climate mitigation through decreasing the level of atmospheric CO_2 by the mere plantation. In this field, agroforestry has its own inherent advantages. Agroforestry promotes growth and development of trees and shrubs and helps to sequester C from the atmosphere which can be channelized through biomass- a mechanism often known as CDM (Clean Development Mechanism). Combination of adaptation and mitigation is an essential need of the hour for the developing countries. But we are in the blink stage as lack of considerable experience at the ground level. This approach often takes the forms of a generalized one and therefore misses to address specific sectors such as agriculture, forestry, animal husbandry, and other allied fields. Regarding policy instrument C credit can be effectively utilized to increase in the income of the agricultural sector further. As per reports C sequestration through agroforestry of about 100 MgC can promote an increase in the income level more than 10.0% (Antle et al., 2007).

A proper relationship between climate change adaptation mitigation needs to be explored in the future. Although the agricultural sector are focusing their R&D towards ecosystem resilience along with the production system. It also helps to increase the capability of the farming community to increase their adaptive capacity (Buchman, 2008). Optimum use of land promoting energy conservation improves the sustainability of the environment by maintaining good environmental and social conditions. Agroforestry due to its diversification is a nature given insurance policy for the subsistence farmers. From future Reducing Emissions from Deforestation and Forest Degradation (REDD) perspective agroforestry has a huge potential towards environmental sustainability (Hare, 2009). Thus, sustainable forestry practices such as agroforestry would lead to an increase in subsidizing products through tree crop management along with reduction in the CO_2 load on the other hand thereby helping to combat climate change.

The farming community of the developing countries of little bit devoid of a technological process and biological know-how about agroforestry and therefore the economic dimension is a significant factor for them. Therefore, proper policy designing is essential for the successful establishment of agroforestry which includes the following:

- Implementation of agroforestry should be devoid of institutional and legal limitations.
- The positive output of agroforestry should always be highlighted in the policy issues of agroforestry.
- Farming community should be provided with subsidy due to delayed output of agroforestry.

Now further as per the reports of FAO (2005) discourages agroforestry if the strict implementation of forest rules were done. If one stops cutting down trees automatically, the need of restoring forest cover goes down. Therefore, the farmers become reluctant to plant trees in farmlands. These strategies of banning the cutting of trees can be effectively utilized for trees planted in agricultural landscapes under agroforestry schemes. Regulated plantation scheme needs to be designed in order to promote success in terms of agroforestry productivity. This would promote the industry to generate an alternate source of wood and its products. As a consequence, this would develop a growing economy for the wood product which will attract the farmers to come under such schemes in India and Kenya.

Malfunctioning of the management system often leads to overexploitation of agricultural products and forests products. Factors such as shifting cultivation, unlawful harvesting of wood is a major problem associated with such system which is totally absent under agroforestry scheme. Also in such conditions of mismanagement, the farmers become reluctant to increase the production. Proper policy formulation needs to be designed to the successful establishment of the agroforestry which includes the following:

- Awareness generation among the farmers about the beneficial role of agroforestry.
- Remove the legal and deleterious constraints for promoting agroforestry.
- Properly designed the land use policy and its aims.
- Develop newer policies that address the integration of agriculture and forestry.
- Go for intersectoral coordination between rural community and government.
- Create an option for eco-friendly practices.
- Organize a stable market structure for tree products of agroforestry.
- Generate a database of community stakeholders.
- Give priority to local need and demand.

1.10.2   NEED FOR AGROFORESTRY POLICY IN INDIA

Agroforestry has a huge potential of developing the country's economic sectors from a rural point of view. It would generate rural employment, values added services and also provides raw materials for industrial setup. As per the study, more than 60.0% of the country's timber demand is fulfilled through agroforestry approaches. Agroforestry also helps to achieve the target of 1/3rd forest cover as per NFP (National Forest

Policy, 1988). Agroforestry helps to conserve natural resources by reducing anthropogenic pressure on forest resources at one hand, and on the other, it helps to mitigate climate change through C sequestration. Agroforestry species are advantageous in terms of sequestering potential cover of the crop and grass systems. Agroforestry can be considered as a revolutionary concept in terms of income generation, maintaining food security, livelihood development and hope for small hold farmers.

1.10.2.1 *LACUNAE OF NATIONAL POLICY AND INSTITUTIONAL MECHANISM ADDRESSING AGROFORESTRY*

Numerous policies at various times were framed aimed towards agroforestry practices. Green India Mission 2010, Farmers Policy 2007, Bamboo Mission 2002, Greening India 2001, Agriculture Policy 2000 and Forest Policy of 1988 have emphasized agroforestry (GoI, 2014). The positive attributes include building up of soil C pool along with improvements in vegetal cover. Agroforestry due to some its inherent problems includes legal restrictions on timber harvesting, use of NTFPs (Non timber/wood forest products), lack of proper extension mechanism, institutional promotion, availability of genetic resource base, insufficient research works on various models of agroforestry under various environmental condition, improper market mechanism and absence of proper postharvest technologies. Such events took place due to the feeble nature of the administrative machinery of our country. Due consideration was not given to agroforestry by the political bureaucrats as they overruled the benefits of agroforestry in terms of lack of public policy. There was a big coordination gap between the grass root level stakeholders and people at the ministry level.

1.10.2.2 *NON-AVAILABILITY OF HOLISTIC APPROACH IN FARMING SYSTEM*

The farming activities of small hold farmers should not be considered as fixed cropping system rather a deviation from the fixed cropping system. Development in such aspect includes a holistic approach which involves agriculture, forestry, conservation of water resource, livestock management, and livelihood maintenance. Such integration is totally lacking in developing countries like India although having a holistic approach seems to be the key process for success under the agro-horticulture scheme of BAIF, NABARD, poplar cultivation system, agro-timber systems in north-western parts of India (GoI, 2014). Such an approach effectively integrates the principles of forestry and agriculture for livelihood generation. As per the previous reports, water stress is a severe factor for maintaining growth and development among tree species as a consequence of that the farmer community becomes reluctant and their morality goes down in terms of implementation of such scheme.

1.10.2.3   LEGAL HINDRANCES

As per the legal instruments marketing of agroforestry, produce are restricted by the state governments. Such restrictions were imposed to save the forests under the hand of government. This is not a convincing practice as fast-growing species with higher economic potential can be effectively grown in private forms under the same climatic conditions. On the other aspect procurement of permission for harvesting, agroforestry produce is often a herculean task and therefore farming community becomes reluctant for agroforestry scheme. Some major problem associated with this phenomenon includes involvement state revenue department in the process, imposement of tax on every processing step limits the on-farm pre-processing, damage the local livelihood, generation opportunities and elevates the transportation cost simply for processing purpose. Such reasons are actually responsible for the gradual shrinking of traditional agroforestry, and that place is replaced by agroforestry produce from other countries. One needs to focus on the improvement of agroforestry produced from Indian perspectives as there is a huge possibility of employ-ment generation (GoI, 2014).

1.10.2.4   IMPROPER APPROACHES IN SIMPLIFICATION OF LEGAL IMPEDIMENTS

Dis-uniformity in the liberalization process among the various state of India is another major challenge in front of successful implementation of agroforestry. The extent of liberalization also remains unknown for the farming community. A miss conception prevails among the farming community that agroforestry leads to changes in their traditional land-use for cultivation and farming. This reflects the lack of awareness among this grass root level people.

1.10.2.5   LACK OF R&D ACTIVITIES

Extension programme related to agroforestry is very much poor in our country as they are incapable of disseminating the research and developmental knowledge to the farmers. It is also associated with improper government machinery and lack of efforts while some other programmes such as institutional setup and support are very much prevalent. Research and developmental activities are also lacking in the direction of various agroforestry models under varied environmental conditions, screening of indigenous agroforestry species and their proper propagation results into an emphasis on specific few species and their limited varieties in certain regions of the country. India has its own technological set back of lacking proper processing technology aiming at fast-growing species.

1.10.2.6   SCARCITY OF IMPROVED PLANTING STOCK

High-quality planting material is also lacking in throughout India system as the quality of the seeds, seedlings or other improved varieties lies on the marketing set up of the country.

1.10.2.7   ECONOMIC LIBERALIZATION AND SECURITY PROVISION

Financial support for any developmental project is a very important aspect for the prosperity of a developmental programme which is reflected in agroforestry. There is a huge gap of knowledge and non-availability of a database regarding the economic output of agroforestry models. The bureaucratic political system of India also does not recognize the financial need for the sustenance of the project. A proper insurance policy is also lacking in agroforestry schemes. Also in the insurance sector lack of awareness regarding the benefits of the insurance, high premium rates, and the inappropriate process of claim settlement often misleads the farming community.

1.10.2.8   IMPROVEMENT FOR OPTIMUM MARKET ACCESS

Development of marketing infrastructure comprises of developing mechanism which includes the development of the proper market for agroforestry produce and proper price tagging for the agroforestry produce. Such a mechanism is totally lacking in the country with few exceptions who have generated their own marketing structure and operating under strict regulation of agricultural produce. Due to such problems, the mediator people associated with agroforestry marketing are having maximum benefits.

1.10.2.9   INDUSTRY INVOLVEMENT AT OPTIMUM LEVEL

In some state of India such as Uttarakhand, Haryana, Punjab has regulated the fate of agroforestry and Indian economy. The rules and regulations behind such development have become harder in due course of time as consequence industries are falling short to work in compliance with the agroforestry prevailing in the country. Sometimes it also demoralizes the growth of the industry. The inherent limitation of processing of agroforest produce increases the transportation cost. As a consequence, there is scarcity of raw material availability among the wood-based industry (WBI). The vital role of industries cannot be ignored, and policy should be framed for the screening of factors inhibiting the growth in the agroforestry sector. As per the record, more than 6 billion wood-based products are being imported per year basis. Due to such mechanism of procuring the raw material from outside as well as lack of availability of raw materials indigenously often hinders the WBI in India. A comprehensive strategy needs to be formulated for developing competitiveness in the field of agroforestry produce which will generate livelihood as well as reduce the burden on imports.

1.10.3   OBJECTIVES/GOAL

A nodal agency should be formulated in order to implement agroforestry successful throughout the country by developing proper coordination among the various segments of agroforestry across various sectors such as forestry, livelihood management, sustainable agriculture, eco-friendly forestry practices under government setup. The major focused should be on the all-round development of small-hold farmers in terms of alternate income generation, livelihood generation and creating employment opportunity through the application of agroforestry. While doing this, one needs to take into consideration that daily needs are fulfilled properly without damaging the environment keeping in view the conservation of natural assets on the one hand and increasing the forest cover throughout the country (GoI, 2014).

To fulfill the aforesaid goals one needs to design certain objective under NAP (National Agroforestry Policy). The objectives include the following:

- Firstly awareness generation among the small hold farmers about the integrated approach of planting trees within farmland which would earn them much better in comparison to only traditional agricultural practices.
- As we are living in the era of climate change; therefore, implementation of climate resilient agricultural practice such as agroforestry helps to reduce the risk due to climatic variability.
- National economy is a major important aspect for the growth and development of our country. Import of timber and non-timber forest products needs to be properly regulated in order to save our country from the burden of foreign exchange.
- Resource dependency of the tribal community people can be minimized by developing such type of alternate income-generating agricultural practices. Also, the daily livelihoods can be meets up through agroforestry practices.
- One of the major objectives of agroforestry should be aimed towards increasing the forest cover to maintain the ecological integrity of the earth ecosystem through considerable research and developmental activity in agroforestry.

1.10.3.1   BENEFITS OF IMPLEMENTING AGROFORESTRY

With due course of time, the landholding capacity of an Indian farmer's is gradually diminishing day by day due to ever-increasing population growth and leading to overuse of forest resources such as fuel, fodder, and other products for maintaining daily livelihoods. Under this context, agroforestry can be effectively implemented to promote conservation of forest. Agroforestry practices are continuous revenue generating processes through producing the product of wide dimension which indirectly promotes sustainable use of varied environmental resources. Agroforestry is also helpful for unskilled rural population as it required comparatively less investments and provides the scope of employment generation. The most beneficial aspect of agro-forestry is that it can be applied under a varied environmental condition in terms of varied landscape, different types of vegetation, varied climatic conditions as well as under mixed ownership condition. By providing employment opportunities among rural people, improvement in their lifestyle and helping in the maintenance of daily livelihood agroforestry addresses towards solving the socio-economic disparity problems among the people of India. Subsequently the health status and nutritional status for the rural people become positive by adopting agroforestry.

1.10.3.2   MANAGEMENT IMPLICATIONS

To minimize the impact of climate change one needs to harmonize between the various components of the environment such as water, land, and biodiversity. As per the reports, the potentiality of agroforestry to sequester up to a level of 2.2 PgC, in case of land ecosystem, with upcoming next 50 years (Solomon et al., 2007). Subsequently, more than 600 million hectares of uncultivable land can be effectively utilized for agroforestry purpose for sequestering more than 5 lacs MgC per year basis on the upcoming next 20 years (Jose, 2009). Scientific citation (Murthy et al., 2013) reveals that higher productivity and biomass happens to be in the case of agroforestry practice than traditional farming.

1.11   AGROFORESTRY AND THEIR ROLE TOWARDS CLIMATE ADAPTATION AND CLIMATE MITIGATION

Adaptations to climatic variability are aimed towards a reduction in the vulnerability of people with respect to changes in the climate. Acceptability towards C sequestration phenomenon among the farming community can be increased if the market economy were grown up in terms of C credit. This can be achieved through revenue generation for farmers for fixing more C or participate in CDM under the emerging C market. Plantation of petro crops in the form of extension forestry can be another way for control in the ambient CO_2 level to combat climate change.

1.11.1   STRATEGIES TOWARDS ADAPTATION AND IMPLEMENTATION FOR CLIMATE RELATED RISKS

- A total agricultural reform in the cultivation practices to maximize production in an environment-friendly way.
- Developments of wind belts, shelterbelts, bund plantation are some of the strategies to regulate the changing microclimatic conditions.
- Promote climate resilient agriculture practice through sustainable diversion of farming practices.
- Screening of effective tree crop combination that would promote soil and water conservation.
- Explore for alternative energy resources with lesser C production.
- Gradual building up of soil nutrient pool in order to maintain the fertility of the soil.
- Technology such as biodiesel, petro crops can be promoted to reduce C emission.

The mechanism involved in C emission reduction can be categorized into three types: C sequestration through sustainable agroforestry practices, C conservation by increasing C concentration in above ground and below ground biomass and C substitution in the form of energy production as an alternate energy resource.

1.11.2   INSTITUTIONAL SETUP BUILDING AT NATIONAL LEVEL

For proper implementation of agroforestry policy throughout the country constitution of a broad or mission is very much essential. This would help to give an organized approach to the agroforestry sector as well as screen key issues and perspective. Agroforestry involves a holistic approach which includes technical and management support, coordination between the scientific community, government official and community stakeholders. Such institutional mechanism would give rise to equal weightages to agroforestry along with other agricultural practices which are yet to be at its optimum level. In this mechanism state, agricultural/forest department could act as the nodal agency to execute the programme. The mission/board through which such programmes are executed can be supported in financial term.

In our country, the Ministry of Agriculture (MoA) happens to be the main driving force for agroforestry. As a consequence Agroforestry Mission or Board will be constituted under the Department of Agriculture & Cooperation (DAC). The national project on agroforestry derives experts from Ministry of Rural Development (MoRD), Ministry of Environment & Forest (MoEF), New and Renewable Energy (MN&RE), National Rainfed Area Authority (NRAA), Planning Commission (PC), Department of Agriculture Research & Education (DARE), International Centre for Research in Agroforestry (ICRAF, South Asia Office), and experts from NABARD, Non-Governmental Organizations (NGOs), Agricultural Universities, State Governments, Industry, etc.

Agroforestry schemes can be integrated with the National Rural Livelihood Mission (NRLM), Integrated Watershed Management Programme (IWMP), National Green India Mission (NGIM), National Bamboo Mission (NBM), Mahatma Gandhi National Rural Employment Guarantee Programme (MGNREGA), National Medicinal Plants Board (NMPB), Mission for Integrated Development of Horticulture (MIDH), CAMPA fund Kisan Mahila Sashatikaran Pariyojana (KMSP), Rashtriya Krishi Vikas Yojana (RKVY), Warehouse Development and Regulation Act 2007 (WDRA). However, the state government may entrust the responsibility of executing agroforestry to a nodal department. In the district, it may be ATMA (Agriculture Technology Management Agency) or other such departments as identified by the state government from time to time. Some examples include VCK (Van Chetna Kendra), KVKs (Krishi Vigyan Kendras), FA (Farmers Associations), FC (Farmers Cooperatives), FPO (Farmer Producer Organizations), SHG (Self-Help-Groups) PRIs (Panchayati Raj Institutions), NGOs, the private sector, etc.

Further, advancement and agroforestry development centers like International Centre for Research in Agroforestry (ICRAF) and Indian Council for Agriculture Research (ICAR) should play a significant role at the field level along with the development of technical expertise among the farming community. ICFRE (Indian Council for Forestry Research & Education) may also be incorporated under this network. Under such a mechanism, key ministries should be identified and effectively used. Such utilization can be stretched up to state, district, and sub-district level. This will promote tree plantation, increase in the green belt and nutrition balance in the forest ecosystem and therefore would become unique and powerful.

1.11.3 SIMPLE REGULATORY MECHANISM

Transportation and marketing is a big issue for the successful establishment of agroforestry in any area. Therefore, processes need to be designed for sustainable harvesting and transportation within and across the state to frame an eco-region for the country. The permission obligation should also be minimized as much as one can do to reduce the time limit and develop transparency. It has been found that rules and regulation imposed by the forest department have a negative implication on agroforestry. Such laws should be simplified for the sake of growth of agroforestry. Modification, relaxation, and appropriation of the database of agroforestry produce needs to be designed in order to avoid the problems and issues in the legal implementation of agroforestry policies. As per 4th Quinquennial Review Team (QRT) of NRCAF and AICRP, agroforestry has screened more than 20 species which has got multipurpose use after consultation with the local community stakeholder. The recommendation of the Bansal committee constituted by MoEF needs to be incorporated within the agroforestry policy.

For proper regulation of agroforestry pioneer organization such as EDC (Eco Development Committees), PRI, JFMCs (Joint Forest Management Committees), Gram Sabhas, or other similar people's institutions, such as those under the Panchayats (Extension to Scheduled Areas) Act 1996 (PESA), Forest Rights Act (FRA), etc. NAP (National Agroforestry Policy) should include FRA, PESA and such Acts (viz. Chotanagpur Tenancy Act) to provide rights for harvesting and to transport agroforest produce for the local community stakeholders. Power delegation to the decentralized organization should be aimed towards capacity building; otherwise, this could lead to possible misuse of power.

1.11.4 BACKGROUND INFORMATION FORMULATIONS

Ownership of land is a critical issue for farmers of developing nations. Under the agroforestry system, the tenure of ownership for the land needs to be much greater as the output is obtained in a much longer time than the normal cultivation practices. This is also a major issue for farmer's willingness to uptake agroforestry. Lack of such rights in landholding often discourages the farmer to invest his labor, cost and other aspects for agroforestry. Long terms ownership may encourage the farmer to go for such strategies. The situation is similar for the tenant farmers who have undergone plantation schemes along with the cropping system. Such rights must be given proper importance. From a financial point of view, trees planted under agroforestry scheme may be considered as collateral security for any financing farms.

From Indian perspective a proper database of agroforestry still lacking till date. Therefore, an efficient system for data generation is required on an urgent basis to meet up these lacunae. Basic information about the plantation schemes, diversity of species planted, their survival rate, harvesting potential of the region and net gain often some times are lacking. In the economic dimension, the role of agroforestry towards national GDP (Gross Domestic Product) is not clear. Such type of mechanism often puts a significant level of challenges in front of us and very interestingly we have no answer to them as no valid database is available to us. Up gradation of our land-use database is very much essential through the application of modern technologies such as remote sensing and geographical information system.

1.11.5   R&D ACTIVITIES AND RELATED SERVICES

To promote research and development in the agroforestry sector under the umbrella of ICAR with joint collaboration of NRCAF (National Research Centre for Agroforestry) several units were started. As per their claim different technologies, models applicable under varied environmental conditions are available. In this aspect, most of the research outputs can be obtained from specific industries such as pulp and timber industries. But the hard realities are that very little is being carried out in ground reality. The lack of proper extension system as well as non-robust technologies involved which leads toward its non-adaptability. Some mini projects aiming towards specific objectives need to be funded for the development of the agroforestry sector.

Agroforestry demands wholehearted participation of the local community stakeholders. It is, therefore, an urgent need to create interest among the local people and also make aware of them about the benefits of the agroforestry. Now under this condition, the education institute should play the key role in transforming knowledge and technology. Training can be undergone through KVKs, Trainers training centers in the form of learning and developing practical knowledge in the field of different agroforestry systems. Extension programmes need to be organized to create awareness among farmers. Proper utilization of mass media, audio-visual aids could be used to make the subject matter simple and trustworthy. Developing a database regarding yield and volume tables for agroforestry species is an essential requirement. Screening of C sequestering potential is another important attributes in this context. Such information needs to be propagated throughout the country for ready access by different stakeholders.

1.11.6   *HIGHER ACCESSIBILITY TO QUALITY PLANTING STOCKS*

A higher level of genetic resource in the form of seed and other planting material needs to be generated through the certification process. Authenticity can be obtained from the proper registration process and subsequent accreditations for the nurseries. It has been observed that private industries play the key role as a supplier of improved quality of planting material such as poplar and clonal eucalyptus species. Therefore, private industries need to encourage acting as a supplier of good quality planting material.

1.11.7   *PROMOTE AGROFORESTRY ORIENTED INDUSTRY*

Specific industries such as petro crops industries and other biomass-based industries can be a leader to promote agroforestry. Identification and encouragement of such industries is the need of the hour. The key sectors which need to be addressed to promote agroforestry include the following:

- Generate elite planting material and supply system.
- R&D activities aiming towards knowledge and technology transfer.
- Extension activities in the form of lab to land programme.
- Sound database on market economy.
- Authentication of the suppliers for good quality planting material.
- Involvement of government through PPP (Public People Participation) establishing a direct relationship between government and community stakeholders.
- Giving more stress upon industry partnership and provide opportunities for more involvement of industries as CSRP (Corporate Social Responsibility Programmes).

1.11.8　HIGHER ACCESSIBILITY OF FARMER FOR TREE PRODUCTS

In agroforestry, there are some shortcomings in terms of improper marketing system for agroforestry produce. As a consequence private organization needs to be integrated to promote the marketing process of agroforestry produce.

1.11.9　FARMER INCENTIVES FOR PROMOTING AGROFORESTRY AND ENERGY NEEDS

The variability of agroforestry scheme on a regional basis should be kept under the minimum for the encouragement of farmers to adopt agroforestry can be done through financial incentive in the form of subsidy and low-interest rate. More active role needs to be played by FPOs (Farmer Producer Organizations) to increase the value of the production chain of agroforestry. Strategies such as plantation of species with high growth rate on degraded lands which can fulfill the energy demands in the form of petro crops would make agroforestry a profitable business. Such energy can be channelized in the process of agroforestry so that the energy demand of the process can be meet up within itself that is to make the agroforestry system self-sufficient in the energy sector. The economic incentive should be there for the promotion of various non-conventional energy resources.

1.11.10　POLICY IMPLEMENTATION MECHANISM

One of the biggest challenges is the incorporation of agroforestry into the policies of agriculture. Schemes such as PPP need to be promoted for extension forestry practices to increase the net economic return as well as to maintain the ecological integrity. Sound database on agroforestry produce needs to be generated through reputed institute such as CSO and NSSO for smooth execution of agroforestry. Agroforestry research should be aimed towards the exploration of MPTs which can serve the dual purpose of providing daily livelihood on the one hand and on the other maintained the soil resource. The market for agroforestry produce on the basis of local needs should be the decisive factor for implementing agroforestry in any area. Cost-effectiveness of each of the agroforestry models needs to be designed involving various sectors of stakeholders for proper dissemination of knowledge. Awareness generation should be initiated at the school level through the incorporation of agroforestry subject matter in the course curriculum. This would also increase the affections towards conserving forest and trees among the youth community. In this area, NBPGR (National Bureau for Plant Genetic Resources) needs to play a significant role in conservation, monitoring and improve germplasm transfer for agroforestry species. For better upliftment of agroforestry scheme practices such as contract farming, PPP, SPV (Special Purpose Vehicles) can be effectively utilized. Extensions activities can be carried out under the PPP scheme. All round development can be mediated thorough quantifying C sequestration and other ecological services along with bringing net economic benefits for farmers. Industries need to be promoted to go for better utilization of agroforestry produce so that the future of agroforestry becomes progressive.

1.12   FUTURE R&D AND WAY FORWARDS

Gradual changes in nature would be playing a key role in the upcoming time for the global world. The policy of sequestering C through the integration of planting practices within the farmland would be an effective strategy to mitigate climate change and would be a major economic reform for the rural small hold farmers. Agro-ecological practices such as agroforestry, development of watersheds, climate resilient agriculture (CRA), not only boost up the production but also build up the soil nutrients pool which comes as co-benefits. The most fruitful aspect of such approaches includes low investment and less emission of GHGs (Keating et al., 2013).

Suitable management practices such as utilization of screened species that would enhance C sequestering potential along with enhancement in the C storage capacity of the ecosystem by biotransformation of C into biomass in the long run. A higher level of agricultural productivity can be achieved in an eco-friendly way by integrating agriculture, forestry, animal husbandry which would promote soil health and productivity. Maintaining soil quality is the answer for a global food security problem.

Due to its varied utility, the scope of agroforestry is widened up throughout the world. It has taken the shape of a different form such as wind and shelterbelts, silviculture scheme, etc. Such findings help the policy and decision makers to understand the global economy from resource perspectives that would lead to sustainable development both in developed and developing nations. In this process, the farmers are the key actors who need a considerable level of training in the process of technological transfer. Appropriate land-use practices automatically solve climate change issues and improve the social wellbeing of the farming community. For the small-hold farmer, agroforestry is a boon due to its wider dimension of products that it produces (Mendez et al., 2010). Secondarily effective utilization of nutrients can be done through agroforestry practices as would it controls nutrient mobilization in the form of runoff and also checks water pollution. Therefore, agroforestry is acting as a fuelling agent for mitigating climate change along with the upliftment of local community stakeholders.

Agricultural expansion, urbanization, population explosion has led to an increase in the agriculture land-use. As a consequence, the integrity of the environment is under question. The impacts include loss of soil quality and biodiversity loss. Farmers are often trying to boost up the agricultural productivity through higher application of chemical fertilizer and pesticides leading to agricultural pollution. Agroforestry plays the role of sustainable agriculture in the form of the betterment of environmental quality. From the various ancient times, trees were very useful for mankind in various forms. Agroforestry has its origin based on traditional knowledge under the theme of integration of principles of agriculture and forestry. Economically and ecologically agroforestry seems to be more dynamic (Lundgren and Raintree, 1982). Agroforestry address three key issues such as productivity, sustainability, and adaptability (Nair, 2008). Agroforestry ultimately leads to the approach of integrated agroecosystem (Jhariya et al., 2015; Singh and Jhariya, 2016). Intentional, intensive, interactive and integrated are the four pillars of agroforestry which make it different from the others. Based upon local need and desire farmers have integrated various species along with cultivation practices which even includes livestock production also. Improper database based on case studies are lacking for agroforestry, and as a consequence, further research and development activity needs to be oriented towards that dimension. These lacunae can be mitigated through systematic research and developmental activities under varied environmental conditions.

1.13 CONCLUSION

From the aforesaid discussion, it is a clear fact that under the grave situation of climate change we have to opt for alternate income generating system in order to maintain the rural livelihood. In this context, one needs to remember this global agenda should be considered under national planning and policies issues. Agroforestry reflects significant promise to combat climate change on the one hand and provide livelihood security on the other. Climate change mitigation through combining the principle of forestry and agriculture should be given top priority as they are doing the side business of alternate income generation for the rural livelihood and helping them to improve their economic dimension.

KEYWORDS

- **agriculture**
- **agroforestry**
- **agroforestry policy**
- **carbon sequestration**
- **carbon stock**
- **climate change**
- **climate change adaptation**
- **climate change mitigation**

REFERENCES

1. Ackerman, F., & Stanton, E. A. (2013). Climate Impacts on Agriculture: A Challenge to Complacency? Global Development and Environment Institute Working Paper No. 13–01.
2. Aggarwal, P. K., & Mall, R. K. (2002). Climate change and rice yields in diverse agro-environments of India. II. Effect of uncertainties in scenarios and crop models on impact assessment. *Climate Change, 52,* 331–343.
3. Ajayi, O. C., Akinnifesi, F. K., Gudeta, S., Chakeredza, S. (2007). Adoption of renewable soil fertility replenishment technologies in the southern African region: lessons learned and the way forward. *Natural Resource Forum, 31*(4), 306–317.
4. Ajayi, O. C., Franzel, S., Kuntashula, E., & Kwesiga, F. (2003). Adoption of improved fallow soil fertility management practices in Zambia: synthesis and emerging issues. *Agroforestry Systems, 59*(3), 317–326.
5. Albrecht, A., & Kandji, S. T. (2003). Carbon sequestration in tropical agroforestry systems. *Agriculture, Ecosystems and Environment, 99,* 15–27.

6. Antle, J. M., & Diagana, B. (2003). Creating incentives for the adoption of sustainable agricultural practices in developing countries: The role of soil carbon sequestration. *American Journal of Agricultural Economics, 85*(5), 1178–1184.

7. Antle, J. M., Stoorvogel, J. J., Valdivia, R. O. (2007). Assessing the economic impacts of agricultural carbon sequestration: Terraces and agroforestry in the Peruvian Andes. *Agriculture, Ecosystems and Environment, 122*, 435–445.

8. Buchmann, N. (2008). Agroforestry for carbon sequestration to improve small farmers' livelihoods, From the North-South Centre Research for development.

9. Chauhan, S. K., Rani, S., & Kumar, R. (2010). Assessing the economic viability of flower seed production under poplar trees. *Asia- Pacific Agroforestry News, 37*, 6–7.

10. ECA (2009). Shaping climate-resilient development: A framework for decisionmaking. A report of the Economics of Climate Adaptation (ECA) working group, Climate Works Foundation, Global Environment Facility, European Commission, McKinsey and Company, The Rockefeller Foundation, Standard Chartered Bank and Swiss Re.

11. Ekpo, F. E., & Asuquo, M. E. (2012). Agroforestry practice as adaptation tools to climate change hazards in Itu Lga, Akwa Ibom State, Nigeria. *Global Journal of Human Social Science Geography & Environmental Geosciences, 12*(11), 27- 36.

12. FAO (2005). Realizing the economic benefits of agroforestry: experiences, lessons, and challenges. *State of the World Forests*. Rome, FAO, pp. 88–97.

13. FAO (2007). Adaptation to climate change in agriculture, forestry, and fisheries: Perspective, framework, and priorities. Interdepartmental Working Group on Climate Change, Food, and Agriculture Organization (FAO) of the United Nations, Rome.

14. GoI (2014). National Agroforestry Policy. Govt. of India, Department of Agriculture & Cooperation, Ministry of Agriculture, New Delhi, India.

15. Hare, W. L. (2009). *A safe landing for the climate*. State of the Wold 2009. Confronting climate change. L. Starke. London, Earth Scan, 13–29.

16. Hepworth, N. D. (2010). Climate change vulnerability and adaptation preparedness in Uganda. Heinrich Boll Foundation, Nairobi, Kenya.

17. Hoffmann, U. (2013). Section B: Agriculture: a key driver and a major victim of global warming, in: Lead Article, in: Chapter 1, in Hoffmann 2013, pp: 3–5.

18. Jhariya, M. K., Bargali, S. S., & Raj, A. (2015). *Possibilities and Perspectives of Agroforestry in Chhattisgarh*. pp. 237–257. In: Precious Forests-Precious Earth, Edited by Miodrag Zlatic (Ed.). ISBN: 978–953–51–2175–6, 286 pages, InTech, Croatia, Europe, DOI: 10.5772/60841.

19. Jose, S. (2009). Agroforestry for ecosystems services and environmental benefits. An overview. *Agroforestry Systems, 76*, 1–10.

20. Keating, B. A., Carberry, P. S., & Dixon, J. (2013). Agricultural intensification and the food security challenge in Sub Saharan Africa. In: *Agro-Ecological Intensification of Agricultural Systems in the African Highlands*. Routledge, US and Canada pp. 20–35 (ISBN: 978-0-415-53273-0).

21. Kumar, B. M. (2006). Agroforestry: the new old paradigm for Asian food security, *Journal of Tropical Agriculture, 44*(1–2), 1–14.

22. Lasco, R., & Pulhin, F. (2009). Agroforestry for climate change adaptation and mitigation. An academic presentation for the College of Forestry and Natural Resources (CFNR), University of the Philippines Los Banos (UPLB), Los Banos, Laguna, Philippines.

23. Leakey, R. R. B. (2010). Should We be Growing More Trees on Farms to Enhance the Sustainability of Agriculture and Increase Resilience to Climate Change?, A Special Report, International Society of Tropical Foresters, USA, pp. 1–12.

24. Lundgren, B. O., & Raintree J. B. (1982). Sustained agroforestry, In: Nestel, B. (eds), Agricultural Research for Development: Potentials and Challenges in Asia, ISNAR, The Hague, The Netherlands, pp. 37–49.

25. Mangala, P. Z., & Makoto, I. (2014). Climate change and agroforestry management in Sri Lanka: Adverse impacts, adaptation strategies, and policy implications. Department of Agricultural Economics, University of Ruhuna, Matara, Sri Lanka and Department of Global Agricultural Sciences, The University of Tokyo, Tokyo, Japan.

26. Mbow, C., Noordwijk, M., Prabhu, R., & Simons, T. (2014a). Knowledge gaps and research needs concerning agroforestry's contribution to Sustainable Development Goals in Africa. *Current Opinion in Environmental Sustainability*, *6*, 162–170.

27. Mbow, C., Smith, P., Skole, D., Duguma, L., & Bustamante, M. (2014b). Achieving mitigation and adaptation to climate change through sustainable agroforestry practices in Africa. *Current Opinion in Environmental Sustainability*, *6*, 8–14.

28. Mendez, V. E., Bacon, C. M., Olson, M., Morris, K. S., & Shattuck, A. K. (2010). Agrobiodiversity and shade coffee smallholder livelihoods: A review and synthesis of ten years of research in Central America. Special Focus Section on Geographic Contributions to Agrobiodiversity Research. *Professional Geographer*, *62*, 357–376.

29. Ministry of Water and Environment (MWE) (2007). Uganda National Adaptation Programme of Action (NAPA), Department of Meteorology, Government of Uganda.

30. Montagnini, F., & Nair, P. K. R. (2004). Carbon sequestration: An underexploited environmental benefit of agroforestry systems. *Agroforestry Systems*, *61*, 281–295.

31. Morgan, J. A., Follett, R. F., Allen, L. H., Del Grosso, S., Derner, J. D., Dijkstra, F., Franzluebbers, A., Fry, R., Paustian, K., & Schoeneberger, M. M. (2010). Carbon sequestration in agricultural lands of the United States. *Journal of Soil and Water Conservation*, *65*(1), 6A–13A.

32. Murthy, I. K., Gupta, M., Tomar, S., Munsi, M., Tiwari, R., Hegde, G. T., & Ravindranath, N. H. (2013). Carbon Sequestration Potential of Agroforestry Systems in India. *J Earth Sci Climate Change*, *4*, 131.

33. Nagarajan, S., & Joshi, L. M. (1978). Epidemiology of brown and yellow rusts of wheat over northern India. II. Associated meteorological conditions. *Plant Dis Rep*, *62*, 186–188

34. Nair, K. R. (2008). Agroecosystem management in 21st century: It is time for a paradigm shift. *Journal of Tropical Agroforestry*, *46*(2), 1–12.

35. NASCO (2006). A Popular Version of National Agroforestry Strategy, NASCO Secretariat, Tanzania Forestry Research Institute, Morogoro, Tanzania.

36. OECD (Organization for Economic Co-operation and Development) (2014). Climate change, water, and agriculture: Towards resilient agricultural and water systems. [http://dx.doi.org. 10.1787/9789264209138-en].

37. Oelbermann, M., & Smith, C. E. (2011). Climate change adaptation using agroforestry practices: A case study from Costa Rica, global warming impacts–case studies on the economy, human health, and on urban and natural environments, Dr. Stefano Casalegno (ed.), In Tech, Available from: http://www.intechopen.com/books/

globalwarming-impacts-case-studies-on-theeconomy-human-healthand-on-urban-andnaturalenvironments/climate-changeadaptation-using-agroforestry-practices-a case study- from-costa-rica.

38. Ofori, D. A., Gyau, A., Dawson, I. K., Asaah, E., Tchoundjeu, Z., & Jamnadass, R. (2014). Developing more productive African agroforestry systems and improving food and nutritional security through tree Domestication. *Current Opinion in Environmental Sustainability, 6,* 123–127.

39. Raj, A., & Jhariya, M. K. (2017). *Sustainable Agriculture with Agroforestry: Adoption to Climate Change.* pp. 287–293. In: Climate Change and Sustainable Agriculture. Edited by Kumar, P. S., Kanwat, M., Meena, P. D., Kumar, V., & Alone, R.A. (Ed.). ISBN: 9789–3855–1672–6. New India Publishing Agency (NIPA), New Delhi, India.

40. Richard, L. C., Munishi, P. K., & Nzunda T., (2013). Agroforestry as Adaptation Strategy under Climate Change in Mwanga District, Kilimanjaro, Tanzania. *International Journal of Environmental Protection, 3*(11), 29–38.

41. Rivest, D., Lorente, M., Olivier, A., & Messier, C. (2013). Soil biochemical properties and microbial resilience in agroforestry systems: Effects on wheat growth under controlled drought and flooding conditions. *Science of the Total Environment Journal, 463–464,* 51–60.

42. Roy, M. M., Tewari, J. C., & Ram, M. (2011). Agroforestry for climate change adaptation and livelihood improvements in India hot arid region. *International Journal of Agriculture and Crop Sciences, 3*(2), 43–54.

43. Saseendran, S. A., Singh, K. K., Rathore, L. S., Singh, S. V., & Sinha, S. K. (2000). Effects of climate change on rice production in the tropical humid climate of Kerala, India. *Climate Change, 44,* 495–514.

44. Schoeneberger, M., Bentrup, G., de Gooijer, H., Soolanayakanahally, R., Sauer, T., Brandle, J., Zhou, X., & Current, D. (2012). Branching out: Agroforestry as a climate change mitigation and adaptation tool for agriculture. *Journal of Soil and Water Conservation, 67*(5), 128A–136A.

45. Shemsanga, C., Omabia, A. N., & Gu, Y. (2010). The cost of climate change in Tanzania: impacts and adaptations. *Journal of American Science, 6*(3), 182–196.

46. Singh, N. R., & Jhariya, M. K. (2016). *Agroforestry and Agri-Horticulture for Higher Income and Resource Conservation.* pp. 125–145. In: Innovative Technology For Sustainable Agriculture Development, Edited by Sarju Narain and Sudhir Kumar Rawat (Ed.). ISBN: 978–81–7622–375–1. Biotech Books, New Delhi, India.

47. Smith, P. E., & Wollenberg (2012). Achieving mitigation through synergies with adaptation Climate Change Mitigation and Agriculture. *ICRAF-CIAT*, London–New York, pp. 50–57.

48. Solomon, S., Qin, D., Manning, M., Chen, Z., Marquis, M., Averyt, K. B., Tignor, M., Miller, H. L. (2007). Climate Change 2007: The Physical Science Basis. Contribution of Working Group I to the Fourth Assessment Report of the Intergovernmental Panel on Climate Change. Cambridge, UK and New York, NY, USA: Cambridge University Press; 2007, (IPCC Fourth Assessment Report (AR4)).

49. Thorlakson, T., & Neufeldt, H. (2012). Reducing subsistence farmers' vulnerability to climate change: evaluating the potential contributions of agroforestry in western Kenya. *Agriculture & Food Security, 1,* 15.

50. Tscharntke, T., Clough, Y., Bhagwat, S. A., Buchori, D., Faust, H., Hertel, D., Holscher, D., Juhrbandt, J., Kessler, M., Perfecto, I., Scherber, S., Schroth, G., Veldkamp, E., & Wanger, T. C. (2011). Multifunctional shade-tree management in tropical agroforestry landscapes–a review. *Journal of Applied Ecology, 48,* 619–629.

51. UNEP (2001). Vulnerability indices. Climate change impacts and adaptation. United Nations Environment Programme (UNEP) Policy Series. UNEP. 91pp.

52. Verchot, L. V., van Noordwijk, M., Kandji, S., Tomich, T., Ong, C., Albrecht, A., Mackensen, J., Bantilan, C., & Palm, C. (2007). Climate change: Linking adaptation and mitigation through agroforestry. *Mitigation and Adaptation Strategies for Global Change, 12,* 902–918.

53. Vuren, D. P. V., Ochola, W. O., Riha, S., Gampietro, M., Ginze, H., et al. (2009). Outlook on agricultural change and its drivers. In: McIntyre, B. D., Herren, H. R., Wakhungu, J., Watson, R. T. (ed.). Agriculture at a Crossroads, Island Press, Washington, DC.

54. Young, A. (2002). Agroforestry for soil management. CAB International, Wallingford, UK.

55. Zomer, R. J., Trabucco, A., Coe, R., & Place, F. (2009). Trees on farm: analysis of global extent and geographical patterns of agroforestry ICRAF (Ed.), vol. Working Paper no. 89, World Agroforestry Centre, Nairobi, Kenya (2009), p. 72.

CHAPTER 2

Vermicomposting Practices: Sustainable Strategy for Agriculture and Climate Change

AMRITA KUMARI PANDA,[1] ROJITA MISHRA,[2] SATPAL SINGH BISHT,[3] ASEEM KERKETTA,[1] and MAHENDRA RANA[4]

[1]Assistant Professor, University Teaching Department, Department of Biotechnology, Sarguja Vishwavidyalaya, Ambikapur–497001 (C.G.), India. Mobile: +00-91-7669347152 (A. K. Panda), +00-91-8871651803 (A. Kerkettu), E-mail: itu.linu@gmail.com (A. K. Panda), aseem.2410@gmail.com (A. Kerketta)

[2]Lecturer, Department of Botany, Polasara Science College, Polasara, Ganjam, Odisha, India, Mobile: +00-91-8018914462, Email: rojitamishra@gmail.com

[3]Professor, Department of Zoology, Kumaun University, Nainital, Uttarakhand–263001, India, Mobile: +00-91-7579138434, E-mail: sps.bisht@gmail.com

[4]Assistant Professor, Department of Pharmacy, Kumaun University, Nainital, Uttarakhand–263001, India, Mobile: +00-91-9639929116, E-mail: rana.mahendra214@gmail.com

ABSTRACT

The remarkable increase in agricultural productivity during the green revolution in developing Asian countries is now witnessing signs of exhaustion in productivity. Intensive agriculture practices with poor concerns of scientific methods and ecological facet led to trouncing of soil health, reduction of freshwater resources and agro-biodiversity. Vermicomposting is considered as an effective technique for sustainable organic waste management that involves both earthworms and microorganisms. Vermicomposting technology has a wider application in the field of sustainable agriculture by altering water, N and P cycles, carbon dynamics and seedbed characteristics. Earthworms are prevalent bio-indicators in soil and act as an eco-toxicological test species in the assessment of the consequences of chemicals in soil. Earthworm gut inhabits a large diversity of microorganisms with varied taxonomical affiliation and function. Greenhouse gas emissions in terms of kg CO_2-eq per tonne of dry matter is less in vermicomposting than thermophilic composting. There are reports that earthworms decrease N_2O emissions during composting; however, role of gut microbiome in denitrification and other physiological process is yet inexplicable. The amount of sequestered C in the soil can be increased by increasing earthworm density led to a reduced rate of atmospheric CO_2 fortification by equalizing fossil fuel emission. This chapter broadly presents the role of earthworm and its gut microbiome in the acceleration of organic matter decomposition and greenhouse gas emission.

2.1   INTRODUCTION

Earthworms are well known for the breakdown of organic matter and serve as bio-tool for organic matter degradation (Moorhead et al., 2012) and consequently hasten decomposition rate and nutrient dynamics. Traditional composting has many disadvantages such as nutrient loss during the long composting process and formation of heterogeneous end products (Nair et al., 2006). Moreover, Vermicomposting is a safe, eco-friendly method and a promising technology for organic waste treatment (Gong et al., 2017). Vermicomposting results in the formation of stabilized, fragmentized and highly porous material with low C:N ratio. Earthworms aerate, grind and blend waste components by secreting enzymes and other chemicals which hasten biochemical degradation processes (Sinha et al. 2002). Recently, interest in the use of vermicomposting, i.e., using earthworms to decompose organic materials has augmented (Ndegwa and Thompson, 2000; Pramanik et al., 2007). The gut microbiome of earthworms has an intricate mutual dependence with the host. Earlier studies focused on classical cultivation and microscopy methods of this microbiome, while current molecular studies insights into both cultivable and uncultivable species in intestines of earthworms (Hong et al., 2011; Prochazkova et al., 2013).

Few functional metagenomic studies revealed biotechnologically important hydrolytic enzymes, i.e., glycosyl hydrolases (GHs), a bifunctional carboxyl/feruloyl esterase and a platelet-activating factor acetylhydrolase (PAF-AH) from earthworm egested matter (Nechitaylo et al., 2015). Vermicomposting decreases methane emissions by 16–32%, reduces N loss and emission of greenhouse gases, therefore practicing vermicomposting could be effective option to reduce greenhouse gas emissions issues comparing the same with conventional composting, specifically in the countries where scientific solutions are either not available at commercial scale or within the means.

2.2 SUSTAINABLE AGRICULTURE AND VERMICOMPOSTING

Sustainable agriculture is a way to pledge food security and can be achieved with the help of organic fertilizers; sustainable development can be redefined as a process of human progress without depleting natural resources for our forthcoming generations in a better way. This path can provide a good and healthy life to the modern world with an inbuilt idea to protect the basic life support system, i.e., global environment (UNDP, 1994; UNEP/GEMS, 1992). The conviction of sustainable development also ensures ample amount and healthy food to all living beings for sustaining a good quality of life. Modern and healthy agricultural technologies are always sought after objectives in the present scenario. Additionally, the agricultural practices have to be sustainable so that the quality and productivity of the ecosystem is retained. The U.S. National Research Council defined it as "those alternative farming methods and technologies integrating natural processes, reducing the use of inputs of off-farm sources, ensuring the long term sustainability of present production levels and conserving soil health, water, energy, and farm biodiversity" (Shinha et al., 2010).

In pursuance of the goal to maintain high and sustainable production crops mainly relies on the use of fertilizers and improved varieties to face the challenge of food security and global climate change. Overuse of chemical fertilizer has been proving to be detrimental for soil health leading to deterioration of soil quality in terms of change in the composition of soil microbial community, alteration in soil pH, change in water retention ability, change in soil texture, etc. Crop productivity has to be increased in a sustainable manner than relying solely on chemical fertilizers would, in the long run, deem us to the status of merely a bystander. Many alternatives such as biomanure, biofertilizers, biocomposites, biopesticides, etc. have been devised to account for such questions. In the same connection, vermicomposting techniques is being widely popularized and being used all over the globe. Vermicomposting is not actually a newly discovered technique but can be considered to have been rediscovered as the essence of vermicomposting, and the role of the earthworm in agriculture has been known to man since time immemorial. Vermicomposting is an applied branch of Life Sciences which utilizes earthworms for the production of value-added organic vermifertilizer and other compost products like vermi-wash. The use of vermifertilizers in the agricultural system is an alternative to chemical fertilizers driving the entire agriculture towards sustainability and food security (Senapati, 1996). Earthworm consumes a wide range of organic waste and capable of converting it to worm cast, on an average earthworm is able to produce 300kg of compost from 1000kg of organic matter (Gunathilagaraj,1994). Vermicomposting plays a key role in sustainable agriculture, and food safety using vermiculture food production can be increased, and the environment can be saved. The sustainability cycle is diagrammatically represented in Figure 2.1.

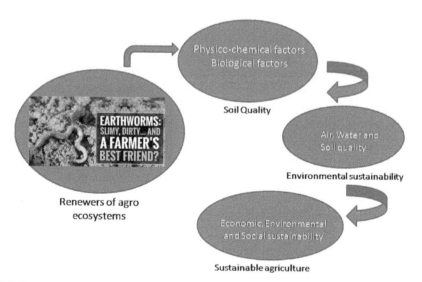

FIGURE 2.1   Schematic depiction of the use of vermicompost for sustainable agriculture.

Vermicompost is acting as a promoter and enhancer of plant growth in comparison to compost and chemical fertilizer. Plant growth promotion is mainly due to an increase in soil fertility and quality parameters of the soil. Comparative account of vermicompost compost and chemical fertilizers towards improving soil quality and health is represented in Table 2.1.

TABLE 2.1 Vermicompost and Chemical Fertilizer's Effect Towards Improving Soil Quality

Soil quality (biological and chemical properties)	Organic farming		Chemical farming
	Compost	Vermicompost	
Availability of N (kg/ha)	256	324	185
Availability of phosphorous (kg/ha)	50.5	72.5	28.5
Availability of potash (kg/ha)	489.5	815.8	426.5
Azatobacter (1000/gm of soil)	11.7	1000	0.8
Phospho bacteria (100,000/kg of soil)	8.8	10000	3.2
Carbonic biomass (mg/kg of soil)	273	394.5	217

2.3 VERMICOMPOSTING AND VERMICOMPOST PROPERTIES

Vermicomposting is an integrated biological process of converting organic waste into vermicompost by employing earthworms and naturally occurring microbial activities under mesophilic environment. The mesophilic composting techniques provide an optimal environment to earthworms and microbes to convert complex substances in a harmonizing means to make them easily accessible to the plants (Pramanik et al., 2011; Song et al., 2014). Vermicompost is generally loaded with 5 times the available nitrogen, 7 times the available potash, and 1½ times more calcium than found in good topsoil, with a low level of heavy metals, low conductivity, high humic acid contents as well as good stability and maturity (Kaushik and Garg, 2004).

Earthworm converts the organic material into soil conditioner (Fig. 2.2) and the process of conversion of organic material into soil conditioner with the help of earthworm is called vermicomposting (Appelhof, 1997). Organic matter along with algae, fungi, protozoa, nematodes, and bacteria was ingested by the worm, and it passes through the digestive tract. Organic waste and bacteria form a cast along with metabolite wastes like ammonium, urea, and other nutrients. The worms also secrete mucus (Bajsa et al., 2003). During important vermicomposting plant, nutrients are released and converted by microbes into soluble forms which are more suitable for plant growth. Vermicomposting is an eco-friendly technique that utilizes earthworms, solid and other agro-waste into vermicast in the form of various vermipellets, involves earthworms and microorganisms that are active at a temperature range of 10 to 32°C (Vermi co, 2001).

2.3.1   *EARTHWORM AND VERMICOMPOST*

Earthworms are the representatives of the phylum Annelida, and till date, more than 3600 types of earthworm are reported worldwide broadly categorized into burrowing and non-burrowing types. Burrowing types exist in the deeper soil layer, and non-burrowing type lives in the upper layer of the soil surface. The burrowing types are pale, 20–30 cm long and live for fifteen years. The non-burrowing types are red or purple and 10–15 cm long, but their life span is only 28 months. The non-burrowing earthworms eat 10% soil and 90% organic waste materials. They can tolerate temperature ranging from 0–40°C and generation capacity is more at 25 to 30°C and 40–45% moisture level in a pile. The burrowing type of earthworms comes onto the soil surface only at night. They produce 5.6 kg casts by ingesting 90% soil and 10% organic waste (Figure 2.2) (Nagavallemma et al., 2004). Earthworm on the basis of their burrowing and feeding activity is divided into three main categories, i.e., epigeics, anecics, and endogeics (Gajalakshmi and Abbasi, 2004). There are reports that epigeics are most suitable for vermicompost. Some species of anecic are also useful in vermicomposting of phytomass and waste paper (Abbasi et al., 2009). Earthworms are well-identified natural tools capable of altering water, N and P cycles, C dynamics of the soil converting waste into vermicompost for organic agriculture (Arancon et al., 2008) in which organic nutrients are released by earthworms (Edwards, 1995). Earthworms are well known voracious feeders, and they excrete a large part of these consumed materials in a semi-digested form accessible to plants. According to Arancon and Edwards (2006), vermicomposts contain plant growth hormones, plant growth regulating substances and humic acids which enhance plant growth and productivity. Therefore, vermicomposts are widely used in organic farming (Atiyeh et al., 2000). Earthworms are well known for their role in nitrogen transformations by initiating the process of nitrogen mineralization. Vermiculture is a cost-effective tool for environmentally sound waste management (Banu et al., 2001; Asha et al., 2008).

SOIL LAYERS BEFORE AND AFTER EARTHWORMS

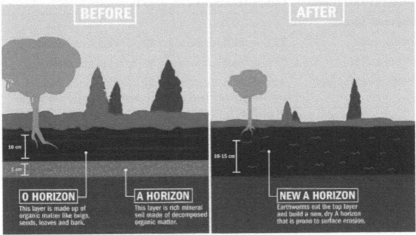

FIGURE 2.2   (See color insert.) Diagram of soil layers before and after vermicompost application (*Source*: Great Lakes Worm Watch). https://www.google.com/search?biw=12 80&bih=609&tbm=isch&sa=1&ei–PwvBXKO6DM27rQGS_hKACA&q=Soil+layer+N ew+A+Horizon+%2B+earthworms&oq=Soil+layer+New+A+Horizon+%2B+earthworm s&gs_l=img.12...202021.207545..209973...0.0..0.233.2829.0j7j7......1....1..gws-wiz-img. KzHfgNMGLRA#imgrc=3bpKWbBwyv7sbM:

2.3.2   EARTHWORM SPECIES USED FOR VERMICOMPOSTING

Epigeic earthworms with their high rate of consumption, excretion, and breakdown of organic waste with leniency to a broad range of environmental factors; short life cycles; high-reproductive rates are very good candidates for vermicomposting. Optimal conditions for growth of a few common species of vermicomposting earthworms are summarized in Table 2.2.

2.3.3   RAW MATERIALS USED FOR VERMICOMPOSTING

Sorghum and rice straw, dry leaves of crops and trees, pigeon pea stalks, groundnut husk, soybean residue, vegetable wastes, weed plants before flowering, fiber from coconut, sugarcane trash are dry wastes used in vermicomposting. Few other raw materials include waste from various industries including animal manure, dairy and poultry wastes, food industry wastes, municipal solid wastes, biogas sludge and bagasse from sugarcane industry (Nagavallemma et al., 2004).

TABLE 2.2   Biology of Some Vermicomposting Earthworm Species

Species of Earthworm	Optimal conditions for growth
Temperate species	
Eisenia fetida	25°C optimal temperature, 80–85% optimal moisture
Eisenia andrei	25°C optimal temperature, 80–85% optimal moisture
Dendrodrilus rubidus	Acid-tolerant, the cocoons are extremely cold tolerant, surviving temperatures lower than -40 °C. However, the adult stage is unable to withstand even slightly negative temperatures.
Dendrobaena veneta	Preference for mild 15–25°C optimal temperature, 75% optimal moisture
Lumbricus rubellus	Breeding takes place mainly in spring and early summer
Tropical species	
Eudrilus eugeniae	Maximum biomass production occurring at 25°C–30°C, while the growth rates were very low at 15°C, the optimum moisture being 80%–82%.
Perionyx excavatus	Unable to withstand low-temperature condition and does not grow during low winter temperature
Polypheretima elongate	Restricted to tropical regions and may not survive severe winters.

2.3.4   VERMICOMPOSTING: ON-FARM AND OFF-FARM SCENARIO

In India, more than 80% of farmers are categorized as small and marginal holders, and suchfarms are hardly economically viable under the existing technological scenario. The structure of Indian agriculture is undergoing transformation by adopting integrated farming systems (IFS). The major characteristics of IFS are gold from the garbage means a waste of one subsystem is utilized for a second subsystem as an input accommodating both the system functional at one place. The major components of IFS in India are crop production, livestock, poultry, apiculture, sericulture, mushroom cultivation, aquaculture, and agroforestry. In addition to these key components small and marginal farmers also practiced vermicomposting, duck farming, rabbit rearing, floriculture, and biogas production as income generating activities. The compatibility of organic agriculture with integrated agricultural approaches can be considered as a development vehicle particularly for developing countries like India (Ramesh et al., 2005).

There are different types of container used to house compost worms for the vermicomposting process. Vermicomposting bins can be a pile of plastic container or an automated unit proficient of processing pounds of organic matter daily. Generally, each square foot of bin area can process 1 pound of food waste per week (Dhoke et al., 2017).

2.3.5 MICROBIOLOGY FOR VERMICOMPOSTING

Studies are there that reported earthworms decreased the soil fungi-to-bacteria ratio, but did not appreciably affect microbial biomass (Lavelle, 1988). Earthworms' increases the bacterial abundance in the soil as the gut conditions are favorable for multiplication of bacteria whereas fungi were suppressed (Brown, 1995). Earthworms increased soil respiration mainly by raising the percentage of bacteria in the total soil microbial biomass (Lv et al., 2016). The low fungi-to-bacteria ratio leads to high soil respiration as bacteria have elevated turnover pace and lesser carbon absorption efficiency compared to fungi (Sakamoto and Oba, 1994; Rousk and Baath, 2011). Soil microorganisms secrete various enzymes such as β-glucosidase to decompose soil organic carbon for their growth and metabolism (Sinsabaugh et al., 2008). Earthworms decreased soil β-glucosidase activity; thereby decreasing the degradation of soil organic carbon by soil microbial flora and the soil microorganisms get more digested C from an earthworm.

2.3.6 VERMICOMPOSTING PROCESS

The process of vermicomposting is a unique process, and it can be divided into four different stages as figured out in the smart chart (Figure 2.3).

2.4 ADVANTAGES OF VERMICOMPOST OVER FERTILIZERS

Food produced by the application of vermicompost is completely organic and chemical free. There are reports that agronomically important traits of agri-products produced in the farms rich in vermicompost are better in quality and quantity. The minerals present in vermicompost are in easily accessible form for plants which is not the case in case of chemical fertilizers and simultaneously vermicompost sizably reduces the water irrigation requirement. Vermicompost helps in building up of natural soil fertility after successive use and produce a large number of earthworms in the soil that reduces the amount of vermicompost required for maturing to attain good yield of crops and productivity, which is reverse in case of chemical fertilizers (Shinha et al., 2010).

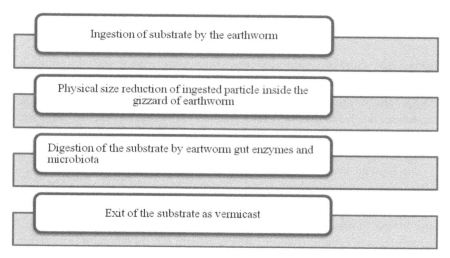

FIGURE 2.3   Schematic representation of vermicomposting process.

2.5   VERMICOMPOST VERSUS PHYSICAL, BIOLOGICAL, CHEMICAL PROPERTIES AND NUTRIENT CONTENT IN SOIL

2.5.1   *VERMICOMPOST AND SOIL PHYSICAL CHARACTERISTICS*

Earthworm secretes gelatinous substances that cast the organic material of the soil and thus stabilizes soil aggregates. A fresh cast is less stable than old casts (Marinissen and Dexter, 1990) and there are reports that soil stability increases as the cast becomes older (Shipitalo and Protz, 1987) and more cast deposited forming layers enhances stability. Vermicompost favors microbial growth and becomes a suitable niche which helps in the growth soil friendly microorganisms; thus stability increases (Emerson et al., 1986). Initially, vermicompost increases the fertility of topsoil later on it helps in the growth of local worms which improves soil aeration and soil turnover (Kale, 1992). Soil stability is maintained by a two-step process: (1) large worms form macroaggregates; and (2) Small worms act on these macro aggregates and convert it into smaller and stable aggregates (Blanchart et al., 1997). Vermicompost increases the number of total cracks at the same time decreases larger cracks this is the indicator of good soil structure (Masciandaro et al., 1997)

A bulk density of soil plays an important role in root development of the plant. Various researchers reported a significant decrease in bulk density when vermicompost is added alone and also with gypsum and chemical fertilizers (Vasanthi and Kumaraswamy, 1999; McColl et al., 1982; Maheswarappa et al., 1999; Raut et al., 2000). Many scientists proved in their study that vermicompost treated soil increases soil porosity and hydraulic conductivity (Maheswarappa et al., 1999; Raut et al., 2000).

2.5.2   VERMICOMPOST AND SOIL BIOLOGICAL PROPERTIES

Vermicompost treatment increases the microbes of the soil like bacteria, spore formers, fungi and nitrogen fixers (Kale et al., 1992). Raut et al. (2000) established five times increase of microbial population in the soil after the augmentation of vermicompost than that of control. Maheswarappa et al. (1999) found the significant increase of fungal and bacterial population of soil by addition of 22 tons of vermicompost per hectare. It was observed a further increase of vermicompost (30 tons per hectare) increases actinomycetes population in the soil. Soil potential to maintain biological fertility is indicated by the increase of enzymatic activities after manuring practices. Masciandaro et al. (1997) and Serra-Wittling et al. (1995) found increased dehydrogenase and hydrolases activity in the vermicompost treated soil, and no difference observed in compost treated soil. Five times increase of enzymes in the soil after treatment of 22 tons per hectare vermicompost to that of control observed by Maheswarappa et al. (1999).

2.5.3   VERMICOMPOST AND NUTRIENT CONTENT OF THE SOIL

The basic chemical composition of the vermicompost is summarized in Table 2.3. The various nutritional parameters of vermicompost and conventional compost proved that vermicompost is having higher nutrients than conventional compost (Table 2.4).

TABLE 2.3   Nutrient Content of Vermicompost and Cattle Dung Compost

S.No.	Nutrients	Cattle dung compost	Vermicompost
1	N	0.4–1%	2.5–3%
2	P	0.4–0.8%	1.8–2.9%
3	K	0.8–1.2%	1.4–2.0%

Source: Modified from Agarwal (1999).

TABLE 2.4   Comparison of a Nutrient Parameter of Vermicompost and Conventional Compost (CNP in % and Other Parameters in mg/100 gm of Compost)

Parameter	Conventional compost	Vermicompost
Total carbon(C)	9.34%	13.5%
Total Nitrogen(N)	1.05%	1.33%
Available P	0.32%	0.47%
Iron	587.87	746.2
Zinc	12.7	16.19
Manganese	35.25	53.86
Copper	4.42	5.16
Magnesium	689.32	832.48

Source: Modified from Jadia and Fulekar (2008).

2.6   VARIOUS SOURCES OF VERMICOMPOST

2.6.1   *VERMICOMPOSTING FROM HOUSEHOLD WASTES*

A wooden box with 45 cm x 30 cm x 45 cm (length × width × height) with a broad base and drainage holes are used for the preparation of vermi-compost. The various layering for vermicomposting is as follows: plastic sheet with small holes in the first layer followed by 3–5 cm soil layer and 5–10 cm coconut coir followed by a compost layer that may vary to the locally available agri-waste and earthworm species. An ideal number of worm for this box size is about 250. Vegetable wastes form the layer above inoculum (earthworm and compost layer), and it has to be added on a daily basis. Top of the box is covered with gunny bags, and when the box is filled completely, it is kept undisturbed for seven days followed by compost harvesting from the upper monolith (Adhikary, 2012).

2.6.2 VERMICOMPOSTING OF FARM WASTES

Adhikary (2012) reported and devised pits of sizes 2.5 m × 1 m × 0.3 m dimensions thatched sheds with opened sides. The sides and bottom part of the pit are usually made with wood. Coconut husk is spread at the bottom with the concave side upward and waste and cow dung (8:1) with a regular sprinkling of water, after a period of 7 to 10 days, the worms are inoculated to the mix (500 to 1000 worms per pit). The pits are covered with gunny bags followed by harvesting the cast along with the worms.

2.7 VERMICOMPOST: MIRACLE PLANT GROWTH PROMOTER AND PROTECTOR

Vermicompost can be used in sustainable agriculture to maintain soil fertility and to combat the food crisis. This process has a fantastic application of conversion of garbage into a gold mine. This biotechnological approach established how vermicomposting helps in plant growth regulation and an increase in crop productivity in comparison to control and chemical fertilizers. The process of transformation or mechanism of action of vermicompost on plants can be dived into two broad categories it affects plant growth directly and indirectly as illustrated in Figures 2.4 and 2.5.

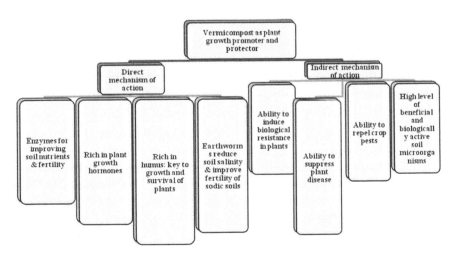

FIGURE 2.4 Vermicomposting: conversion of garbage to goldmine by earthworm.

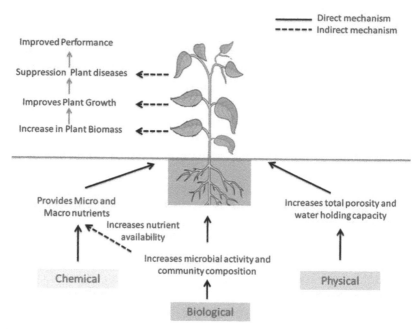

FIGURE 2.5   (See color insert.) Vermicompost and plant growth regulating mechanisms (direct and indirect) (*Source*: Modified from Lazcano and Domínguez, 2011).

Vermicompost is used for the growth of various types of crops as depicted in Table 2.5. The consequences of vermicompost on plant growth include the induction of seed germination to overall vegetative growth. Vermicompost also helps in altering the seedling morphology like an increase in leaf surface area and root branching, plant flowering, the biomass of flower, the nutritional quality of fruits and fruit yield. The raw material used for vermicomposting, rate and application time of vermicompost, etc. determines the crop response mainly. Agronomic value of vermicompost on crops is tested in field trial either alone or along with chemical fertilizers.

Vermicomposts are daintily alienated peat-like matter with excellent structure, aeration, porosity, drainage, enhancing the moisture holding capacity and with these properties vermicomposts enhanced the plant growth (Ali et al., 2007). The plant parameters such as height, total weight, shoot and root weight are increased in the vermicompost treated plants than those treated with the control compost. There are similar reports that vermicomposted coir pith enhances the growth of *Andrographis paniculata* (Vijaya et al., 2008), decomposed coir pith increases the yield of tomato (Baskaran and Saravanan 1997). According to Vinodhini et al. (2005) biodegraded coir pith compost enhanced the growth of two valuable medicinal plants. Vermicompost of degraded coir pith and cow dung used for organic farming of *C. tetragonaloba*, a nutritionally important crop (Dube et al., 2011).

TABLE 2.5 Scientific Application of Vermicompost on Various Crops

Sl No	Types of crop	Name of the crop	References
1	Horticultural crops	Tomato	Atiyeh et al. (1999 & 2001); Atiyeh et al. (2000a & b); Hashemimajd et al. (2004); Gutierrez-Miceli et al. (2007)
		Pepper	Arancon et al. (2004a & 2005)
		Garlic	Arguello et al. (2006)
		Aubergine	Gajalakshmi and Abbasi (2004)
		Strawberry	Arancon et al. (2004b)
		Green gram	Karmegam et al. (1999)
		Sweet corn	Lazcano et al. (2011)
2	Cereals	Sorghum	Bhattacharjee et al. (2001); Reddy and Ohkura (2004); Sunil et al. (2005)
3		Rice	Bhattacharjee et al. (2001); Reddy and Ohkura (2004); Sunil et al. (2005)
4	Fruit crops	Banana	Cabanas-Echevarria et al. (2005); Acevedo and Pire (2004)
		Papaya	Cabanas-Echevarria et al. (2005); Acevedo and Pire (2004)
5	Ornamentals	Geranium	Chand et al. (2007)
		Marigolds	Atiyeh et al. (2002)
		Petunia	Arancon et al. (2008)
		Chrysanthemum	Hidalgo and Harkess (2002a)
		Poinsettia	Hidalgo and Harkess (2002b)
6	Forestry species	Acacia	Donald and Visser (1989); Lazcano et al. 2010a & b).
		Eucalyptus	Donald and Visser (1989); Lazcano et al. 2010a & b).
		Pine tree	Donald and Visser (1989); Lazcano et al. 2010a & b).
7	Medicinal and aromatic plants		Anwar et al. (2005); Prabha et al. (2007)

2.8   CLIMATE CHANGE SCENARIO OF WORLD AND INDIA

The whole planet is experiencing the cost of climate change resulting high temperature, melting of glaciers, shifting rain and snowfall patterns, unexpected drought, sea level rise leading to the issues linked to agriculture, food, water, and energy security. Invasion of sea water and mounting temperature would be a serious threat to the crop yields directly affecting the food security. Crop diversification, climate-smart agriculture, more efficient water use, changes in cropping pattern and improved soil management practices together with the development of drought-resistant crops would be very helpful to the pessimistic impacts of climate change. It is worldwide accepted that the international trade and security in terms of food implies to address effects at global scale. A 2008 study published in Science suggests that Southern Africa could lose more than 30% of maize, by 2030, South Asia losses many regional staples, such as rice, millet and maize up to 10% due to climate change (Lobell et al., 2008).

India is a major greenhouse gas evolver and one of the most susceptible developing countries in the world in terms of climate change protuberance and policy framework for the mitigation of climate change. India may trade in double the amount of food grain under 2°C warming by the 2050s (The World Bank group). The country is already experiencing the impacts of climate change, including the change in seasons, frost-free season, vegetation shift, severe storms and flooding, water stress and associated negative consequences on health and livelihoods. With an escalating population and dependence on agriculture, India probably will be badly impacted by continuing climate change.

2.8.1   *ROLE OF MODERN AGRICULTURAL PRACTICES IN CLIMATE CHANGE*

Agriculture is responsible for over a quarter of total global greenhouse gas emissions (IPCC, 2007), share about 4% in the global gross domestic product (GDP) suggests that agriculture is highly greenhouse gas intensive. Leading edge agricultural practices and forefront technologies can play a role in climate change mitigation, moderation, and alteration (Basak, 2016). Cutting edge agriculture has been providing tools and techniques to reduce carbon emission as well as to lower carbon footprint. There are many emerging innovations to drive climate-smart agriculture, i.e., release of the carbon-neutral crop production model, integrated pest management practices, improved seed varieties, applications of digital tools and data science in agronomy, etc.

2.8.2 VERMICOMPOSTING VIS-A-VIS CLIMATE CHANGE

Waste is a reusable resource, and the national asset can act as a rich source of food and energy if utilized properly. The accumulation of waste in the environment creates health and aesthetic problems. Policy decisions on waste management and recycling of waste for energy and manure making will only help India to be a developed nation in the next few years. Vermicomposting is an economically and environmentally adopted technology over the traditional composting technology as it is a speedy and nearly odorless process, reducing the composting time and producing the nutritive end product. Earthworms eat the sludge components, not only convert them into vermicompost but also reduce the pathogens and absorb the heavy metals. Earthworms significantly bioaccumulate the heavy metals mercury (Hg), cadmium (Cd), copper (Cu), Manganese (Mn) and lead (Pb) from the digested sludge and produce pathogen-free fortified biofertilizer.

2.8.3 EFFECT OF EARTHWORM ON GREENHOUSE GAS EMISSION

Earthworms are generally assigned as ecological engineers, differ in their capability to ingest materials as a carbon source, and their gut microbiome leads to increase in carbon assimilation efficiencies ranging from 1 to 60% (Edwards. 2004).

The global warming potentials of CH_4 and N_2O, in a 100-year time frame, are 25 and 298 times higher than that of CO_2, respectively (IPCC, 2014). Yang et al. (2017) reported that earthworm activities in soil create aerobic conditions leads to a reduction in CH_4 production by methanogenic microflora thereby decreases the CH_4-C loss in comparison to thermophilic compost. The biological activity of earthworm improved aeration of the composting piles, thus vermicompost released much less GHG (CH4 and N2O) and NH3. The research of Fukumoto et al. (2003) and Szanto et al. (2007) showed a similar N_2O-N and NH_3-N loss level and observed in thermophilic compost the N loss slightly higher than that of vermicompost. They found that earthworm activities, different temperature, and moisture content are the primary reasons for this decrease. Yang et al. (2017) reported lesser GHG emissions in terms of kg CO_2-eq per tonne of dry matter in vermicompost than thermophilic compost and concluded that vermicompost is an effective compost method to reduce GHG emissions (Table 2.6).

TABLE 2.6 Comparison of Vermicomposting and Composting on the Basis of Greenhouse Gas Emission

Greenhouse gas emissions (kg CO_2-eq per tonne of dry matter)	Vermicomposting	Thermophilic composting
CH_4	2.28	10.52
N_2O	5.76	12.29
Total GHG	8.1	22.8

Source: Modified from Yang et al. (2017).

Vermicomposting sizably reduces total N loss, CH_4 and N_2O emissions in comparison to conventional composting practices, irrespective of the nature of the vermicomposting bed and other parameters, i.e., moisture content, C:N ratio and labile C pool. The higher worm load decreases CH_4 emissions and gears up the decomposition activity. The addition of labile C sources increases CO_2 and CH_4 emissions during composting. Many declarations have been made for reducing N losses and non-CO_2 GHG emissions from composting, however, the existing technologies are difficult to apply due to economic viability, simultaneously vermicomposting is a cost-effective practice that makes it farmer friendly.

In post-modern (after merging current policies and practices in farming) agricultural systems that do not use deep tillage, earthworms are vital for piercing packed in soil layers and providing appropriate physical conditions, such as aeration and infiltration, which are essential for healthy soils. Earthworm burrows into soil results in vertical mixing and pore size distribution followed by soil ingestion thereby biodegrading recalcitrant organic matter and released nutrient-rich castings (Fonte et al., 2007; Bossuyt et al., 2004). Therefore, earthworms are one of the important indicator and causative factors of biological soil health (Doran and Zeiss, 2000).

A recent analysis reported that earthworms increased soil CO_2 and N_2O emissions by 33% and 42% respectively (Lubbers et al., 2013). There are many factors and interactions during the process of vermicomposting which can affect CO_2 and N_2O emissions. They are as follows:

1. Interactions between earthworms and mesofauna and their role in climate change

During the past decades, several studies have shown that soil fauna influences soil CO_2 and N_2O emissions (Kuiper et al., 2013; Porre et al., 2016; Thakur et al., 2014; Wu et al., 2015). Zhu et al. (2017) reported that addition of predators does not increase soil CO_2 and N_2O emissions and also confirmed that the interactions of soil fauna could affect the availability of soil C and N, which can decrease or increase soil CO_2 and N_2O emissions.

2. Influence of earthworm size on the emission of nitrogenous gases and contribution to greenhouse effect

Wust et al. (2009) studied on the ability of large earthworms on greenhouse gas emission and concluded that earthworm size might influence the in vivo emission of this greenhouse gas (Wust et al., 2009). But later on, Peter et al. (2013) reported the emission of nitrogenous gases is independent on the size of the earthworm rather ingested soil-derived members of Rhizobiales, gut passage time and competing redox processes are important factors for emission of nitrous oxide (Figure 2.6).

2.8.4 EFFECT OF FEEDING RATIO ON N_2O EMISSION AND ITS IMPACT ON CLIMATE CHANGE

Substrate utilization is an important measure for assessing the earthworms' effect on the organic material utilization and emission of greenhouse gases. Earthworms reduce N_2O emissions by 23–48% (Nigussie et al., 2017). The higher feeding ratio had an insignificant effect on N_2O emissions as compared to non-earthworm treatments. The feeding ratio is broadly defined as the ratio of substrate added to earthworm biomass (Ndegwa et al., 2000). A high feeding ratio reduces the breakdown speed of fresh matter to vermicompost. Many other investigations made known that higher substrate supply reduces the biomass and reproduction

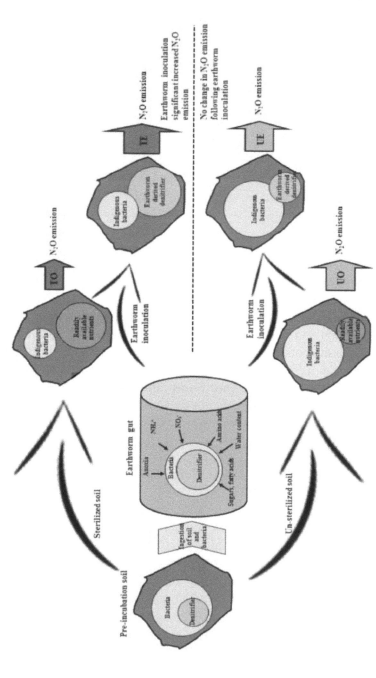

FIGURE 2.6 **(See color insert.)** Hypothetical model demonstrating the stimulation of growth of denitrifiers in earthworm gut and evolution of nitrogen gas after vermicompost treatment (*Source:* Reproduced from Wu, H., Lu, H., Lu, M., Lu, X., Guan, Q., & He, X. (2015a). Interactions between earthworms and mesofauna has no significant effect on emissions of CO2 and N2O from soil. *Soil Biol. Biochem, 88,* 294–297. Copyright © 2015 Elsevier Masson SAS. All rights reserved.).

potential of earthworms (Luth et al., 2011). However, reduction feeding ratio enhances the process of mineralization of nitrogen in comparison to high feeding ratio (Ndegwa et al., 2000). High feeding ratio increases temperature and checks air circulation in a pile (Luth et al., 2011), influencing GHG emissions. The highly optimal feeding ratio per unit of earthworm biomass, increase in pile temperature leads to earthworm mortality and higher GHG emissions.

2.8.5 EFFECT OF EARTHWORM IN SOIL CARBON SEQUESTRATION: A KEY FACTOR IN CLIMATE CHANGE

Carbon sequestration is a process to fix, store atmospheric carbon dioxide that results in mitigation of global warming. Soil organic matter (SOM) can be formed due to litter decomposition and therefore plays an indispensable role in soil carbon (C) sequestration (Angst et al., 2017). Soil physico-chemical properties, soil fauna, and various management practices, the presence of earthworms strongly influence the environment of soil and litter decomposition (Garcia-Palacios et al., 2014). Earthworms also positively control soil physical and chemical properties such as infiltration, air, and water penetration, organic C content and total polysaccharides (Kladivko et al., 1997; Bossuyt et al., 2005). The amount of sequestered C in the soil can be increased by increasing earthworm density. The ingested soil passes through the earthworm gut creates casts that lead to enhanced stabilization of organic matter (Angst et al., 2017). Zhang et al. (2013) reported that earthworm affects net C sequestration due to unequal amplification of carbon stabilization in comparison to C mineralization and generates an earthworm mediated carbon trap (Figure 2.7). The earthworm treated soil organic matter over a longer period of time reduces C loss and increases C sequestration. Earthworms make a significant difference in C stabilization in macro and micro-aggregates those formed in their castings (Bossuyt et al., 2005; Pulleman et al., 2005a, 2005b). Similar kind of observations is made by Fonte et al. (2007) that 35% of new C augmented in biogenic aggregates, compared to a conventional system. The scale and way of C dynamics in an agroecosystem is greatly influenced by the earthworms (Hedde et al., 2013). The feeding habit of earthworms can differentially change integration of fresh organic material (OM) into biogenic aggregates. This might have significant consequences for C protection and extended soil organic carbon (SOC) storage (Bossuyt et al., 2006).

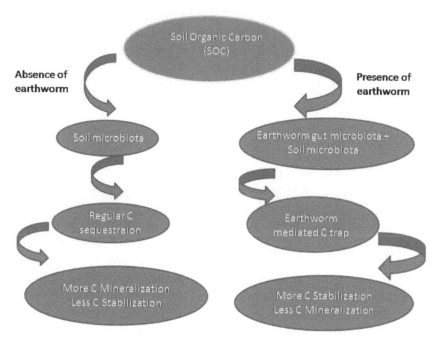

FIGURE 2.7   Earthworm intervened carbon entrap (Modified from Zhang et al., 2013).

2.8.6   *REDUCTION IN SOIL CARBON NITROGEN RATIO*

The recycling of waste material, i.e., conversion of waste into new material is one of the current trends about waste management strategy. Household wastes are accumulating day-by-day to a significant amount, and vermicomposting of these waste products not only handle waste management but also will be useful information for policymakers, state, and central government to explore different options for meeting the 2050 emissions reduction targets. Earthworms eat on the organic material and convert into casting rich in two, seven, fifteen fold more available magnesium, potassium, and nitrogen respectively compared to the surrounding soil (Majlessi et al., 2012). The biological activity associated with vermicomposting play an important role to determine the suitability of vermicompost as a soil conditioner. The suitability of the compost for plant growth determined by the maturity of the compost which in turn explains low C:N ratio, slow biological activity and less phytotoxic substances (Aslam et al., 2008; Benito et al., 2005; Brewer and Sullivan, 2003). The application of immature vermicompost gives rise to N-deficiency in crops; blocks plant growth and even kill the plants by decreasing the concentration of O_2 around the root system (Chikae et al., 2006; Komilis and Tziouvaras, 2009) whereas the mature vermicompost increases the surface area for microbial exposure and lower C/N ratio.

2.9 CONCLUSION

New agriculture may not be able to sustain without eco-friendly technologies and innovations. Vermicomposting is amongst best strategy for waste management and organic manure which increases biological, physiochemical and nutrient status of soil to increase crop yield and production which can increase soil self-life and biological fertility of the soil. There are undeniable facts that humic substances of vermicompost promote plant growth. Further attention is required to expand the potential of vermicompost as an eco- friendly, sustainable, non-toxic alternative to chemical pesticides. Earthworms are really justifying the beliefs as *"unheralded soldiers' of mankind"* and *"friends of farmers."* Agriculture causes around 10% of the world's GHG emissions; the use of vermiculture practices will be an effort to reduce GHG emissions from agricultural practices. To understand the effect of earthworms on N losses and GHG emissions by optimizing different parameters such as substrate quality, earthworm species, and feeding frequency need further research. The outcomes of vermicomposting technology not only reduce organic waste accumulation also slowing the contamination of waterways, improve the risk of flooding and provide income-earning potentials through the sale of compost.

KEYWORDS

- agricultural productivity
- agriculture practices
- carbon sequestration
- carbon stabilization
- climate change
- greenhouse gases
- gut microbiome
- mesofauna
- nutrient dynamics
- organic compound

- **organic carbon**
- **organic matter**
- **productivity gain**
- **soil fertility**

- **soil management practices**
- **soil quality**
- **soil respiration**
- **vermicomposting**

REFERENCES

1. Abbasi, T., Gajalakshmi, S., & Abbasi, S. A. (2009). Towards modeling and design of vermicomposting system: Mechanism of composting/vermicomposting and their implications. *Indian Journal of Biotechnology, 8,* 177–182.

2. Acevedo, I. C., & Pire, R. (2004). Effects of vermicompost as substrate amendment on the growth of papaya (*Carica papaya* L.). *Interciencia, 29*(5), 274–279.

3. Adhikary, S. (2012). Vermicompost, the story of organic gold: A review. *Agricultural Sciences, 3*(7), 905–917.

4. Agarwal, S. (1999). Study of vermicomposting of domestic waste and the effects of vermicompost on growth of some vegetable crops. Ph.D. Thesis, University of Rajasthan, Jaipur.

5. Ali, M., Griffiths, A. J., Williams, K. P., & Jones. D. L. (2007). Evaluating the growth characteristics of lettuce in vermicompost and green waste compost. *Eur. J. Soil Biol., 43,* 316–319.

6. Angst, S., Mueller, C. W., Cajthaml, T., Angst, G., Lhotakova, Z., Bartuska, M., & Frouz, J. (2017). Stabilization of soil organic matter by earthworms is connected with physical protection rather than with chemical changes of organic matter. *Geoderma, 289,* 29–35.

7. Anwar, M., Patra, D. D., Chand, S., Kumar, A., Naqvi, A. A., & Khanuja, S. P. S. (2005). Effect of organic manures and inorganic fertilizer on growth, herb and oil yield, nutrient accumulation, and oil quality of French basil. *Communications in Soil Science and Plant Analysis, 36*(13–14), 1737–1746.

8. Appelhof, M. (1997). Worms Eat My Garbage; 2nd ed, Flower Press, Kalamazoo, Michigan.

9. Arancon, N. Q., & Edwards, C. A. (2006). Effects of vermicomposts on plant growth. In: Proceedings of the Vermi-Technologies Symposium for Developing Countries, Department of Science and Technology-Phillippine Council for Aquatic and Marine Research and Development, Los Banos, Philippines.

10. Arancon, N. Q., Edwards, C. A., Babenko, A., Cannon, J., Galvis, P., & Metzger, J. D. (2008). Influences of vermicomposts, produced by earthworms and microorganisms from cattle manure, food waste, and paper waste, on the germination, growth, and flowering of *Petunias* in the greenhouse, *Applied Soil Ecology, 39,* 91–99.

11. Arancon, N. Q., Edwards, C. A., Bierman, P., Metzger, J. D., & Lucht, C. (2005a). Effects of vermicomposts produced from cattle manure, food waste and paper waste on the growth and yield of peppers in the field. *Pedobiologia, 49,* 297–306.

12. Arancon, N. Q., Edwards, C. A., Bierman, P., Welch, C., & Metzger, J. D. (2004b). Influences of vermicomposts on field strawberries: 1. Effects on growth and yields. *Bioresource Technology, 93,* 145–153.
13. Arancon, N. Q., Edwards, C. E., Atiyeh, R. M., & Metzger, J. D. (2004a). Effects of vermicompost produced from food waste on the growth and yields of greenhouse peppers. *Bioresource Technology, 93,* 139–144.
14. Arancon, N. Q., Galvis, P. A., & Edwards, C. A. (2005b). Suppression of insect pest populations and damage to plants by vermicompost. *Bioresource Technology, 96,* 1137–1142.
15. Arancon, N. M., Edwards, C. A., Babenko, A., Cannon, J., Galvis, P., & Metzger J. D. (2008). Influences of vermicomposts, produced by earthworms and microorganisms from cattle manure, food waste, and paper waste, on the germination, growth and flowering of petunias in the greenhouse. *Appl. Soil Ecol., 25,* 26–28.
16. Arguello, J. A., Ledesma, A., Nunez, S. B., Rodriguez, C. H., & Diaz Goldfarb, M. D. C. (2006). Vermicompost effects on bulbing dynamics nonstructural carbohydrate content, yield, and quality of 'Rosado Paraguayo' garlic bulbs. *Hortscience, 41*(3), 589–592.
17. Asha, A., Tripathi, A. K., & Soni, P. (2008). Vermicomposting: a better option for organic solid waste management. *J Human Ecology, 24,* 59–64.
18. Aslam, D. N., Horwath, W., VanderGheynst, J. S. (2008). Comparison of several maturity indicators for estimating phytotoxicity in compost-amended soil. *Waste Management, 28,* 2070–2076.
19. Atiyeh, R. M., Arancon, N., Edwards, C. A., & Metzger, J. D. (2002). The influence of earthworm-processed pig manure on the growth and productivity of marigolds. *Bioresource Technology, 81,* 103–108.
20. Atiyeh, R. M., Arancon, N. Q., Edwards, C. A., & Metzger, J. D. (2000). Influence of earthworm- processed pig manure on the growth and yield of greenhouse tomatoes. *Bioresource Technology, 75,* 175–180.
21. Atiyeh, R. M., Edwards, C. A., Subler, S., & Metzger, J. D. (2001). Pig manure vermicompost as a component of a horticultural bedding plant medium: Effects on physicochemical properties and plant growth. *Bioresource Technology, 78,* 11–20.
22. Atiyeh, R. M., Subler, S., Edwards, C. A., & Metzger, J. (1999). Growth of tomato plants in horticultural media amended with vermicompost. *Pedobiologia, 43,* 724–728.
23. Atiyeh, R. M., Subler, S., Edwards, C. A., Bachman, G., Metzger, J. D., & Shuster, W. (2000a). Effects of vermicomposts and compost on plant growth in horticultural container media and soil. *Pedobiologia, 44,* 579–590.
24. Atiyeh, R. M., Dominguez, J., Subler, S., & Edwards C. A. (2000b). Changes in biochemical properties of cow manure during processing by earthworms (*Eisenia Andrei Bouche*) and the effects on seedling growth. *Pedobiologia, 44,* 709–724.
25. Bajsa, O., Nair, J., Mathew, K., & Ho, G. E. (2003). Vermiculture as a tool for Domestic Wastewater Management. *Water Science and Technology, 48*(11–12), 125–132.
26. Banu, J. R., Logakanthi, S., & Vijayalakshmi, G. S. (2001). Biomanagement of paper mill sludge using an indigenous (*Lampito mauritii*) and two exotic (*Eudrilus eugineae* and *Eisenia foetida*) earthworms. *J Environ Biol., 22,* 181–185.
27. Basak, R. (2016). Benefits and costs of climate change mitigation technologies in paddy rice: Focus on Bangladesh and Vietnam. CCAFS Working Paper no. 160.

Copenhagen, Denmark: CGIAR Research Program on Climate Change, Agriculture and Food Security (CCAFS).

28. Baskaran, M., & Saravanan, A. (1997). Effects of coir pith based potting mix and methods of fertilizer application on tomato. *Madras Agric. J., 84,* 476–480.

29. Benito, M., Masaguer, A., Moliner, A., Arrigo, N., Palma, R. M., & Effron, D. (2005). Evaluation of maturity and stability of pruning waste compost and their effect on carbon and nitrogen mineralization in soil. *Soil Sci., 5,* 360–370.

30. Bhattacharjee, G., Chaudhuri, P. S., & Datta, M. (2001). Response of paddy (Var. TRC-87- 251) crop on amendment of the field with different levels of vermicompost. Asian *Journal of Microbiology, Biotechnology, and Environmental Sciences, 3*(3), 191–196.

31. Blanchart, E., Lavelle, P., Braudeau, E., Le Bissonnais, Y., & Valentin, C. (1997). Regulation of soil structure by geophagous earthworm activities in humid savannas of Cˆote d'Ivoire. *Soil Biol. Biochem., 29,* 431–439.

32. Bossuyt, H., Six, J., & Hendrix, P. F. (2006). Interactive effects of functionally different earthworm species on aggregation and incorporation and decomposition of newly added residue carbon. *Geoderma, 130,* 14–25.

33. Bossuyt, H., Six, J., & Hendrix, P. F. (2005). Protection of soil carbon by microaggregates within earthworm cast. *Soil Biol. Biochem., 37,* 251–258.

34. Bossuyt, H., Six, J., & Hendrix, P. F. (2004). Rapid incorporation of carbon from fresh residues into newly formed stable microaggregates within earthworm casts. *Eur. J. Soil Sci., 55,* 393–399.

35. Brewer, L. J., & Sullivan, D. M. (2003). Maturity and stability evaluation of composted yard trimmings. *Compost Science and Utilization, 11,* 96–112.

36. Brown, G. G. (1995). How do earthworms affect microfloral and faunal community diversity? *Plant Soil, 170,* 209–231.

37. Cabanas-Echevarria, M., Torres –García, A., Diaz-Rodriguez, B., Ardisana, E. F. H., & Creme-Ramos, Y. (2005). Influence of three bioproducts of organic origin on the production of two banana clones (*Musa* spp AAB.) obtained by tissue cultures. *Alimentaria, 369,* 111–116.

38. Chand, S., Pande, P., Prasad, A., Anwar, M., & Patra, D. D. (2007). Influence of integrated supply of vermicompost and zinc-enriched compost with two graded levels of iron and zinc on the productivity of geranium. *Communications in Soil Science and Plant Analysis, 38,* 2581–2599.

39. Chikae, M., Ikeda, R., Kerman, K., Morita, Y., & Tamiya, E. (2006). Estimation of maturity of compost from food wastes and agro-residues by multiple regression analysis. *Bioresour Technol., 97,* 1979–1985.

40. Dhoke, S., Dalal, P., & Srivastava, J. K. (2017). Municipal solid waste management of Ujjain city by on-site vermicomposting technique: A review. *IJAR, 3*(6), 106–111.

41. Donald, D. G. M., & Visser, L. B. (1989). Vermicompost as a possible growth medium for the production of commercial forest nursery stock. *Appl. Plant Sci., 3,* 110–113.

42. Doran, J. W., & Zeiss, M. R. (2000). Soil health and sustainability managing the biotic component of soil quality. *Appl. Soil Ecol., 15,* 3–11.

43. Dube, K. G., Bajaj, A. S., & Gawande, A. M. (2011). Mutagenic efficiency and effectiveness of gamma rays and EMS in *Cyamopsis tetragonoloba* (L.) var. Sharada. *Asiatic J. Biotech. Res., 2*(4), 436–440.

44. Edwards, C. A. (2004). *Earthworm Ecology.* 2nd ed. Boca Raton, FL: CRC Press; p. 441.
45. Edwards, C. A (1995). Historical overview of vermicomposting. BioCycle, pp. 56–58.
46. Emerson, W. W. et al. (1986). In: *Interactions 0/Soil Minerals with Natural Organics and Microbes.* (Huang, P. M., & Schnitzer, M. OOs.), Soil Science Society of America Special Publication No. 17. pp. 521–547.
47. Fonte, S. J., Kong, A. Y. Y., Van Kessel, C., Hendrix, P. F., & Six, J. (2007). Influence of earthworm activity on aggregate-associated carbon and nitrogen dynamics differs with agroecosystem management. *Soil Biology & Biochemistry, 39,* 1014–1022.
48. Fukumoto, Y., Osada, T., Hanajima, D., & Haga, K. (2003). Patterns and quantities of NH_3, N_2O and CH_4 emissions during swine manure composting without forced aeration effect of compost pile scale. *Bioresource Technol., 89*(2), 109–114.
49. Gajalakshmi, S., & Abbasi, S. A. (2004). Neem leaves as a source of fertilizer-cum-pesticide vermicompost. *Bioresour Technol., 92*(3), 291–296.
50. Garcia-Palacios, P., Maestre, F. T., Bradford, M. A., & Reynolds, J. F. (2014). Earthworms modify plant biomass and nitrogen capture under conditions of soil nutrient heterogeneity and elevated atmospheric CO_2 concentrations. *Soil Biol. Biochem., 78,* 182–188.
51. Gong, X., Wei, L., Yu, X., Li, S., Sun, X., & Wang, X. (2017). Effects of Rhamnolipid and Microbial Inoculants on the Vermicomposting of Green Waste with *Eisenia fetida. PLoS ONE, 12*(1), e0170820.
52. Gunathilagaraj, K. (1994). Vermicompost for sustainable agriculture. The Hindu, dated March 6, 1994.
53. Gutierrez-Miceli, F. A., Santiago-Borraz, J., Montes Molina, J. A., Nafate, C. C., Abdud- Archila, M., Oliva Llaven, M. A., Rincón-Rosales, R., & Deendoven, L. (2007). Vermicompost as a soil supplement to improve growth, yield and fruit quality of tomato (*Lycopersicum esculentum*). *Bioresource Technology, 98,* 2781–2786.
54. Hashemimajd, K., Kalbasi, M., Golchin, A., & Shariatmadari, H. (2004). Comparison of vermicompost and composts as potting media for growth of tomatoes. *Journal of Plant Nutrition, 6,* 1107–1123.
55. Hedde, M., Bureau, F., Delporte, P., Cecillon, L., & Decaens, T. (2013). The effects of earthworm species on soil behavior depend on land use. *Soil Biology and Biochemistry, 65,* 264e273.
56. Hidalgo, P. R., & Harkess, R. L. (2002a). Earthworm casting as a substrate amendment for Chrysanthemum Production. *Hortscience, 37*(7), 1035–1039.
57. Hidalgo, P. R., & Harkess, R. L. (2002b). Earthworm casting as a substrate for Poinsettia production. *Hortscience, 37*(2), 304–308.
58. Hong, S. W., Lee, J. S., & Chung, K. S. (2011). Effect of enzyme producing microorganisms on the biomass of epigeic earthworms (*Eisenia fetida*) in vermicompost. *Bioresource Technology, 102*(10), 6344–6347.
59. IPCC (2014). Climate Change 2014: Synthesis Report. Contribution of Working Groups I, II and III to the Fifth Assessment Report of the Intergovernmental Panel on Climate Change [Core Writing Team, R. K. Pachauri & L.A. Meyer (eds.)]. IPCC, Geneva, Switzerland, 151 pp.

60. IPCC (2007). Climate Change 2007: Synthesis Report. Contributions of Working Groups I, Ii, and Iiito the Fourth Assessment Report of the Intergovernmental Panel on Climate Change. IPCC, Geneva, Switzerland.

61. Jadia, C. D., & Fulekar, M. H. (2008). Vermicomposting of vegetable wastes: A biophysiochemical process based on hydro-operating bioreactor. *African Journal of Biotechnology, 7*(20), 3723–3730.

62. Kale, R. D. (1992). Earthworms: nature's quiet and biggest gift. *Spice India, 5*(9), 2–9.

63. Karmegam, N., Alagumalai, K., & Daniel, T. (1999). Effect of vermicompost on the growth and yield of green gram (*Phaseolus aureus* Roxb.). *Tropical Agriculture, 76,* 143–146.

64. Kaushik, P., & Garg, V. K. (2004). Dynamics of biological and chemical parameters during vermicomposting of solid textile mill sludge mixed with cow dung and agricultural residues. *Bioresour. Technol., 94,* 203–209.

65. Kladivko, E. J., Akhouri, N. M., & Weesies, G. (1997). Earthworm populations and species distributions under no-till and conventional tillage in Indiana and Illinois. *Soil Biol. Biochem., 29,* 613–615.

66. Komilis, D. P., & Tziouvaras, I. S. (2009). A statistical analysis to assess the maturity and stability of six composts. *Waste Manage, 29,* 1504–1513.

67. Kuiper, I., Deyn, G. B., Thakur, M. P., & Groenigen, J. W. (2013). Soil invertebrate fauna affect N_2O emissions from soil. *Global Change Biology, 19*(9), 2814–2825.

68. Lavelle, P. (1988). Earthworm activities and the soil system. *Biology and Fertility of Soils, 6*(3), 237–251.

69. Lazcano, C., & Dominguez, J. (2011). The use of vermicompost in sustainable agriculture: impact on plant growth and soil fertility, In: Soil Nutrients, Chapter *10,* Editor: Mohammad Miransari, Nova Science Publishers, Inc.

70. Lazcano, C., Revilla, P., Malvar, R. A., & Dominguez, J. (2011). Yield and fruit quality of four sweet corn hybrids (*Zea mays*) under conventional and integrated fertilization with vermicompost. *Journal of the Science of Food and Agriculture, 91*(7), 1244–1253.

71. Lazcano, C., Sampedro, L., Zas, R., & Domínguez, J. (2010a). Vermicompost enhances germination of the maritime pine (*Pinus pinaster* Ait.). *New Forest, 39,* 387–400.

72. Lazcano, C., Sampedro, L., Zas, R., & Domínguez, J. (2010b). Assessment of plant growth promotion by vermicompost in different progenies of maritime pine (*Pinus pinaster* Ait.). *Compost Science and Utilization, 18,* 111–118.

73. Lobell, D. B., Burke, M. B., Tebaldi, C., Mastrandrea, M. D., Falcon, W. P., & Naylor, R. L. (2008). "Prioritizing climate change adaptation needs for food security in 2030". *Science, 319*(5863), 607–610.

74. Lubbers, I., Gonzalez, E. L., Hummelink, E., Van Groenigen, J. (2013). Earthworms can increase nitrous oxide emissions from managed grassland: a field study. *Agric. Ecosyst. Environ., 174,* 40–48.

75. Luth, R. P., Germain, P., Lecomte, M., Landrain, B., Li, Y., & Cluzeau, D. (2011). Earthworm effects on gaseous emissions during vermifiltration of pig fresh slurry. *Bioresour. Technol., 102,* 3679–3686.

76. Lv, M., Shao, Y., Lin, Y., Liang, C., Dai, J., Liu, Y., & Fu, S. (2016). Plants modify the effects of earthworms on the soil microbial community and its activity in a subtropical ecosystem. *Soil Biology and Biochemistry, 103,* 446–451.

77. Maheswarappa, H. P., Nanjappa, H. V., & Hegde, M. R. (1999). Influence of organic manures on yield of arrowroot, soil physicochemical and biological properties when grown as intercrop in coconut garden. *Ann. Agric. Res., 20*(3): 318–323.

78. Maity, S., Bhattacharya, S., & Chaudhury, S. (2009). Metallothionein response in earthworms *Lampito mauritii* (Kinberg) exposed to fly ash. *Chemosphere, 77,* 319–324.

79. Majlessi, M., Eslami, A., Saleh, H. N., Mirshafieean, S., & Babaii, S. (2012). Vermicomposting of food waste: assessing the stability and maturity. *Iranian Journal of Environmental Health Science & Engineering, 9*(1), 25.

80. Marinissen, J. C. Y., & Dexter, A. R. (1990). Mechanisms of stabilization of earthworm casts and artificial casts. *Biology and Fertility of Soils, 9*(2), 163–167.

81. Masciandaro, G., Ceccanti, B., & Garcia, C. (1997). Soil agro-ecological management: fertigation and vermicompost treatments. *Bioresource Technology, 59,* 199–206.

82. McColl, H. P., Hart, P. B. S., & Cook, F. J. (1982). Influence of earthworms on some soil chemical and physical properties, and the growth of ryegrass on a soil after topsoil strippingŰa pot experiment. *New Zealand Journal of agricultural research, 25*(2), 239–243.

83. Moorhead, D. L., Lashermes, G., & Sinsabaugh, R. L. (2012). A theoretical model of C- and N-acquiring exoenzyme activities, which balances microbial demands during decomposition. *Soil Biol. Biochem.,* 53, 133–141.

84. Nagavallemma, K. P., Wani, S. P., Lacroix, S., Padmaja, V. V., Vineela, C., Babu Rao, M., & Sahrawat, K. L. (2004). Vermicomposting: Recycling wastes, into valuable organic fertilizer. Global Theme on Agroecosystems Report no.8. Patancheru 502324, Andhra Pradesh, India: International Crops Research Institute for the Semi and Tropics, 20pp.

85. Ndegwa, P., Thompson, S., & Das, K. (2000). Effects of stocking density and feeding rate on vermicomposting of biosolids. *Bioresour. Technol., 71*(1), 5–12.

86. Ndegwa, P. M., Thompson, S. A., & Das, K. C. (2000). Effects of stocking density and feeding rate on vermicomposting of biosolids. *Bioresour. Technol., 71,* 5–12.

87. Nigussie, A., Bruun, S., de Neergaard, A., & Kuyper, T. W. (2017). Earthworms change the quantity and composition of dissolved organic carbon and reduce greenhouse gas emissions during composting. *Waste Management, 62,* 43–51.

88. Peter, S. Depkat-Jakob, George G. Brown, Siu, M. Tsai, Marcus A. Horn and Harold L. Drake. (2013). Emission of nitrous oxide and dinitrogen by diverse earthworm families from Brazil and resolution of associated denitrifying and nitrate-dissimilating taxa. *FEMS Microbiology Ecology, 83*(2), 375–391.

89. Porre, R. J., van Groenigen, J. W., De Deyn, G. B., de Goede, R. G. M., & Lubbers, I. M. (2016). Exploring the relationship between soil mesofauna, soil structure and N_2O emissions. *Soil Biology and Biochemistry, 96,* 55–64.

90. Prabha, M. L., Jayraay, I. A., Jayraay, R., & Rao, D. S. (2007). Effect of vermicompost on growth parameters of selected vegetable and medicinal plants. *Asian Journal of Microbiology, Biotechnology and Environmental Sciences, 9*(2), 321–326.

91. Pramanik, P., & Chung, Y. R. (2011). Changes in fungal population of fly ash and vinassee mixture during vermicomposting by *Eudrilus eugeniae* and Eisenia fetida: Documentation of cellulose isozymes in vermicompost. *Waste Management, 31*(6), 1169–1175.

92. Pramanik, P., Ghosh, G., Ghosal, P., & Banik, P. (2007). Changes in organic-C, N, P and K and enzyme activities in vermicompost of biodegradable organic wastes under liming and microbial inoculants. *Bioresour. Technol, 98*(13), 2485–2494.

93. Prochazkova, P., Sustr, V., Dvorak, J., Roubalova, R., Skanta, F., Pizl, V., & Bilej, M. (2013). Correlation between the activity of digestive enzymes and nonself recognition in the gut of *Eisenia andrei* earthworms. *Journal of Invertebrate Pathology, 114*(3), 217–221.

94. Pulleman, M. M., Six, J., Uyl, A., Marinissen, J. C. Y., & Jongmans, A. G. (2005a). Earthworms and management affect organic matter incorporation and microaggregate formation in agricultural soils. *Appl. Soil Ecol, 29,* 1–15.

95. Pulleman, M. M., Six, J., Van Breemen, N., & Jongmans, A. G. (2005b). Soil organic matter distribution and microaggregate characteristics as affected by agricultural management and earthworm activity. *European Journal of Soil Science, 56,* 453–467.

96. Ramesh, P., Singh, M., & Rao, A. S. (2005). Organic farming: Its relevance to the Indian context. *Current Science, 88*(4), 561–568.

97. Raut, R. S. *et al.* (2000). In: *International Conference on Managing Natural Resources,* New Delhi. *Extended Summaries, 2,* 375–376.

98. Reddy, M. V., & Ohkura, K. (2004). Vermicomposting of rice-straw and its effects on sorghum growth. *Tropical Ecology, 45*(2), 327–331.

99. Rousk, J., & Baath, E. (2011). Growth of saprotrophic fungi and bacteria in soil. *FEMS Microbiology Ecology, 78*(1), 17–30.

100. Sakamoto, K., & Oba, Y. (1994). Effect of fungal to bacterial biomass ratio on the relationship between CO_2 evolution and total soil microbial biomass. *Biology and Fertility of Soils, 17*(1), 39–44.

101. Senapati, B. K. (1996). In: Proceedings and Recommendations. National Workshop on Organic Farming for Sustainable Agriculture: 187–189.

102. Serra-Wittling, C., Houot, S., & Barriuso, E. (1995). Soil enzymatic response to addition of municipal solid waste compost. *Biology and Fertility of Soils, 20*(4), 226–236.

103. Shipitalo, M. J., & Protz, R. (1987). Comparison of morphology and porosity of a soil under conventional and zero tillage. *Can. J. Soil Sci., 67,* 445–456.

104. Sinha, R. K., Agarwal, S., Chauhan, K., Chandran, V., & Soni, B. K. (2010). Vermiculture Technology: Reviving the Dreams of Sir Charles Darwin for Scientific Use of Earthworms in Sustainable Development Programs. *Technology and Investment, 1*(03), 155–172.

105. Sinha, R. K., Valani, D., Chauhan, K., & Agarwal, S. (2010). Embarking on a second green revolution for sustainable agriculture by vermiculture biotechnology using earthworms: Reviving the dreams of Sir Charles Darwin. *Journal of Agricultural Biotechnology and Sustainable Development, 2*(7), 113–128.

106. Sinha, R. K., Herat, Agarwal, S. Asadi, R., & Carretero, E. (2002). Vermiculture technology for environmental management: study of action of earthworms *Eisenia foetida, Eudrilus euginae* and *Perionyx excavatus* on biodegradation of some community wastes in India and Australia. *The Environmentalist, 22*(2), 261–268.

107. Sinha, R. K., Agarwal, S., Chauhan, K., & Valani, D. (2010). The wonders of earthworms & its vermicompost in farm production: Charles Darwin's 'friends of farmers,' with potential to replace destructive chemical fertilizers from agriculture. *Agricultural Sciences, 1*(2), 76–94.

108. Sinsabaugh, R. L., Lauber, C. L., Weintraub, M. N., Ahmed, B., Allison, S. D., Crenshaw, C. L., Contosta, A. R., Cusack, D., Frey, S., Gallo, M. E., Gartner, T. B., Hobbie, S. E., Holland, K., Keeler, B. L., Powers, J. S., Stursova, M., Takacs-Vesbach, C., Waldrop, M. P., Wallenstein, M. D., Zak, D. R., & Zeglin, L. H. (2008). Stoichiometry of soil enzyme activity at global scale. *Ecology Letters, 11,* 1252–1264.

109. Song, X., Liu, M., Wu, D., Qi, L., Ye, C., Jiao, J., & Hu, F. (2014). Heavy metal and nutrient changes during vermicomposting animal manure spiked with mushroom residues. *Waste management, 34*(11), 1977–1983.

110. Sturzenbaum, S. R. et al. (2012) Biosynthesis of luminescent quantum dots in an earthworm. *Nature Nanotechnology, 8,* 57–60.

111. Sunil, K., Rawat, C. R., Shiva, D., & Suchit, K. R. (2005). Dry matter accumulation, nutrient uptake and changes in soil fertility status as influenced by different organic and inorganic sources of nutrients to forage sorghum (*Sorghum bicolor*). *Indian Journal of Agricultural Science, 75*(6), 340–342.

112. Szanto, G. L., Hamelers, H. V. M., Rulkens, W. H., & Veeken, A. H. M. (2007). NH_3, N_2O and CH_4 emissions during passively aerated composting of straw-rich pig manure. *Bioresource Technology, 98,* 2659–70.

113. Thakur, M. P., van Groenigen, J. W., Kuiper, I., & De Deyn, G. B. (2014). Interactions between microbial-feeding and predatory soil fauna trigger N_2O emissions. *Soil Biology and Biochemistry, 70,* 256–262.

114. UNDP (1994). Sustainable Human Development and Agriculture; UNDP Guide Book Series, NY.

115. UNEP/GEMS (1992). The Contamination of Food. UNEP/GEMS Environment Library No. 5, Nairobi, Kenya.

116. Vasanthi, D., & Kurriaraswamy, K. (1999). Efficacy of vermicompost to improve soil fertility and rice yield. *J. Indian Soc. Soil Sci., 47,* 268–272.

117. Vermi Co. (2001) Vermicomposting technology for waste management and agriculture: An executive summary. Vermi Co., Grants Pass.

118. Vijaya, D., Padmadevi, S. N., Vasandha, S., Meerabhai, R. S., & Chellapandi, P. (2008). Effect of vermicomposted coir pith on the growth of *Andrographis paniculata*. *J. Org. Syst., 3*(2), 52–56.

119. Vinodhini, S., Gnanambal, V. S., Sasikumar, J. M., & Padmadevi, S. N. (2005). Growth of two medicinal plants using biodegraded coir pith. *Plant Archives, 5,* 277–280.

120. Wu, H., Lu, M., Lu, X., Guan, Q., & He, X. (2015a). Interactions between earthworms and mesofauna has no significant effect on emissions of CO_2 and N_2O from soil. *Soil Biol. Biochem, 88,* 294–297.

121. Wu, Y., Shaaban, M., Zhao, J., Hao, R., & Hu, R. (2015). Effect of the earthworm gut-stimulated denitrifiers on soil nitrous oxide emissions. *European Journal of Soil Biology, 70,* 104–110.

122. Wust, P. K., Horn, M. A., Henderson, G., Janssen, P. H., Rehm, B. H., & Drake, H. L. (2009b). Gut-associated denitrification and in vivo emission of nitrous oxide by the earthworm families Megascolecidae and Lumbricidae in New Zealand. *Applied and Environmental Microbiology, 75*(11), 3430–3436.

123. Yang, F., Li, G., Zang, B., & Zhang, Z. (2017). The Maturity and CH4, N2O, NH3 Emissions from Vermicomposting with Agricultural Waste. *Compost Science & Utilization,* 1–10.

124. Zhang, H., & Schrader, S. (1993). Earthworm effects on selected physical and chemical properties of soil aggregates. *Biology and Fertility of Soils, 15*(3), 229–234.

125. Zhang, W., Hendrix, P. F., Dame, L. E., Burke, R. A., Wu, J., Neher, D. A., Li, J., Shao, Y., & Fu, S. (2013). Earthworms facilitate carbon sequestration through unequal amplification of carbon stabilization compared with mineralization. *Nat. Commun., 4,* 2576.

126. Zhu, X., Chang, L., Li, J., Liu, J., Feng, L., & Wu, D. (2017). Interactions between earthworms and mesofauna affect CO_2 and N_2O emissions from soils under long-term conservation tillage. *Geoderma*, DOI: 10.1016/j.geoderma.2017.09.007.

CHAPTER 3

Agroforestry with Horticulture: A New Strategy Toward a Climate-Resilient Forestry Approach

ABHISHEK RAJ,[1] M. K. JHARIYA,[2] D. K. YADAV,[2] and A. BANERJEE[2]

[1]PhD Scholar, Department of Forestry, College of Agriculture, I.G.K.V., Raipur–492012 (C.G.), India, Mobile: +00-91-8269718066, Email: ranger0392@gmail.com

[2]Assistant Professor, University Teaching Department, Department of Farm Forestry, Sarguja Vishwavidyalaya, Ambikapur–497001 (C.G.), India, Mobile: +00-91-9407004814 (M. K. Jhariya), +00-91-9926615061 (D. K. Yadav), +00-91-9926470656 (A. Banerjee), E-mail: manu9589@gmail.com (M. K. Jhariya), dheeraj_forestry@yahoo.com (D. K. Yadav), arnabenvsc@yahoo.co.in (A. Banerjee)

ABSTRACT

Horticulture-based agroforestry (AF) has played a multifarious role in production (timber, fuelwood, nutritive fruits, pasture/fodder, etc.) and protection of the environment through carbon sequestration to mitigate ongoing climate change problem for better well-being of mankind. Different type of models under horticulture-based agroforestry viz., agri-horticultural system, agrihortisilviculture, hortipasture, etc. are playing a very diverse role in production potential, upliftments of marginal and degraded land (through the incorporation of the hortipasture system), economic benefits and national food and nutritional security. Growth and productivity of AF components vary due to nature and arrangement of trees (including fruit trees), agricultural crops and pastures/animals. Moreover, horticulture-based agroforestry plays a pivotal role in potential carbon storage in tree biomass and in the soil in addition to soil nutrient pool. Overall, horticulture-based agroforestry is very profitable in the context of productivity, improvement of soil physicochemical properties, nutrition management, food security, and environmental sustainability and there should be further research and developmental activities in this direction for the betterment of the whole agroecosystem.

3.1   INTRODUCTION

Agroforestry is an ecologically-based sustainable land use farming practices, new repute for an ancient land use practice which comprises agriculture and forestry to maintain the need of forest cover up to 33% as per India's national forest policy (Singh et al., 2013; Raj et al., 2014; Jhariya et al., 2015; Raj et al., 2016a&b; Singh and Jhariya, 2016; Toppo et al., 2016). As per earlier reports agroforestry has the potential to produce multiple products and other tangible products in a sustainable way (Raj and Jhariya, 2016). Fruits tree of tropical, subtropical, temperate region, etc. are considered under agroforestry systems, and the major fruit trees in different agro-climatic region and states of India is depicted in Table 3.1 (Singh and Jhariya, 2016). As we know, fruits and vegetables are the major sources of minerals and vitamins and the minimum dietary requirement of fruits and vegetables per day per capita is 85 g and 220 g/head/day but availability is only 60 g and 80 g, respectively (Roy, 2011). Fruit trees like *Carica papaya, Citrus* spp., *Mangifera indica, Psidium guajava, Annona squamosa, Emblica officinalis, Malus domestica, Prunus armeniaca, Prunus domestica,* etc. are very common and popular in India. In the rainfed area, integration of horticulture crops offers exclusive advantages to food and nutritional security (Korwar et al., 1988; Singh and Malhotra, 2011) and provides nutritive fruits, income, employment and livelihood securities (Samara, 2010). Moreover, the outcomes of growth, yield, and productivity depend upon tree spacing, density, type and nature of plantation and their shading effects which include effects on morphology (internodes length, leaf area) and effects on flower initiation/fruit-set of associated crops (Raj et al., 2016b). From an economic perspective, fruit-based systems are economically worthwhile with high benefit-to-cost ratios and total system-productivity in fruit tree + annual mixed systems remaining high although individual yields of annual components are commonly reduced relative to sole-cropping (Ashour et al., 1997).

TABLE 3.1 Various Fruit Trees Used in Agroforestry of Different Agroclimatic Regions and States in India

Agroclimatic Regions	Fruit trees used	States
Western Himalayan region	Almond, apple, apricot, cherry, peach, plum, strawberry and walnut	Uttarakhand, Himachal Pradesh, Jammu, and Kashmir
Eastern Himalayan region	Orange, banana, cherry, lemon, and papaya	West Bengal, Arunachal Pradesh, Assam, Meghalaya, Mizoram, Nagaland, Sikkim, Tripura
Lower Gangetic plain region	Mango, guava, and litchi	West Bengal
Middle Gangetic plain region	Mango, papaya Guava, jamun and litchi	Bihar and Uttar Pradesh
Upper Gangetic plain region	Mango, guava, jamun, papaya, and peach	Uttar Pradesh
Trans Gangetic plain region	Mango, aonla, guava, and kinnow	Punjab, Rajasthan, Chandigarh, Delhi, Haryana
Eastern plateau and hills region	Mango, guava, apple, aonla, custard, lemon, pomegranate and papaya	Chandigarh, Jharkhand, Madhya Pradesh, Maharastra, Odisha, West Bengal
Central plateau and hills region	Mango, Aonla, ber, and mandarin	Uttar Pradesh, Madhya Pradesh, Rajasthan
Western plateau and hills region	Mango, papaya, Banana, grapes, lemon, malta, mandarin, and pomegranate	Maharashtra and Madhya Pradesh.
Southern plateau and hills region	Mango, guava, banana, citrus, grapes and sapota	Karnataka, Tamil Nadu and Andhra Pradesh
East coast plains and hills region	Mango, apple, banana, custard apple and sapota	Odisha, Tamil Nadu, Andhra Pradesh, and Pondicherry.
West coast plains and ghat region	Mango and citrus	Kerala, Maharashtra, Tamil Nadu Goa, and Karnataka
Gujarat plains and hills region	Mango, banana, dates, grapes, guava, and sapota	Gujarat, Daman and Diu, Dadra and Nagar Haveli
Western dry region	Ber, Mosambi, pomegranate, and Kinnow	Rajasthan
Island region	Mango, papaya, and sapota	Lakshadweep, Andaman and Nicobar Islands

Source: Adapted and modified followed by NRCAF (2007); Singh and Jhariya (2016).

Although, agroforestry could sequester atmospheric carbon through carbon sequestration process and recognized as a possible solution for the mitigation of climate change (Alavalapati and Nair, 2001). As per Nair et al. (2011), carbon storage potential of agroforestry system ranged from 0.3 to 15.2 Mg C/ha/yr and was reported to be highest in humid tropics receiving high rainfall. Horticulture-based agroforestry plays an inevitable role in this context with greater stability of atmospheric carbon and receiving wider appreciation in agricultural sustainability, profitability, and context of carbon sequestration or climate change mitigation. In the view of above points, this chapter review about different models of horticulture-based agroforestry systems, their role in overall productivity enhancement, livelihood security through economic benefits of farmers and environment benefits from pollution reduction and climate change mitigation through carbon sequestration.

3.2   CONCEPT OF AGROFORESTRY AND HORTICULTURE AND THEIR INTEGRATION

Agroforestry is found to be a practical, low cost means of integrated land management especially for small-scale producers which involve cultivation of woody perennial trees in farming systems (Leakey, 2010), as effective means of targeting sustainable production (Padmavathy and Poyyamoli, 2013) with conservation and preservation of natural resources without any environmental destruction (Ajayi et al., 2005). Agroforestry is an integrated unit of cultivating agricultural crops along with tree species, economic crop with domesticated livestock rearing. Horticulture similarly facilitates diversified product and therefore address food and nutritional security for the rural farming community. Such an approach can be integrated for sustainable production of agro-products. Under such circumstances, agri-horticulture proves significant promise. The major advantage of such practice occurs in the form of fruits as a secondary crop in the agro-ecosystem. Fruit-based agroforestry system is an alternative land use system that integrates the cultivation of herbaceous crops, perennial fruit trees and trees other than fruits. Likewise, fruit trees are integrated with pasture or grasses to form a hortipasture system which is prevalent in the rainfed area. Therefore, trees and their products such as timber, and non-timber forest products, etc. are the major source and satisfying the seasonal food needs with maintaining household food. Fruit trees like jack fruit, mango, banana, guava, coconut, papaya, etc. from home garden play a major role in household consumption and cash earn from these tree and tree products help farmers to the household consumption of fruits like jack fruit, mango, banana, guava, coconut, papaya, etc. from the homestead trees and the cash earned through the sale of fruits and other tree products help the poor farmers to conquer the lean periods. Moreover, the availability of homestead vegetables is year round while tree fruits are available during one lean period (Hassan and Mazumdar, 1990). Worldwide, fruit-based agroforestry receives high popularity among other agroforestry practices due to market value with the economically sound return from fruits trees and their products and high dietary and nutritional value (Bellow et al., 2008).

3.3 DIFFERENT FORMS OF AGROFORESTRY SYSTEMS

Fruit-based agroforestry models like agri-horticulture (Crops + Fruit trees), hortipastoral (fruit trees + pasture/animals), and agrihortisilviculture (crops + fruit trees + tree other than fruits) are prevalent in humid, semiarid and arid climatic regions.

3.3.1 AGRIHORTICULTURAL SYSTEM

It is the practice of cultivating agricultural crops of different nature with horticultural species (Pant et al., 2014) and widely practiced on the dry and marginal areas of the country of different agroclimatic zones (Table 3.2). Multiple products are harvested along with food production which is widely practiced in different developing nations of Asian sub-continent. Agri-horticultural system covers 0.5 M ha area with total employment/yr (millions of man-days) of 20.3 (NRCAF, 2007). Therefore, the agri-horticultural system is more preferable practice by farmers than agrisilvicultural system because of guaranteed returns and notional security (Kareemulla et al., 2002).

3.3.2 AGRIHORTISILVICULTURE

It could be considered a combination of the above two approaches, involving the growing of fruit as well as forest tree species, along with

TABLE 3.2 List of Agrihorticulture Systems for Various Agro-Climatic Zones of India

Agro-climatic Zone	Agroforestry System	Tree component	Crop/Grasses
Western Himalayan Region	Agri-horticulture	*Malus pumila*	Wheat, Millets
	Agri-horticulture	*Prunus persica*	Soybean, Millets
Eastern Himalayan Region	Agri-horticulture	*Alnus nepalensis*	Large cardamom/Coffee
Lower Gangetic Plains	Agri-horticulture (Irri.)	*Mango/Banana/Litchi*	Paddy, Wheat, Maize
Middle Gangetic Plains	Agri-horticulture (Irri.)	*Mango/Citrus spp.*	Rice-Wheat
Trans Gangetic Plains	Agri-horticulture (Irri.)	*Emblica officinalis*	Black gram/Green gram
Central Plateau and Hills	Agri-horticulture (Irri.)	*Psidium gujava*	Bengal gram/Groundnut
	Agri-horticulture (RF)	*Emblica officinalis*	Black gram/Green gram
Western Plateau and Hills	Agrihortisilviculture (Irri.)	*Tectona grandis, Achras sapota,*	Paddy, Maize
	Agri-horticulture	*Areca catechu*	Black pepper, cardamom
Southern Plateau and Hills	Agri-horticulture	*Tamarindus indica*	Chilli
West Coast Plains and Hills	Agri-horticulture (RF)	*Artocarpus heterophyllus*	Black pepper
	Agri-horticulture	*Cocus nucifera/Areca catechu*	Paddy
Island Region	Agrohorticulture	*Cocus nucifera*	Paddy

*Note: Irri–Irrigated; RF–Rainfed.

Source: Adapted and modified as given by NRCAF (2007).

agricultural crops. The underlying principle for the adoption of agrihor-tisilviculture is that mature fruit trees can provide income to poor families on sustainable basis unlike agricultural crop cultivation, which is highly dependent on monsoons. Moreover, family's food requirements are satisfied by incorporation of agricultural crops in alleys and spaces between trees, till fruit trees begin to yield. Along the boundaries, integration of multipurpose tree species meet the family's fuel, wood or fodder needs and acts as a source of secondary income. Thus, agro-horti-forestry meets a wider spectrum of key basic needs of poor families. A number of agroforestry models have been developed and evaluated for their productivity and sustainability (Dhyani and Chauhan, 1995) but the agri-horticulture model amongst various agroforestry practices have been found most suitable in terms of productivity, sustainability, productivity and tree-crop compatibility (Bhatt et al., 2001).

3.3.3 HORTIPASTURE

It is another type of model which integrates fruit trees with pasture (either legumes or grasses) and plays an important role to bridge the gap of fruit and fodder demand and supply through proper utilization of degraded lands and secured income, livelihood and nutrition (Kumar et al., 2011). Kumar et al. (2009) has reported various type of fruit-based hortipasture systems namely Ber-based hortipasture system, Kinnow-based hortipastoral system under partial irrigation, Aonla-based hortipastoral system, Aonla-based hortipastoral system with soil and water conservation, Annona (custard apple)-based hortipastoral system, Bael-based hortipastoral system, Tamarind-based hortipastoral system, etc. under rainfed area. Such diverse nature of systems meets the need of the people as well as socio-economic upliftment of the local community stakeholders.

3.4 MONOCROPPING SYSTEM VERSUS HORTICULTURE-BASED AGROFORESTRY SYSTEMS

Horticulture-based agroforestry is ecologically and economically more diverse than Monocropping system. As per Kassa (2015), the combination of fruit trees with perennials like Enset (*Ensete ventricosum*) and Coffee (*Coffea arabica*) is economically more profitable, less risky investment and labor-saving with diversified income sources than monocrop land use system. Also the value of net present value (NPV), benefit-cost ratio (BCR) and annual equivalent value (AEV) is more in fruit-based agroforestry as compared to monocrop system (Table 3.3). Higher NPV were recorded for the agri-horticulture system in comparison to other agroforestry systems. The BCR values of fruit crops reflect higher output values in comparison to crop species, respectively. The value of AEV of fruit-tree-based agroforestry system was 80,600.28 ETB per annum, that is more in compared to 52,089.97 ETB per annum (sugarcane monocrop), 36,445.68 ETB per annum (sequential monocrop of tomato with maize) and 20,625.17 ETB per annum (monocrop of potato with maize), respectively.

TABLE 3.3 Value of NPV, BCR, and AEV per Hectare in Ethiopian Birr (ETB) of Fruit Tree-Based Agroforestry SYSTEM COMPARED OTHER MONOCROPPING SYSTEM

Economical variables	Fruit tree-based Agroforestry system	Monocropping system		
		Sugarcane	Tomato + maize	Potato + maize
Net present value (NPV)	731,608.35	472,820.76	330,817.59	187,214.76
Benefit-cost ratio (BCR)	80,600.28	52,089.97	36,445.68	20,625.17
Annual equivalent value (AEV)	3.43	2.43	1.90	1.44

1 USD = 17.80 ETB.
Source: Adapted and modified as given by Kassa (2015).

3.5 WORLDWIDE SCENARIO OF IMPLEMENTATION OF VARIOUS AGROFORESTRY SYSTEMS

Around 1,023 million ha area has been estimated under agroforestry system in the world (Nair et al., 2009) which comparatively higher than India alone covers 25.32 Mha (8.2% of the total land area) (Dhyani et al., 2013). As per Rizvi et al. (2014) area under agroforestry is 13.75 m ha which is-based on data from CAFRI, Jhansi and Bhuvan LISS III. However, Forest Survey of India (FSI, 2013) estimated the same as 11.54 m ha, which is 3.39% of India's total land area. Nair and Nair (2003) estimated the extent of alley-cropping, silvo-pastoral, windbreaks and riparian buffers in the USA as 235.2 m ha. Kumar (2006) estimated the area of the home garden in South and Southeast Asian home garden as 8.0 m ha.

3.6 DRYLAND HORTICULTURE-BASED PRODUCTION SYSTEM

Dryland area is located in the arid, and semiarid region comprises very low productivity and not fit for crop farming. Therefore, cultivation of fruit crops-based agroforestry practices under horticulture-based farming system plays a significant role in these areas. The horticulture-based agro-forestry systems in the arid and semiarid region of India are described in the following subsections.

3.6.1 HORTICULTURE-BASED AGROFORESTRY SYSTEMS IN INDIAN ARID REGION

The Indian arid zone covers states like Rajasthan, Gujarat, Punjab, Haryana, Andhra Pradesh, Karnataka and Maharashtra and which occupying land of 31.8 million ha (12%) area of country's geographical area (Figure 3.1). The region experiencing sandy soil with high wind velocity and having low and erratic rainfall (100 and 500 mm annually) with high temperature ranged from 1 to 48°C (Bhandari et al., 2014). Along with cultivation of some important annual crops in this arid region, the importance of horticulture could not be underestimated because this sector makes land productive, employment generation, improve the socio-economic condition of poor farmers with income security, crop diversification and nutritional security to desert dwellers, etc. Some horticulture crops such as Indian Jujube (*Ziziphus mauritiana*), Indian Cherry (*Cordia myxa*), Indian Gooseberry (*Emblica officinalis*), Indian Mesquite (*Prosopis cineraria*), Pome-granate (*Punica granatum*), Kair (*Capparis deciduas*), Karonda (*Carissa carandas*), Kinnow (*Citrus reticulate*) and Bael (*Aegle marmelos*), etc. are important crops in this region which makes not only productive land but also economically viable. In this region, horticulture-based agroforestry systems provide food and nutrition to rural peoples (Chundawat, 1993; Chadha, 2002) and offer multifarious outputs and provide benefits in the term of production and income (Osman, 2003).

3.6.1.1 AGRIHORTICULTURE SYSTEM

Integration of agricultural crops with horticultural trees are very common practices that would surely be profitable and make productive land in the arid region. But the suitability of arable crops intercropping with the horti-cultural tree depends on tree crop interaction, their competitiveness, soil and climatic condition in this region. This can be justified through the study of Krishnamurthy (1959) and Naik (1963) and according to them, some crops such as maize, wheat, sugarcane, and cotton, etc. are considered as exhaustive crops that are incompatible with horticultural crops. Therefore, the recommended combination of agriculture crops with horticulture trees are depicted in Figure 3.2 (Bhandari et al., 2014). Several authors have been carried out research in this context. As per Singh (1997), the higher yield was seen in both jujube and the intercrops than in monoculture (sole crop). Similarly, intercropping of green gram, sesame and guar with jujube enhanced the yield of fruits (14.8 to 5.2 kg tree^{-1}) and seeds (782 kg).

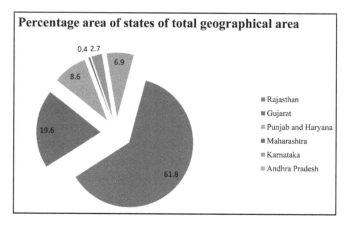

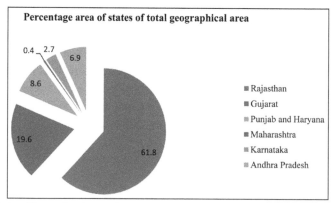

FIGURE 3.1   (See color insert.) Area of arid region in different state of India (Bhandari et al., 2014).

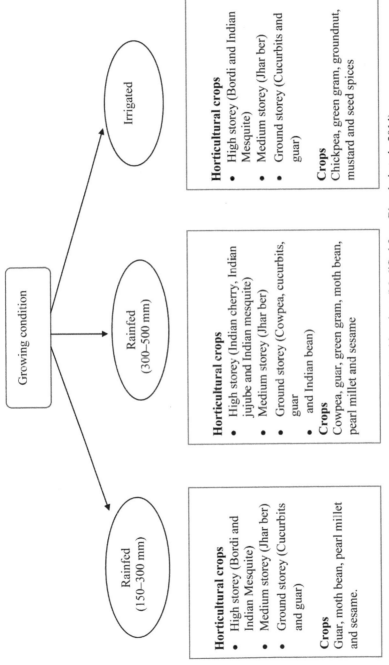

FIGURE 3.2 Suitable agri-horti crops components for Indian arid regions (Modified from Bhandari et al., 2014).

3.6.1.2   AGRIHORTISILVICULTURE SYSTEM

As we know that the incorporation of the agri-horticulture system is highly successful in the arid region. But the major problem prevailing in that region is the frequent occurrence of high wind speed that results wind erosion which does erode the topmost productive soil, leads to high water evaporation and impair the health of cultivated crops and fruit plants. Therefore, scientifically it is advisable to use some perennial trees as a shelterbelt in or around the field. Such practice is termed as agrihortisilviculture which is a more diverse and economically sound system that produce timber, food, fruits, fodder, fuelwood, and non-timber forest products.

3.6.1.3   HORTIPASTURE

This is another type of agroforestry practices which involve cultivation of some grasses for rearing cattle in the fruits orchard. In this system, farmers can grow some fruit crops like mango, citrus, guava, pomegranate, Indian gooseberry, jujube, etc. in the first tier with the integration of pasture in the ground tier for feeding the livestock and cattle. Some crops like *Z. nummularia* (jhar ber) works as the rescue of the farmers in another crop failed condition in a drought area and provides fruits, fodder, and fuelwood. Earlier reports revealed the output of horticulture-based agroforestry system. As per one estimates, around 80 and 60% of mortality is found in *Z. nummularia* and *Z. rotundifolia* during establishment with buffelgrass pasture which is incompatible with above shrubs due to the vigorous root system of grass (Sharma and Vashishtha, 1985). Other experiment reported the additional production of dry forage of 2.5-ton ha^{-1}, grass seed of 25–30 kg ha^{-1} and jujube fruit of 7.5-ton ha^{-1} under hortipasture system comprised *Cenchrus ciliaris* which is intercropped with jujube orchard (Vashishtha and Prasad, 1997).

3.6.1.4   HORTI-SILVIPASTURE SYSTEM

In an arid region, the practice of horti-silvi-pastoral has several advantages such as the inclusion of tree and fruit crop makes productive soil through the addition of organic matter which can help in proficient nutrient cycle, prevent soil erosion and runoff, improve microclimate, maintain quantity and quality of grass as forage which results healthy livestock's population. Therefore, this system can maintain sustainability in different land use system in the arid region of India. Although, the types and nature of species for feeding maintain the health of livestock. As per one estimates, species like *Acacia senegal, Acacia tortilis, Albizia lebbeck, Anogeissus pendula, Anogeissus rotundifolia, Azadirachta indica, Calligonum polygonoides, Grewia tenax, Gymnosporia spinosa, Prosopis cineraria, Prosopis juliflora, Salvadora oleoides, Ziziphus nummularia,* and *Tecomella undulata* are considered top feeding species and characterized by high palatability of leaves which remain green still during hot summer months (Ganguli et al., 1964). Moreover, tree, fruit crops, and pasture should be arranged in favor of better utilization of resources along with the reduction of competition for nutrient, light, and water.

3.7 HORTICULTURE-BASED AGROFORESTRY SYSTEM IN SEMIARID REGION

Globally, semiarid tropics covered 2.1 billion ha area which supports a large population of which up to 99% population in 90% area are located in developing countries. In these developing countries, India covered 10% of semiarid (SAT) area of 56% population (Burford and Virmani, 1983) with Vertisols and associated soils of 72 million ha (Murthy et al., 1982). In this harsh condition of the semiarid region of India, horticulture crops including fruit trees such as Aonla, Ber, Pomegranate, Datepalm, Mango, Guava, Custard apple, Sapota, Tamarind, Wood apple, Chironji, Mahua, Jamun, etc. play an inevitable role due to nutritive and medicinal value. Moreover, identification of high yielding and superior varieties of such crops and fruit trees can potentially increase the productivity, soil health, microclimate and improved the socioeconomic condition of poor farmers in the semiarid area or even in marginal degraded land. Therefore, productivity and economic evaluation in term of yield (q/ha) and net return (Rs/ha) are analyzed for some fruit trees in semiarid region compared to traditional cropping system (Table 3.4) In this table, fruit-bearing event start from second years in some trees like ber, custard apple and pomegranate, guava and aonla as compared to mango and sapota, i.e., from third year. Similarly, highest productivity in q/ha was recorded in ber (176.40) due high population of plant per hectare followed by guava (98.80), Aonla (94.5), Custard apple (80.80), Mango (79.5), Sapota (73.8) but least in pomegranate (70.40) with second highest net return amounting

TABLE 3.4 Yield (q/ha) and Net Return (Rs/ha) Fruit Crops in the Semiarid Region Compared to Traditional Cropping System

Crop Year	Ber	Guava	Pomegranate	Aonla	Custard apple	Sapota	Mango	Maize + pigeon pea (Traditional method)
Second	18.8*(2,640**)	16.80 (5,900)	12.72 (9,720)	7.9 (1,450)	6.92 (960)	–	–	15.96
Third	50.80 (1,990)	26.00 (10,250)	29.48 (26,230)	25.8 (10,150)	17.84 (6,170)	20.6 (7,800)	7.3 (4,800)	3.57
Fourth	83.00 (21,400)	46.80 (20,400)	41.00 (37,750)	39.3 (20,400)	21.08 (7,540)	27.4 (10,450)	16.7 (13,700)	–
Fifth	107.60 (28,530)	59.60 (26,300)	46.00 (42,750)	61.7 (16,550)	32.32 (12,910)	38.2 (15,600)	27.8 (24,500)	–
Sixth	126.00 (33,800)	71.20 (31,850)	54.89 (51,640)	75.4 (27,150)	45.20 (18,850)	47.9 (20,200)	35.4 (31,650)	–
Seventh	134.8 (35,940)	80.80 (36,400)	59.20 (54,750)	83.3 (33,700)	61.60 (26,800)	56.4 (24,200)	48.7 (44,700)	–
Eight	166.00 (45,050)	91.20 (41,350)	61.60 (56,850)	79.4 (37,400)	71.60 (31,550)	64.3 (26,950)	63.8 (56,550)	–
Ninth	176.40 (47,920)	98.80 (44,900)	70.40 (65,400)	94.5 (42,750)	80.80 (35,900)	73.8 (32,400)	79.5 (75,000)	10,050

*Yield q/ha and **Net return (Rs/ha).

Source: Adapted and modified as given by Mutanal et al. (2001).

to Rs. 65,400 per hectare due to high market rate followed by mango (Rs. 75,000) in which highest net return was recorded the 9th year. Moreover, net return was recorded least (10,050) in traditional cropping system comprised maize and pigeon pea due to less productivity (Mutanal et al., 2001). In semiarid tropics, tree component is deciduous in nature and produces a large quantity of biomass and can be used to add leaf litter to the soil (Hiwale, 2004). Moreover, some cropping system like agri-horti, horti-silvi, and horti-pasture can improve the sustainability of degraded lands and maintain soil health.

3.8   EFFECT ON YIELD AND PRODUCTIVITY

In the agri-horticultural model, associated crops are affected by associated fruit tree species in terms of yield, productivity, and growth. Under varied agro-ecosystem, it has been observed that climate always plays a significant role in productive function. Bijalwan (2012) reported the range of productivity output of agri-horticulture system happens to be near about 12 quintal yearly per hectare basis on both northern and southern region compare to traditional farming. Reduction in productivity in agro-ecosystem combining agriculture and fruit crops is usually supplemented by fruit cultivation by local community stakeholders in the valley region. Combinations of silvicultural species with orchards have proven to be positive compared to isolated cultivation practice. Dhara and Sharma (2015) reported the best results in terms of growth and production of eucalyptus and mango as a comparison to other systems of silviculture or individual orchard plantation. Similarly, as per Singh ct al. (2014), combined plantation of Ber (*Ziziphus mauritiana*) and Khejri (*Prosopis cineraria*) leads to higher economic outputs in comparison to wheat-based silviculture system and horticulture system. Up to one-fourth times of yield have reported to be reduced under agroforestry in comparison to solo plantation of crops. Also, crop yield reduced by 5% to 23% in the agroforestry systems than the yield in sole crop plot. Under the Indian system of agroforestry, spinach recorded higher biological efficiency as per KS green < Local < KS red sequence under Litchi-Papaya plantation. The growth attributes such as sprouting in KS green and local variety reflected the value of 2.97 and 2.89 which is higher than the KS red variety of 2.57. The highest yield was reported in KS red as 36.32 t ha^{-1} followed by local (34.61 t/ha) and KS green (34.00 t/ha). Growth attributes of litchi plant reflected 21.39%

(height) and 44.94% (stem diameter) increment during the cultivation period. Similarly, papaya gave the yield biomass of approximately 24 tons per hectare with more than 40 fruits per plant.

3.9   EFFECT ON PHYSICO-CHEMICAL PROPERTIES OF SOIL

The agri-horticulture and allied system help to develop soil fertility and productivity by efficient nutrient cycling into the plant-soil system. This is justified by Dhara and Sharma (2015) as soil fertility improved by available N, P, K and higher soil organic carbon under different agroforestry system components as compared to the system of fruit trees and trees alone. The increased value of soil N, P, K and organic carbon under AFS model comprises, T1 (*E. tereticornis*+ Mango + Pigeon pea) and T2 (*E. tereticornis*+ Mango+ Black gram) are 34.3% and 27%, 35% and 27%, 18% and 13%, 29%, and 24%, respectively, after 2 year. This was due to a valuable interaction between tree species and legume crop which helped to improve the fertility status of the soil. These findings corroborate with earlier findings of Biswas et al. (2003). Organic manure tends to change soil physicochemical attributes. Under peach plantation 25 tons per hectare FYM application increases the soil pH values in comparison to other organic sources (Tripathi, 2017). The application of 6 tons per hectare vermicompost application increased soil conductivity value and 25 tons per hectare FYM application reduced soil bulk density. Under varied soil depth higher level of soil organic carbon were recorded with a gradual increment of FYM doze in comparison to other organic inputs. A higher level of soil macronutrients was recorded under peach plantation with 6 tons per hectare vermicompost application. A higher level of soil potassium was reported in case of 25 tons per hectare FYM applications. Peach plantation reported better soil condition in comparison to plots where there is no peach plantation. Integration of horticulture system within the agroforestry may help to improve the soil structure along with higher productivity and yield in case of woody and nonwoody components.

3.10   ROLE TOWARDS LIVELIHOOD SECURITY

Model of agroforestry with the incorporation of horticultural crops can play an inevitable role towards socioeconomic development and livelihood security. *Wadi* programme is a good example for in the context of agro-horti-forestry which works very efficiently towards socioeconomic development and livelihood security. It is implemented in family-owned plots of under-utilized lands and spread over Gujarat, Maharashtra, Rajasthan, Uttar Pradesh, Uttarakhand, Madhya Pradesh, Chhattisgarh, Bihar, Karnataka, Andhra Pradesh, and Jharkhand. For example, horticulture crops such as amla, mango, and cashew with intercrops grown in these spaces and trees like *Gliricidia* and Subabul planted in closer spacing in the border in an area of 0.4–1.0 ha. Integrated nutrient management and eco-friendly farming practices lead to socio-economic upliftment of the local community stakeholders which forms the basis of such programme (Chavan et al., 2015). Moreover, the value addition of the products would further enhance income and maintain the livelihood security of poor farmers (Malik et al., 2013). Therefore, horticulture-based production system assures the improvement in the socioeconomic condition of poor farmers in resource deficient area.

3.11 AGROFORESTRY AND HORTICULTURE-BASED LAND-USE ECONOMY

From an economic standpoint agroforestry and horticulture seems to be a very positive approach for large land holding farmers. Under such scheme crop production is taking place along with fruits which gives a considerable amount of secondary income. Such positivity in terms of economic return has gained interest for the farmers to go for agri-horticulture practice on a priority basis (Bijalwan, 2012). As per Kassa (2015), some factors such as family labor, farming experience, total land holding, and gross annual income affect the practice of fruit tree-based agroforestry. In the other side, Dhara and Sharma (2015) have analyzed the economic return from Different Mango + *E. tereticornis*-based Agroforestry model. The income from different intercrops and fruit and gross income under each AFS model during the study for two consecutive years are presented in Table 3.5. The results showed that the integration of cropping with components of tree and fruit gave higher gross income than tree and fruit tree alone. The model T4: *E. tereticornis* + Mango + Lady's finger followed by Mustard was found to be superior with maximum gross income of Rs. 1.682 and 2.237 lakh ha^{-1} during the 4th and 5th year of the ongoing

TABLE 3.5　Effect of Different Agroforestry System on Economics (Rs. Lakh ha^{-1} year^{-1}) (Adapted from Dhara and Sharma, 2015)

Mango-based AFS model	2011–12					2012–13				
	Tree*	Fruit	Kharif crops	Rabi crop	Total	Tree*	Fruit	Kharif crops	Rabi crop	Total
T1	0.518	0.182	0.625	-	1.324	0.626	0.485	0.615	-	1.725
T2	0.406	0.170	0.520	0.248	1.343	0.550	0.470	0.536	0.252	1.808
T3	0.346	0.168	0.588	0.240	1.342	0.515	0.422	0.612	0.248	1.796
T4	0.309	0.174	0.968	0.232	1.682	0.477	0.479	1.038	0.244	2.237
T5	0.297	0.173	0.565	0.248	1.282	0.446	0.426	0.584	0.252	1.708
T6	0.137	0.098	-	-	0.235	0.140	0.171	-	-	0.311
T7	0.115	-	-	-	0.115	0.105	-	-	-	0.105
T8	-	0.095	-	-	0.095	0.000	0.108	-	-	0.108

Treatments comprised T1: *E. tereticornis* + Mango + Pigeon pea; T2: *E. tereticornis*+ Mango+Black gram followed by Mustard; T3: *E.tereticornis* + Mango + Bottle gourd followed by Mustard; T4: *E. tereticornis* + Mango + Lady's finger followed by Mustard; T5: *E.tereticornis* + Mango + Maize followed by Mustard; T6: *E. tereticornis* + Mango; T7: *E. Tereticornis*; T8 = Mango.

experiment, followed by model T2: *E. tereticornis*+ Mango+ Black gram followed by Mustard Rs. 1.808 and 1.343 lakh ha^{-1}. This was followed by model T3, T1, T5 and T6 which ranked third, fourth, fifth and sixth position in both years, respectively. The model T8 gave the minimum gross income of Rs. 0.095 lakh ha^{-1} among all models during 2011–12, whereas, T7 generated minimum gross return during 2012–13. Because the production of mango was low in the first year, Prasad et al. (2010) stated that the net return and benefit/cost ratio of intercropping in eucalyptus-based agroforestry systems were significantly higher than for sole tree system. Similar findings were also reported in the case of eucalyptus and poplar, respectively (Singh et al., 1997; Dube et al., 2003). Therefore, income is prime factor that positively affects the practice of and incorporation of fruit-based agroforestry systems by farmers (Kham and Thuy, 1999) and higher income of farmer results maximum probability of adopting this fruit-based AF system. Similarly, Bhatt and Misra (2003) has conducted research on economic analysis in agri-horticulture AFS in North-east India and found the average monetary input was Rs. 25,440 (US$ 553.00), Rs. 24,650 (US$ 536.00) and Rs. 21,440 (US$ 466.00) to guava block, Assam lemon block and control plots, respectively. The net monetary benefit was significantly higher in tree-based systems than the control. Guava-based system significantly reflects positive economic output in comparison to a lemon-based system in Assam. Overall, the net monetary benefit was Rs. 20,610/ha (US$ 448.00) to guava-based AFS, followed by Assam lemon-based AFS (Rs. 13,787.60/ha or US$ 300.00), irrespective of rice varieties. Thus the guava and Assam lemon-based agri-horticultural AFS exhibited, respectively, 2.96 and 1.98 fold higher net returns as compared to control. In Uttarakhand Lal et al. (2014) studied different Litchi-based agri-horticultural models comprised litchi with cowpea–toria, sesame–toria, pigeon pea, black gram–toria, okra–toria, turmeric and colocasia and found all the litchi-based models performed well and indicated that the practice was economically viable and profitable in respect of Net Present Value (NPV), BCR (Benefit-Cost Ratio) and Pay Back Periods (PBP). Similarly, all these models have shown higher NPV, BCR and less PBP comparative to sole litchi plantation. On the whole, intercropping with litchi is successful up to 15 years with suitable crops. The intercropped received proper light under litchi due to slow growing behavior, and thus intercrops provide additional returns, contribute to reduce PBP and also improve soil health on degraded lands.

3.12 ENTREPRENEURSHIP

Horticulture-based agroforestry system plays an important role in the establishment of small cottage industries in the village for making pickle, jam, jelly, and squash. This will generate employment opportunities and create an income source for poor farmers and villagers. Some industry use horticulture crop like tomato and chili for making sauce and pickles in a lean period of under irrigated conditions. The dried and fermented pulp of jujube is used for the production of cakes in both countries of western Sudan and Zambia (Dalziel, 1937; Kalikiti, 1998). Therefore, this programme of generating newer products in the village can surely develop entrepreneurship and create additional employment. In arid region, tree of *Ziziphus mauritiana* commonly known as jujube trees, utilized as the host plant for the both insects of *Kerria lacca* and *K. sindica* (Li and Hu, 1994) considered the best plant for rearing lac insects (Hussain and Khan, 1962) and their leaves contain high protein (Dalziel, 1937; Dastur, 1952; Ngwa et al., 2000; Arndt, 2001). Likewise, flowers of *Ziziphus jujube* are the important source of nectar for honeybees in India and Queensland (Dash et al., 1992) and attract honeybees in Pakistan (Fatima and Ramanujam, 1989; Chemas and Gray, 1991; Muzaffar, 1998).

3.13 ENVIRONMENTAL BENEFIT

Horticulture-based agroforestry system contributes various advantages in the context of the better environment such as improvement in soil physicochemical properties for examples better soil health through good soil aeration and porosity, moisture holding capacity, reclamation of degraded land and barrier to soil and wind erosion. Intercropping of turmeric with papaya; guar, pearl millet or cowpea with peach; *Chrysopogon fulvus* or Napier grass with aonla or ber (Annon, 2010); turmeric with kinnow or peach; and cowpea, black gram, and toria with mango and litchi (Dhyani, 2011) have been identified as economically viable for conserving the natural soil resources.

Microclimate amelioration is another benefit that has been seen in horticulture-based production system. For example, shade effect plays an amazing effect in the dropping of excess soil temperature and helps to improve soil moisture content that results in better growth and productivity of intercrop. In Sahelian climate *Z. mauritiana* shows significant promise in soil conservation and protection due to their fibrous rooting architecture

(Depommier, 1988; Arndt, 2001). Some hardy and deep-rooted fruit crops such as mango (*Mangifera indica*), ber (*Ziziphus mauritiana*) and guava (*Psidium guajava*) are major components of horti-based agroforestry systems and recommend for rehabilitation of degraded wasteland. Likewise, agroforestry and horticulture-based agroforestry techniques could enable the rehabilitation and economic exploitation of such salt affected wastelands. As per Sharma (2009) native jharber shrubs (*Ziziphus nummularia*) were budded with improved cultivars (Umran and Banarasi Kadaka) of ber. The system proved to be a successful low input hortipastoral system for degraded lands. Therefore, agroforestry, agri-horticultural and agripastoral systems have the potential to reduce erosion, runoff and to maintain soil organic matter, improve soil physical properties, augment nitrogen fixation and promote efficient nutrient cycling (Nair, 1984).

3.14 CARBON SEQUESTRATION AND CLIMATE CHANGE MITIGATION

Agroforestry systems which may possibly help by buffering crops from the effects of temperature and precipitation variation, agro-ecosystem to adapt to greater climate variability as well as mitigate greenhouse gases through carbon sequestration and storage (Verchot et al., 2007; Chauhan et al., 2012; Nair and Nair, 2014). It plays a great role in overall biomass productivity, enhances carbon sequestration, maintains soil health and fertility, check soil erosion, improve nutrient cycling and potentially works as bio-drainage and provides bioenergy and biofuel (Fanish and Priya, 2013). Out of all valuable contribution, carbon sequestration is a natural or artificial process by which carbon dioxide is removed from the atmosphere and stored in woody vegetation as biomass. Besides wood biomass production and farmers income generation, agroforestry practices are a viable option for climate change mitigation through sequestration of carbon in vegetation and soil, produces (Sudha et al., 2007; Parihaar, 2016). Moreover, overall biomass productivity increased from 2 to 10 t ha^{-1}y^{-1} in rainfed, arid and semiarid regions through the adoption of agroforestry practices (Dhyani et al., 2013).

Although, carbon sequestration capacity under agroforestry systems generally depends upon the nature of tree species and their associated crops with their growing condition and management practices. Therefore, carbon stock (Mg ha) in the different agrihortisilviculture model in Rajasthan is depicted in Table 3.6 (Singh and Singh, 2015). In this table, carbon stock in tree biomass and soil organic carbon is range from 1.47 Mg ha^{-1} and 3.60 Mg ha^{-1} (*Cordia myxa* + *P. cineraria*) to 0.05 Mg ha^{-1} and 0.10 mg ha^{-1} (Sole *A. excelsa*). This concludes that *Cordia myxa* + *P. cineraria-based* agroforestry has a greater potential of carbon storage and sequestration in both soil and biomass that result in total maximum carbon value of 5.07 Mg ha^{-1}. Similarly, Yadav et al. (2015) reported the carbon dynamics of the agri-horticulture ecosystem. According to this, aboveground biomass, carbon stock, and carbon stock equivalent CO_2 ranged from 10.8 to 37.8 Mg ha^{-1}, 4.8 to 17.0 Mg ha^{-1}, 17.6 to 62.3 Mg ha^{-1}, respectively. The biomass, carbon stock, and carbon stock equivalent CO_2 were recorded significantly (<0.05) higher (37.8, 17, and 62.3 Mg ha^{-1}) under the pear + wheat and the lowest was observed in wheat mono-cropping. Also, the more biomass, carbon and CO_2 buildup were reported for pear and wheat (12.0, 5.3, 19.6 Mg ha^{-1} annum^{-1}) after apricot and wheat (11.5, 5.2, and 18.9 Mg ha^{-1} annum^{-1}) and varied with diverse fruit tree species. Therefore, fruit tree biomass showed a significant and positive relationship with total biomass, total carbon, and total CO_2 mitigation. This suggests the benefits of agroforestry in carbon mitigation through storage and sequestration which helps to mitigate global climate change issue and furthermore study should be done in this direction.

3.15 AGROFORESTRY AND HORTICULTURE: ECOLOGICAL PERSPECTIVES

Agroforestry is known as good for maintenance of healthy agroecological functions because of this farming system can play a diverse role in production and potential and contains mature tree which sequesters carbon from the atmosphere and stored as biomass like a natural forest. Agroforestry is ecologically more dynamic and natural resources management-based systems which play a viable role in diversification through the incorporation of the tree in farm and sustain production that leads to maximizing socioeconomic and environmental benefits (Leakey, 1996). Likewise, the integration of fruit trees in agroforestry is socio-economically viable and also ecologically acceptable. There are normally both ecological and economic interactions between the woody (fruits trees and silvi) and non-woody components (crops, pastures, grasses, etc.) in agroforestry systems. Therefore, the diverse multistoried and multicomponent systems of agroforestry or horticulture-based agroforestry are more complex than single-species cropping systems, exhibit positive and negative interactions between different components and display wider variety in temporal and spatial ecological interactions. This promoted an increased number of scientific investigations that explored the underlying ecological principles of horticulture-based agroforestry practices. Thus, a study of ecological interactions, both above- and below-ground, became the heart of the agroforestry scientific community.

TABLE 3.6 Carbon Stock (Mg ha) in Different Agri-Horti-Silvi Agroforestry Systems in Arid Region of Rajasthan (Adapted from Singh and Singh, 2015)

Species combination under AFS	Biomass (Mg ha⁻¹)	Biomass carbon (Mg ha⁻¹)	Soil carbon (Mg ha⁻¹)	Total carbon (Mg ha⁻¹)
Ziziphus mauritiana + Prosopis cineraria	2.85	1.25	2.25	3.50
Ziziphus mauritiana + A. excelsa	1.03	0.46	2.70	3.16
Ziziphus mauritiana + C. mopane	2.54	1.16	1.35	2.51
Cordia myxa + Prosopis cineraria	3.36	1.47	3.60	5.07
Cordia myxa + A. excelsa	1.51	0.66	0.91	1.57
Cordia myxa + C. mopane	3.16	1.43	3.15	4.58
E. officinalis + Prosopis cineraria	2.55	1.12	1.80	2.92
E. officinalis + A. excelsa	0.81	0.36	2.26	2.61
E. officinalis + C. mopane	2.31	1.06	0.90	1.96
Sole *A. excelsa*	0.10	0.05	0.10	0.15
Sole *C. mopane*	0.30	0.13	0.31	0.44
Sole *Prosopis cineraria*	0.36	0.15	0.36	0.51
Sole *Cordia myxa*	0.20	0.08	0.20	0.28
Sole *Ziziphus mauritiana*	0.24	0.10	0.24	0.34
Sole crop	-	-	0.55	0.55

3.16 FUTURE RESEARCH AGENDA RELATED TO AGROFORESTRY AND HORTICULTURE

Horticulture-based agroforestry options for the development of salt-affected lands and other degraded lands found in various agroclimatic zones of India are needed to be explored. Crop rotation practices need to be altered to obtain a higher yield and maintain the sustainability of agroecosystem. As India has a land area of diverse nature comprising of 7 soil type, therefore, each model of agroforestry needs to be practiced in each soil zone and agroclimatic zone to find out the effective different agroforestry systems under variable environmental conditions. Scientific exploration needs to be designed in the field of agroforestry aiming towards mitigating climate change and other environmental consequences. This can be achieved through proper screening of potential agroforestry species having more yield and productivity under varied agroecological conditions.

3.17   CONCLUSION

Horticulture-based agroforestry system maintains higher productivity, enhances the socio-economic status and provides livelihood security of the farmers and mitigates the burning issue of climate change through carbon sequestration. It has a great potential to protect the environment by reducing soil erosion, creating wildlife habitat and rehabilitate the degraded wasteland. Due to diversity and nutritional value of fruits, the farmer should encourage incorporating the fruit trees on their farm under horticulture-based agroforestry system. There should be a need for more research in this context to explore the benefits of horti-based agroforestry systems in the coming era.

KEYWORDS

- agri-horticulture
- agroforestry
- climate-resilient
- climatic change
- forestry
- fruit-based agroforestry

- global climate
- horticulture sectors
- mitigation
- natural resources
- tree-crop compatibility

REFERENCES

1. Ajayi, O., Akinnifesi, F., Mullila-Mitti, J., Dewolf, J., Matakala, P., & Kwesiga, F. (2005). Adoption, Profitability, Impacts, and Scaling-Up of Agroforestry

Technologies in Southern African Countries. World Agroforestry Centre, pp. 1–37. http://worldagroforestry.org/downloads/Publications/PDFS/BC08031.pdf.

2. Alavalapati, J. R. R., & Nair, P. K. R. (2001). Socioeconomic and institutional perspectives of agroforestry. In M. Palao and J. Uusivuori (eds.) *World Forest Society and Environment: Markets and Policies*. Kluwer Academic Publishers, Dordrecht, pp. 71–81.

3. Anonymous (2010). Annual Report, CSWCRTI, Dehradun, Uttarakhand, India.

4. Arndt, S. K. (2001). http://chemsrv0.pph.univie.ac.at/ska/ziplant.htm

5. Ashour, N. I., Saad, A. O. M., Tayel, M. Y., & Kabesh, M. O. (1997). Lentil and wheat production in peach orchards under rainfed marginal conditions of North Sinai. *Egypt J. Soils. Sci., 37,* 549–565.

6. Bellow, G., Hudson, R. F., & Nair, P. K. R. (2008). Adoption potential of fruit-tree-based agroforestry on small farms in the subtropical highlands. *Agroforestry System, 73,* 23–36.

7. Bhandari, D. C., Meghwal, P. R., & Lodha, S. (2014). Horticulture-Based Production Systems in Indian Arid Regions. *Sustainable Horticultural Systems*, Sustainable Development and Biodiversity *2,* pp. 19–49, D. Nandwani (Ed.), ISBN 978-3-319-06903-6, © Springer International Publishing Switzerland, DOI 10.1007/978-3-319-06904-3_2.

8. Bhatt, B. P., & Misra, L. K. (2003). Production Potential and Cost-Benefit Analysis of Agri-horticulture. Agroforestry Systems in Northeast India. *Journal of Sustainable Agriculture, 22*(2), 99–108.

9. Bhatt, B. P., Singh, R., Misra, L. K., Tomar, J. M. S., Singh, M., Chauhan, D. S., Dhyani, S. K., Singh, K. A., Dhiman, K. R., & Datta, M. (2001). Agroforestry research and practices: An overview. In: *Steps towards Modernization of Agriculture in NEH Region* (eds. N. D. Verma and B. P. Bhatt), pp. 365 392, ICAR Publication, Meghalaya, India.

10. Bijalwan, A. (2012). Structure, Composition, and Diversity of Horticulture Trees and Agricultural Crops Productivity under Traditional Agri-Horticulture System in Mid Hill Situation of Garhwal Himalaya, India. *American Journal of Plant Sciences, 3,* 480–488.

11. Biswas, S., Ghosal, S. K., Sahoo, S. S., & Mukherjee, D. (2003). Some soil properties under agroforestry in Gangetic alluvial tract of West Bengal. *Environment and Ecology, 21,* 562–567.

12. Burford, J. R., & Virmani, S. M. (1983). Improved farming systems for vertisols in semiarid tropics. Annual report, Farming systems research ICRISAT, Hyderabad, pp. 17–21.

13. Chadha, K. L. (2002). Diversification to horticulture for food, nutrition, and economic security. *Indian J Hortc, 59,* 209–229.

14. Chauhan, S. K., Sharma, R., Sharma, S. C., Gupta, N., & Ritu (2012). Evaluation of Poplar (*Populus deltoides* Bartr. Ex Marsh) Boundary plantation-based Agri-Silvicultural System for Wheat-Paddy Yield and C storage. *Int. J Agric., & For.,* 2(5), 239–246.

15. Chavan, S. B., Keerthika, A., Dhyani, S. K., Handa, A. K., Ram Newaj and Rajarajan, K. (2015). National Agroforestry Policy in India: a low hanging fruit. *Current Science, 108*(10), 1826–1834. www.baifwadi.org/

16. Chemas, A., & Gray, V. R. (1991). Apiculture and management of associated vegetation of the Maya of Taxicacaltuyub, Yucatan, Mexico. In: Iqbal I, Hamayun M (eds.) Studies on the traditional uses of plants of Malam Jabba Valley, District Swat, Pakistan. *Agroforest Syst., 13*(1), 13–26.

17. Chundawat, B. S. (1993). Intercropping in orchards. *Adv Hortic, 2,* 763–775.

18. Dalziel, J. M. (1937). The useful plants of west tropical Africa. Crown Agents. 2nd Reprint, London 1955.

19. Dash, A. K., Nayak, B. K., & Dash, M. C. (1992). The effect of different food plants on cocoon crop performance in the Indian Tasar silkworm, *Autheraea Mylitta* Drury (Lepidoptera: Saturniidae). *J Res Lepidoptera., 31*(1–2), 127–131.

20. Dastur, J. F. (1952). Useful plants of India and Pakistan. A popular handbook of trees and plants of industrial, economic and commercial utility. D. B. Tarporewala Sons & Co, Bombay, pp. 252.

21. Depommier, D. (1988). *Ziziphus mauritiana*: cultivation and use in Kapsiki country, northern Cameroon. *Bois et Forêts des Tropiques* No. *218,* 57–62, CTFT-CIRAD, ICRAF, Nairobi, Kenya.

22. Dhara, P. K., & Sharma, B. (2015). Evaluation of Mango-Based Agroforestry is an Ideal Model for Sustainable Agriculture in Red & Laterite Soil. *Journal of Pure and Applied Microbiology, 9*(2), 265–272.

23. Dhyani, S. K. (2011). Alternative land use and agroforestry system for resource conservation and enhanced productivity in hills. Proceeding of workshop on mountain agriculture in Himalayan region: Status, constraints, and potentials, April 2–3, 2011, pp. 107–132.

24. Dhyani, S. K., & Chauhan, D. S. (1995). Agroforestry interventions for sustained productivity in northeastern region of India. *J. Range Management & Agroforestry, 16*(1), 79–85.

25. Dhyani, S. K., Handa, A. K., & Uma. (2013). Area under agroforestry in India: An assessment for present status and future perspective. *Indian Journal of Agroforestry,* 15 (1), 1–11.

26. Dube, F., Couto, L., Silva, M. L., Leite, H. G., Garcia, R., & Araujo, G. A. A. (2003). A simulation model for evaluating technical and economic aspects of an industrial eucalyptus-based agroforestry system in Minas Gerais, Brazil. *Agroforestry Systems, 55,* 73–80.

27. Fanish, S. A., & Priya, R. S. (2013). Review on benefits of agroforestry system. *International Journal of Education and Research, 1*(1), 1–12.

28. Fatima, K., & Ramanujam, C. G. K. (1989). Pollen analysis of two multi-floral honeys from Hyderabad, India. *J Phytol Res, 2*(2), 167–172.

29. FSI (2013). India State of Forest Report 2013. Forest Survey of India, (Ministry of Environment & Forests), Dehradun, India.

30. Ganguli, B. N., Kaul, R. N., & Nambiar, T. N. (1964). Preliminary studies on a few top feed species. *Ann Arid Zone, 3*(2), 33–37.

31. Hassan, M. M., & Mazumdar, A. H. (1990). An exploratory survey of trees on homestead and wasteland of Bangladesh. ADAB News, pp. 26–32.

32. Hiwale, S. S. (2004). Technical bulletin on "Develop sustainable Agri-Horti production system under rainfed conditions on marginal lands, pp. 1–60.

33. Hussain, A. K. M., & Khan, T. A. (1962). Lac cultivation in East Pakistan. *Pak J Forest.*, *12*(3), 220–222.

34. Jhariya, M. K., Bargali, S. S., & Raj, A. (2015). Possibilities and perspectives of agroforestry in Chhattisgarh, Precious Forests–Precious Earth, pp. 237–257, Dr. Miodrag Zlatic (Ed.), ISBN:978–953–51–2175–6, In-Tech, DOI: 10.5772/60841.

35. Kalikiti, F. (1998). *Ziziphus mauritiana* in Siavonga district, Zambia. In: International Workshop on *Ziziphus mauritiana*, Harare, Zimbabwe, 13–16 July 1998.

36. Kareemulla, K., Rizvi, R. H., Singh, R., & Dwivedi, R. P. (2002). Trees in rainfed agro-ecosystem: A socio-economic investigation in Bundelkhand region. *Indian Journal of Agroforestry*, *4*(1), 53–56.

37. Kassa, G. (2015). Profitability analysis and determinants of fruit tree-based agroforestry system in Wondo District, Ethiopia. *African Journal of Agricultural Research*, *10*(11), 1273–1280.

38. Kham, T., & Thuy, L. (1999). An economic analysis of agroforestry systems in Central Vietnam. In: Francisco F, Glover D (eds.), Economy and Environment case studies in Vietnam.

39. Korwar, G. R., Osman, Mohd., Tomar, D. S., & Singh, R. P. (1988). *Dryland Horticulture*, CRIDA Hyderabad, 34 pp.

40. Krishnamurthy, S. (1959). Effect of intercrops on the citrus health. *Indian J Hortic*, *16*(4), 221–227.

41. Kumar, B. M. (2006). Carbon sequestration potential of Tropical homegardens. In: Tropical Homegardens: A time tested example of sustainable agroforestry. Advance in Agroforestry 3. Edited by Kumar, B. M., & Nair, P. K. R. (eds.), Springer, Dordrecht, the Netherlands, pp. 185–204.

42. Kumar, S., Satyapria, Singh, H. V., & Singh, K. A. (2009). Horti-pasture for nutritional security and economic stability in rainfed area. *Progressive Horticulture*, *41*(2), 187–195.

43. Kumar, S., Shukla, A. K., Satyapria, Singh, H. V., & Singh, K. A. (2011). Horti-pasture for nutritional security and economic stability in rainfed area. *Paper presented in National Conference on Horti Business Linking farmers with Market held at Dehradun* from 28–31 May 2011 pp. 18–19.

44. Lal, H., Sharma, N. K., Mehta, H., Jayaprakash, J., & Chaturvedi, O. P. (2014). Livelihood security through Litchi (*Litchi chinensis*)-based agri-horticultural models for resource-poor communities of Indian Sub-Himalayan. *Current Science, 106*(11), 1481–1484.

45. Leakey, R. (1996). Definition of Agroforestry Revisited. *Agroforestry Today, 8*, 1.

46. Leakey, R. (2010). Should we be growing more trees on farms to enhance the sustainability of agriculture and increase resilience to climate change? Special Report, ISTF News, USA.

47. Li, J., & Hu, X. (1994). Rejuvenation and utilization of wild *Ziziphus mauritiana*. *Forest Res.*, *7*(2), 224–226.

48. Malik, S. K., Bhandari, D. C., Kumar, S., & Dhariwal, O. P. (2013). Conservation of multipurpose tree species to ensure ecosystem sustainability and farmers livelihood in Indian arid zone. In: Nautiyal, S., Rao, K. S., Harald, K., Raju, K. V., Ruediger, S. (eds) Knowledge systems of societies for adaptation and mitigation of impacts

of climate change, Environment Science and Engineering. doi:10.1007/978-3-642-36143-2-16. Springer, Berlin.

49. Murthy, R. S., Bhattacharjee, T. C., Lande, R. J., & Pofali, R. M. (1982). Distribution, characterization, and classification of vertisol. In: 2th International Congress of Soil Science, New Delhi, 8–16 February 1982, pp. 3–22.

50. Mutanal, S. M., Nadagoudar, B. S., & Patil, S. J. (2001). Economic evaluation of an Agroforestry system in hill zone of Karnataka. *Ind J Agric Sci*, *71*(3), 163–165.

51. Muzaffar, N. (1998). Beekeeping in Islamabad area. Islamabad Horticultural Society 16th Annual Spring flowers, fruit and vegetable show. In: Iqbal L, Hamayun M (eds) Studies on the traditional uses of plants of Malam Jabba valley. district Swat, Pakistan pp. 44

52. Naik, K. C. (1963). South Indian fruits and their culture. P Verdachari & Co, Madras.

53. Nair, P. K. R. (1984). Role of trees in soil productivity and conservation. Soil productivity aspects of agro-forestry. The International Council for Research in Agro-Forestry, Nairobi, p. 85.

54. Nair, P. K. R., & Nair, V. D. (2003). Carbon storage in North American Agroforestry systems. In: The potential of U.S. forest soils to sequester carbon and mitigate the greenhouse effect, Boca Raton, FL. Edited by Kimble, J., Heath, L. S., Birdsey, R. A., & Lal, R. (eds), CRC Press LLC. pp. 333–346.

55. Nair, P. K. R., & Nair, V. D. (2014). Solid-fluid-gas: the state of knowledge on carbon sequestration potential of agroforestry systems in Africa. *Current Opinion in Environ. Sustain.*, *6*, 22–27.

56. Nair, P. K. R., Kumar, B. M., & Nair, V. D. (2009). Agroforestry as a strategy for carbon sequestration. *J. Plant Nutr. Soil Sci.*, *172*, 10–23.

57. Nair, P. K. R., Vimala, D. N., Kumar, B. M., & Showalter, J. M. (2011). Carbon sequestration in agroforestry systems. *Advances in Agronomy*, *108*, 237–307.

58. Ngwa, A. T., Pane, D. K., & Mafeni, J. M. (2000). Feed selection and dietary preferences of forage by small ruminants grazing natural pastures in the Sahelian zone of Cameroon. *Anim Feed Sci Technol*, *88*, 253–266.

59. NRCAF (2007). Perspective plan: Vision 2025, National Research Centre for Agroforestry (NCRAF), Jhansi, India.

60. Osman, M. (2003). *Alternate land use systems for sustainable production in rainfed areas*. In: Pathak, P.S., & Newaj, R. (eds) Agroforestry-Potential and Opportunities. Agrobios, Jodhpur, pp. 177–181.

61. Padmavathy, A., & Poyyamoli, G. (2013). Role of agro-forestry on organic and conventional farmers' livelihood in Bahour, Pondicherry, India. *Int. J. Agric. Sci.*, *2*(12), 400–409.

62. Pant, K. S., Yewale, A. G., & Prakash, P. (2014). *Fruit Trees-based Agroforestry Systems*. In: Agroforestry Theory and Practices. Edited by Raj, A. J., & Lal, S. B., Scientific Publishers, India, 564–588.

63. Parihaar, R. S. (2016). Carbon Stock and Carbon Sequestration Potential of different Land-use Systems in Hills and Bhabhar belt of Kumaun Himalaya. PhD. Thesis, Kumaun University, Nainital, pp. 350.

64. Prasad, J. V. N. S., Korwar, G. R., Rao, K. V., Mandal, U. K., Rao, C. A. R., Rao, G. R., Ramakrishna, Y. S., Venkateswarlu, B., Rao, S. N., Kulkarni, H. D., & Rao M. R. (2010). Tree row spacing affected agronomic and economic performance of

Eucalyptus-based agroforestry in Andhra Pradesh, Southern India. *Agroforestry Systems, 78,* 253–267.

65. Raj, A., & Jhariya, M. K. (2016). Joint forest management: A program to conserve forest and environment. *Van Sangyan, 3*(6), 38–42.

66. Raj, A., Jhariya, M. K., & Bargali, S. S. (2016b). Bund-Based Agroforestry Using *Eucalyptus* Species: A Review. *Current Agriculture Research Journal, 4*(2), 148–158.

67. Raj, A., Jhariya, M. K., & Pithoura, F. (2014). Need of agroforestry and impact on ecosystem. *Journal of Plant Development Sciences, 6*(4), 577–581.

68. Raj, A., Jhariya, M. K., & Toppo, P. (2016a). Scope and potential of agroforestry in Chhattisgarh state, India. *Van Sangyan, 3*(2), 12–17.

69. Rizvi, R. H., Dhyani, S. K., Ram Newaj, Karmakar, P. S., & Saxena, A. (2014). Mapping agroforestry area in India through remote sensing and preliminary estimates. *Indian Farm., 63*(11), 62–64.

70. Roy, A. (2011). "Requirement of vegetables and fruit" The Daily Star (A English Newspaper). 24/03/2011.

71. Samara, J. S. (2010). Horticulture opportunities in rainfed areas. *Indian J. Hort., 67*(1), 1–7.

72. Sharma, S. K. (2009). Development of fruit tree-based agroforestry systems for degraded lands—a review. *Range Management and Agroforestry, 30*(2), 98–103.

73. Sharma, S. K., & Vashishtha, B. B. (1985). Evaluation of jujube-buffelgrass hortipastoral system under arid environment. *Ann Arid Zone, 24*(3), 303–309.

74. Singh, B., & Singh, G. (2015). Biomass Production and Carbon Stock in a Silvi-Horti-based Agroforestry System in Arid Region of Rajasthan. *Indian Forester, 141*(12), 1237–1243.

75. Singh, B., Bishnoi, M., Baloch, M. R., & Singh, G. (2014). Tree biomass, resource use, and crop productivity in agri-horti-silvicultural systems in the dry region of Rajasthan, India. *Archives of Agronomy and Soil Science, 60*(8), 1031–1049.

76. Singh, G., Singh, N. T., Dagar, J. C., Singh, S., & Sharma, V. P. (1997). An evaluation of agriculture, forestry and agroforestry practices in a moderately alkali soil in northwestern India. *Agroforestry Systems, 37,* 279–295.

77. Singh, H. P., & Malhotra, S. K. (2011). Horticulture for food, nutrition, and healthcare: a new paradigm. *Indian Horticulture, 56*(2), 3–11.

78. Singh, N. R., & Jhariya, M. K. (2016). Agroforestry and agrihorticulture for higher income and resource conservation. pp. 125–145. In: Innovative Technology for Sustainable Agriculture Development, Edited by Sarju Narain and Sudhir Kumar Rawat (Ed.). ISBN: 978-81-7622-375-1. Biotech Books, New Delhi, India.

79. Singh, N. R., Jhariya, M. K., & Raj, A. (2013). Tree crop interaction in agroforestry system. *Readers Shelf, 10*(3), 15–16.

80. Singh, R. S. (1997). Note on the effect of intercropping on growth and yield of ber (*Z. mauritiana*) in semi-arid region. *Curr Agric, 21*(1–2), 117–118.

81. Sudha, P., Ramprasad, V., Nagendra, M. D. V., Kulkarni, H. D., & Ravindranath, N. H. (2007). Development of an agroforestry carbon sequestration project in Khammam district, India. *Mitigation and Adaptation Strategies for Climate Change, 12,* 1131–1152.

82. Toppo, P., Raj, A., & Jhariya, M. K. (2016). Agroforestry systems practiced in Dhamtari district of Chhattisgarh, India. *Journal of Applied and Natural Science, 8*(4), 1850–1854.

83. Tripathi, P., Kashyap, S. D., Shah, S., & Pala, N. A. (2017). Effect of organic manure on soil physicochemical properties under fruit-based agroforestry system. *Indian Forester, 143*(1), 48–55.

84. Uddin, H., & Chowhan, S. (2016). Performance of Indian spinach and papaya in litchi-based agroforestry system. *Int. J. Agril. Res. Innov. & Tech.*, *6*(1), 34–40.

85. Vashishtha, B. B., & Prasad, R. N. (1997). Horti-pastoral systems for arid zone. In: Yadav, M. S., Singh, M., Sharma, S. K., Tewari, J. C., Burman, U. (eds) Silvipastoral systems in arid and semi-arid ecosystems. CAZRI, Jodhpur, pp. 277–288.

86. Verchot, L. V., Noordwijk, M. V., Kandji, S., Tomich, T., Ong, C., Albrecht, A., Mackensen, J., Bantilan, C., Anupama, K. V., & Palm, C. (2007). Climate change: linking adaptation and mitigation through agroforestry. *Mitg. Adapt. Strat. Glob. Change, 12,* 901–918.

87. Yadav, R. P., Bisht., J. K., & Pandey, B. M. (2015). Above ground biomass and carbon stock of fruit tree-based land use systems in Indian Himalaya. *The Ecoscan, 9*(3&4), 779–783.

CHAPTER 4

Role of Agroforestry in Carbon Sequestration: An Indian Perspective

VISHNU K. SOLANKI[1] and AMIT SEN[2]

[1]*Assistant Professor, Department of Agroforestry, College of Agriculture, Ganjbasoda, JNKVV, Jabalpur, Madhya Pradesh, India, Mobile: +91-9509536620, E-mail: drvishnu@hotmail.com*

[2]*Assistant Professor, Department of Biotechnology, Mewar University, Chattisgarh, Rajasthan, India, Mobile: +91-7357770161, E-mail: amitsenbhl28@gmail.com*

ABSTRACT

Agroforestry is an alternative land use practice which supports the diversification of crop, conservation of natural resources, efficient functioning of the ecological system and CO_2 assimilation in plant and soil ecosystem. Conjoint cultivation of tree and crop makes the land more productive with profitable revenue generation and also increases the green cover for better carbon sequestration. By deliberately integrating trees with the crop production system, agroforestry systems offer a promising avenue for carbon storage. Trees that provide multiple products and services, the proportion of total photosynthesis that is removed from the system is minimized, thereby facilitating storage of more carbon in the agroforestry system as compared to monocropping systems. Carbon sequestration has been suggested as a means to help mitigate the increase in atmospheric carbon dioxide concentration. Agroforestry systems can better sequester carbon in soil and biomass and help to improve soil conditions. The soil organic carbon and net carbon sequestered were greater in the agroforestry system.

4.1   INTRODUCTION

An integrated system of agroforestry helps to maintain the natural resource base, proper management towards socio-economic and environmental output through efficient land use practices. To optimize the profit from the biological relation formed when trees and any other crops are deliberately shared with crops and livestock in a concentrated land management system (Nair, 1993). At present time agroforestry is an integrated land management system. Many different agroforestry systems are found under the tropics, partially because of their favorable climate situation and partially because of the socioeconomic reason such as higher human population, more labor availability and smaller size of land-holding, complex land tenure and nearer closeness to markets (Nair, 2007; Nair et al., 2008). The practices of growing forest trees and other crops in interacts the mixture of sustainable land use. Higher capture and higher utilization of the light, water and nutrient in agroforestry system as compared to monocropping so that agroforestry system is more capable to carbon sequestration. Under various agroforestry systems, it has been observed that higher carbon accumulation takes place in deep soil surface surrounding the rhizospheric tree zone in comparison to open area without trees. The advantage of agroforestry systems toward mitigating climate change over the oceanic components in terms of secondary benefits that help to maintain the daily livelihood, food security in developing countries. Further such approaches also sustain biodiversity, maintains forest corridor, help to keep the balance of hydrological features of the watershed and above all conservation of soil resource. Simultaneously agroforestry also reduces the pressure over forest due to demand for wood. Encourage woodcarving industry to ease long-term looking-up of carbon in carved wood and latest sequestration through intensifying plantation.

4.2   CARBON SEQUESTRATION

Carbon sequestration has been recommended as a way to assist mitigate enhance carbon dioxide concentration in atmospheric. Carbon dioxide concentration and other greenhouse gases (GHGs) in the atmosphere have been extremely raised since the industrial revolution. The main causes credited to global warming and related climatic changes are improved concentration of greenhouse gases in the atmosphere. Present approaches for handling with global warming consist of reducing fossil fuel burning as well as restriction emission of other greenhouse gases and rising carbon sequestration. Atmospheric carbon can be sequestered in long-lasting carbon pools of plant biomass both above ground and below ground or unmanageable organic and inorganic carbon in soils and deeper subsurface atmospheres. Apart from counterbalance CO_2 emissions and global warming sequestration of carbon in soils also assist to develop soil value and yield by improving several physical, chemical, and biological characteristics of soils like infiltration rate, aeration, bulk density, nutrient availability, cation exchange capacity, buffer capacity, etc. Soil organic carbon sequestration is more significant in arid regions where soils are essentially small in organic carbon content. In arid atmosphere trees, pastures and agroforestry systems are essential for carbon sequestration approaches. Systems relating trees act as carbon sinks due to their capability to sequester atmospheric carbon in deep soil profiles and different tree components.

4.3 FOREST, AGROFORESTRY, AND CARBON SEQUESTRATION

To decrease CO_2 emissions and increasing carbon sinks by means of forestry is a standard way. The role of forest in the carbon cycle is well accepted and is a great sink of carbon. Agroforestry, i.e., agriculture along with forestry, offers many objectives *viz.* control the CO_2 concentration, reduce the CO_2 emissions, climate change adaptation, mitigation, etc. For dealing with the unfavorable impact of climate change, it has the potential to increase the flexibility of the system. The purpose of the agroforestry system also include significant chances for generating synergies among both adaptation and mitigation strategies, i.e., it delivers important mitigation selections including correct management that influence the quantity of carbon sequestered. Agroforestry as a diverse land use helps to promote carbon sink. To increase the carbon storage potential of tree species in land ecosystem, various plantation schemes including agroforestry have been promoted. The principle behind agroforestry for incorporating tree plantation along with agricultural system is the most demanding approach for maintaining the livelihood of people and overall ecological sustainability. Higher carbon content through such alternate land use system in comparison to indigenous vegetation improves the carbon accumulation rate. Carbon storage can be increased through conversion of lesser biomass generating land use to higher biomass generating land use practices.

Agroforestry practice is not aimed towards carbon sequestration process, but research reports reveal such approaches can promote increment in the level of carbon in the above-ground biomass, soil component and underground soil ecosystem. The two beneficial roles of agroforestry system include increment in the terrestrial carbon sink and reduce GHGs level in the ambient atmosphere.

4.4 AGROFORESTRY IN INDIA

India has a long history and traditional practices for agroforestry system. In India, the agroforestry systems consist of tree plantation on farms, a range of management strategies by local peoples, ethno-forestry exercises by tribal peoples and community forestry on community land. Here the exercise of planting trees on farmlands is relatively old, and the trees used for this purpose are versatile, *e.g.*, for shade, fodder for livestock, for fuelwood, fruit, and vegetables for food purpose and medicinal uses. *Eucalyptus* and *Populus* plantation has been reported in farm boundaries of an agricultural field in Punjab and Haryana. The predominance of shifting cultivation has been reported from the northeast region. On the contrary taungya systems have been found in Uttar Pradesh, Kerala, and West Bengal along with parts of South India. Home gardens, large cardamom cultivated area, alder based agricultural system has been found in Eastern Himalayas and Northeast India, respectively (Table 4.1). The main agroforestry practices of India are found in four essential regions.

4.4.1 NORTHEAST REGION IN INDIA

Agroforestry in the form of shifting cultivation has been represented in northeast India. Shifting cultivation is also known as *Jhum* cultivation in the northeast region. Farmers form the resource base for maintaining biodiversity and socio-economic status of rural India. Under the present scenario reduction in the fallow period has led to unsustainability in agro-ecosystem coupled with forest degradation. Restoration and preservation of soil fertility and enhancement in agricultural productivity are possible by agroforestry. Some of the prevalent agroforestry system in this region includes agrohorticulture, agrisilvicultural, kitchen gardens, silvihorticulture, and Silvo-pastoral.

4.4.2 NORTHWEST REGION IN INDIA

Very little or irregular rainfall and thereby failure of the crop is a normal phenomenon observed in the states of Gujarat and Rajasthan. The potential to reduce the impact of extreme climatic situation and availability of alternate income source to the people is achievable by agroforestry exercises. Agricultural crops species along with *Prosopis cineraria* are the main prevalent agroforestry practice in this region.

4.4.3 THE WESTERN GHATS REGION IN INDIA

The Western Ghats agroforestry practice is an integrated approach of crop cultivation along with tree plantation and animal husbandry. Some farming systems dominated the area in plantation agriculture include crops likes spices and other beverage crops (coffee and tea) along with woody species and rubber, paddy-based farming, home garden/kitchen garden, and coconut.

4.4.4 SOUTHERN REGION IN INDIA

The southern part of India represents the diverse nature of agroforestry models. For example, Kerala represents kitchen garden and multistate agroforestry, plantation crops, coastal Karnataka tree and spice based agroforestry practices, petro crops in Tamil Nadu and Andhra Pradesh. In the case of Tamil Nadu farming is practiced along with intercrops such fruit trees and silk cotton tree (Table 4.1).

4.4.5 AREA UNDER AGROFORESTRY IN INDIA

Presently agroforestry covers an area of 11.54 m ha, which is 3.39% of the geographical area of the country (FSI, 2013). Maharashtra, Gujarat, and Rajasthan rank high in the state-wise area under agroforestry. National Mission for Green India aims towards the conversion of degraded lands comprising of agricultural land and fallow land (1.5 Mha) into the land area for plantation of agroforestry species by the National Climate Change Action Plan. From Indian perspective carbon storage via afforestation can be done through agroforestry species in cultivation area as it is the most

TABLE 4.1 Scenarios of Agroforestry Systems in India

Agroforestry Models Agrisilvi practice	Components (crops with woody species)	Nature of components (a–livestock, w–woody species, h–herb, f–forage)	Agro-ecological need
Advance fallow system	Trees are grown and left in fallow land	w–Legume-based short rotation practice, h–indigenous crops	In areas of shifting cultivation
Taungya method	During initial phase trees grown with agricultural crops	w–Economic tree species, h–common indigenous crops	Diverse ecological zone
Alley cropping practice or hedgerow intercropping system	Timber species are planted in hedges, alley and strip plantation of agricultural crops within hedges	w–Short duration legume with coppicing potential, h–traditional crops	Humid tropics
Multistrada trees based gardens	Multi-layered and diverse plant species	w–Tree species with variable growth and form, h–occasional crop	Fertile soil region
MPTS (Multipurpose trees species) based agroforestry	Tree plantation in the extension areas of farmlands	w–MPTS and other Orchards crops, h–traditional crops	Entire ecological zone comprising of subsistence farming along animal husbandry
Plantation crop-based system	Multi-story crops along with plantation crops beneath shade plants besides agricultural crops	w–plantation crops like beverages, orchards trees, firewood, forage crops, h–shade bearers crop	Humid, sub-humid tropics
Kitchen garden	Diversified crops species (herb, shrubs, climber and trees)	w–Orchards tree species with other woody vegetation.	Entire ecological zones
Restorative plants based practice	Extension plantation with mulching and nitrogen fixers	w–MPTs and orchard plantation, h–indigenous crops	Undulating region, & difficult sites recovery
Shelter-belts, wind-breaks, hedgerows practice	Timber species near agricultural plots	w–Woody plants, h–traditional crops	Windy region
Energy blocks	Fuelwood species with agriculture	w–Fuelwood timber species, h–Indigenous crop	All ecological zone

TABLE 4.1 (Continued)

Agroforestry Models Agrisilvi practice	Components (crops with woody species)	Nature of components (a-livestock, w-woody species, h-herb, f-forage)	Agro-ecological need
Silvipastoral models			
Trees on rangeland or pastures	Timber species planted on irregular or systematic manner	w–multipurpose timber and forage species with livestock	Prevalent grazing zones
Protein banks	Inclusion of protein-rich forage species in farm or rangelands	w–legume-based forage species with livestock	Area with high man: land ratio
Plantation crops along with pastures and livestock	Livestock with coconuts plantation in South Pacific and South-east Asia	w–plantation crops, with forage and livestock	Area with less plantation crop
Home gardens involving animals	close, multi-store and mixture of different trees and crops and animals near homesteads	w–fruit trees dominate and also other woody species, f - present	All ecological areas with high human population
MPTs hedgerows system	Palatable woody hedges, mulch and green manuring crops for conservation.	w–palatable fast-growing timber with coppicing ability, h–herbaceous crops for conservation of soil	Humid tropics, valley and steep terrain regions
Supplementary models			
Apisilvi	Woody species planted for the production of honey	w–Woody species preferably for honeybee foraging	Feasible region for apiculture
Aqua-forestry	Woody species are maintained beside fish ponds, palatable tree species by fish are preferred for food supply	w–Woody plants and shrubs ideal for fisheries	Lowlands regions
MPTs lots	Woody species are incorporated based on different principles	w–Region-specific MPTs	Wide regions

Source: Modified from Chundawat and Gautam (2006).

acceptable approach. One of the problems in preparation for capitalizing on the possible profits of agroforestry has been the lack of availability of a clear-cut approximation of the area at a particular period of time. The main complexity in estimating the area under agroforestry is the lack of corrective measures for delineating the area influenced by trees in a various stand of trees and crops. In simultaneous systems, the whole area engaged by multi-strata systems like home gardens system and shaded perennial tree systems and the intensive tree-intercropping condition can be listed as agroforestry. However, most of the agroforestry systems are relatively extensive where the components mainly trees are not planted at regular spacing or density. The problem is more hard in the case of practices like windbreaks planting and boundary plantation where although the tree species are planted at large distances between rows (windbreaks) or around agricultural (boundary planting), the influence of trees expand over larger than the simply perceivable level of areas. The problem has a diverse measurement of complexity when it comes to sequential systems like improved fallows system and shifting cultivation system. In such conditions, the valuable effect of trees and other woody vegetation on the crops that follow them is supposed to last for a changeable length of time. Approximately 1023 million hectares total area under agroforestry in the world (Nair et al., 2009). The multitude of agroforestry systems that they practice in the tropics, or temperate area is tightly grounded on physically powerful ecological values and through the condition of various basic needs and ecosystem services they supply to the ability of several regional developmental goals. Thirty years ago agroforestry began to draw the consideration of the international development and scientific community mainly as a means for satisfying agricultural production in marginal lands and remote areas of the tropics that were not profited by the Green Revolution. Today, agroforestry has been accepted as having the possibility to present much more toward making sure not only food safety in poor countries but also environmental truth in poor and rich nations equally.

4.4.6   AREA UNDER TRADITIONAL AGROFORESTRY SYSTEMS

Under humid tropical regions shifting cultivation and home gardens are the main farming practices which are plantation based cropping system. In India, Kitchen gardens have their origin in the states of Kerala and Andaman and the Nicobar Islands. Throughout India range land trees, live hedges, taungya plantation and boundary plantations are practiced. Bihar, Gujarat, Haryana, Himachal, Orissa, Uttar Pradesh and Nilgiri hills of south India represents the boundary plantation. Wind-prone areas like coastal belts represent shelterbelts, and Andhra Pradesh, Assam, Gujarat, Haryana, Karnataka, Punjab, Tamil Nadu, and Orissa states represent woodlots plantation. Arid and semiarid regions represent scattered tree plantation on farmlands. In semiarid regions, agroSilvo-pastoral exercises are found in India (Table 4.2).

4.5   BENEFITS OF AGROFORESTRY SYSTEMS

To supply both economic and environmental profits has the potential of Agroforestry systems. In maintaining agroforestry systems, a few of the main profits are discussed in the following subsections.

4.5.1   TO INCREASE CARBON STOCK

Mitigation policy towards the shifting climate and possibility to sequester carbon using several plant species and soil has an enormous potential of agroforestry systems. Accurate designing of the agroforestry can be an effective carbon sink. Active absorption of atmospheric carbon dioxide by green plants through the process of photosynthesis and subsequent storage of carbon as biomass has been defined as carbon sequestration (Baes et al., 1977). Agroforestry practice provides supplementary wood for the same products which are ruthlessly harvested from forest plantation in an unsustainable way. Agroforestry also becomes a good income source for the farming community and simultaneously can be adopted as a strategy for climate change mitigation through sequestering carbon in soil and vegetation (Sudha et al., 2007). Pruning is done within the tree component of agroforestry system in order to reduce ecological competition named species interaction. The pruned materials are used as Non-timber forest products which when returns back to soil add to the soil carbon pool (Alavalapati and Nair, 2001).

4.5.2   TO IMPROVE SOIL FERTILITY

Food security, decreasing poverty, conserving environment and sustainability for all that it is very important to increasing and maintaining soil

TABLE 4.2 Area Expansions of Traditional and Commercial Agroforestry Systems (AFS) in India

Type of species	State	Area (ha)	Species	Reference
Traditional agroforestry system				
Alder-cardamom agroforestry	North-East India	34	*Alnus nepalensis*	Srinivasa et al., (2006)
Kangayam agroforestry	Tamil Nadu	384	*Acacia leucophloea*	Kumar et al., (2011)
Homegardens	Kerala	1330	Mix tree species	Kumar et al., (2006)
Khejri-based agroforestry	Rajasthan	1586	*Prosopis cineraria*	Tewari et al., (2007)
Commercial agroforestry system				
Pulpwood agroforestry (paper)	Punjab, Haryana, Uttar Pradesh, Andhra Pradesh, Gujarat, and Tamil Nadu	657	Eucalyptus, Poplar, Casuarina, Subabul	Kulkarni (2013)
Timber-based agroforestry (furniture)	Kerala, Maharastra, Tamil Nadu, and Madhya Pradesh	1700	*Tectona grandis*	Luukkanen & Appiah (2013)
Willow-based agroforestry (bat industry)	Jammu & Kashmir, Himachal Pradesh, Uttarakhand and Punjab	137	Salix species	ICFRE (2011)

fertility. Many agroforestry land use systems, e.g., agrohorticulture, agro-pastoral system, agroSilvo-pastoral system, etc. are capable ways of restore soil organic material. More yield with enhanced crop rotation than with uninterrupted cropping. Humification process through microbial decomposition is mediated through the contribution of leaf litter from agroforestry species. Agroforestry practice contributes towards the check of soil erosion and runoff loss. This therefore significantly influences a reduction in the water loss, soil materials, organic matter content and soil nutrients leading to improvement of soil texture. Further, it also helps to combat the problem of soil salinization and soil acidification which thus helps to renovate the soil.

4.5.3 TO INCREASE INCOME

The various constituent of agroforestry offers several harvests at diverse times of the years. Agroforestry system produces firewood, forage for livestock population and helps to improve the fertility of the soil. In this way, it reduces the risk of crop failure and provides an alternate income source for farmers.

4.5.4 TO INCREASE PRODUCTIVITY

Soils under forest have the potentiality to provide more yields in comparison to normal soils. Intercropping of fodder grasses with fodder trees improves the fodder yield in comparison to mono-cropping of fodder grass. Food crops intercropping systems are more prevalent in South India and some states like Gujarat. Haryana, Punjab and Uttar Pradesh.

4.5.5 TO IMPROVE AESTHETIC VALUE

A vigorous environment can be generated by aesthetic value. Improvement in environmental conditions and farmland can be done through agroforestry practices. A small area of land involved under agroforestry systems promotes recovery of biodiversity through habitat improvement with simultaneous presence of beneficial bird and insect species which regulates the pest population. Diversity to the landscape is developing aesthetics by the use of tree biodiversity. Improved nitrogen inputs because of nitrogen-fixing trees and shrubs, reduced insect pests and connected diseases and consumption of solar energy more resourcefully as compared to mono-cultural systems, and it can also moderate microclimates.

4.6   ECOLOGICAL SUSTAINABILITY THROUGH AGROFORESTRY

The idea of agroforestry stems from the normal function of on-farm and off-farm forest tree production in supporting sustainable land use and natural resource management. Structurally and functionally added land-use systems are much more resource efficient than monocultures of the tree. For example, nutrient, light, water, etc. capture and consumption and better structural diversity that involve tighter nutrient cycle. Above and below-ground diversity exhibits more system stability and flexibility at the site level. Such systems are connected through the forest and other larger area of landscapes such as landscape and watershed stages (Nair et al., 2008). The integration of tree component into cultivation systems incorporates significant interspecific interaction and intraspecific competition for abiotic factors thereby generating a more complex agro-ecosystem (Rao et al., 1998). Both positive, e.g., improved productivity, nutrients cycling, fertility of soil and improve microclimate, etc. and negative, e.g., allelopathic effect, pest and disease affected, etc. and proper management in agroforestry systems may alter the negative and positive consequences (Gordon et al., 2009; Jose et al., 2009). Ecologically combining capability underlies the method of selecting an arrangement of species with different growth characteristics to certify better complementarity in resource utilization and supply better ecosystem services like water-quality amelioration (Menalled et al., 1998). Growth features of trees vary significantly among species, e.g., shade tolerant species versus shade intolerant species, evergreen species versus deciduous species, rapid height growth versus slow juvenile height growth, deep rooting versus shallow rooting.

4.7   CARBON SEQUESTRATION IN DIFFERENT AGROFORESTRY SYSTEMS

The concept of carbon sequestration is similar across diverse land-use systems. While the fundamental mechanisms too are related, they may obvious themselves in a different way in diverse systems depending on system-specific description. Atmospheric carbon was removed and added into reservoir pool of soil and forest for sequestering carbon. The atmospheric carbons are transferring particularly CO_2 and its safe storage in long-lived pools (UNFCCC, 2007). The long-term total carbon cycle explains that the biogeochemical cycling of carbon with surface systems consisting of oceans, the atmosphere, biosphere, and soil manage the atmospheric CO_2 concentration above geological time scales of more than 100,000 years (Berner, 2003). The short-term carbon cycle above decades and centuries is of larger significance than the long-term cycle in the forest, agroforestry systems, and agricultural ecosystems. The significant processes of this cycle are the complex of atmospheric CO_2 in plants during photosynthesis and come back of part of that carbon to the atmosphere through the plant, animal and microbial respiration as CO_2 in aerobic and CH_4 in anaerobic conditions. Vegetation fires, burning and land clearing for farming of agricultural and forestry purposes can also discharge important quantities of CO_2 and CH_4 to the atmosphere, but more of this carbon is recaptured in consequent re-growth of vegetation (Lorenz and Lal, 2010; Nair et al., 2011). Carbon pools in such terrestrial systems consist of the aboveground plant biomass, durable produce derived from biomass like timber and belowground biomass like roots, soil microorganisms and the relatively constant forms of organic and inorganic carbon in soils and deeper subsurface environments. The uptake of atmospheric CO_2 during photosynthesis and transfer of fixed carbon into vegetation, detritus and soil pools for safe storage are the main sources of carbon sequestration in the agroforestry. There are direct and indirect modes of sequestration of carbon (SSSA, 2001). Inorganic chemical reaction helps to sequester C in soil directly through the transformation of atmospheric CO_2 into carbonate salts of calcium and magnesium. Indirect mode of C sequestration takes place during photosynthetic conversion of C into plant biomass. Further decomposition of plant biomass sequesters C into the soil. Amount of carbon sequestered in a site maintains the balance of input and output of carbon in a particular ecosystem. The rate of such processes tends to be higher, and modification of such processes in the form of changing land-use and land-cover in the form of agroforestry practices may significantly influence the global C cycle and climatic features of the earth (Table 4.3).

TABLE 4.3   Categorization of Biomes and Their Carbon Sequestration Potential

Biomes	Primary method to increase C sequestration	Potential C sequestration (Pg C year-1)
Agricultural lands	Management	0.85–0.90
Biomass-based croplands	Alteration/Manipulation	0.50–0.80
Grasslands	Systematized/Management	0.50
Pastureland/Rangelands	Systematized/Management	1.20
Woodland/Forests	Systematized/Management	1–3
Deserts and degraded lands	Alteration/Manipulation	0.80–1.30

Source: Adapted from DOE (1999).

4.8   CARBON ACCUMULATION IN AGROFORESTRY SYSTEMS

Improvement of soil carbon as well as root biomass in belowground and carbon accumulates in standing biomass in aboveground both occurs of carbon sequestration in varied agroforestry systems. On the other hand on the level of process and the last use of wood, are factors that the degree of carbon sequestration from forestry performances would depend. It is clear that carbon sequestration takes place in the two most significant segments of the agroforestry ecosystem aboveground and belowground. The former into exact plant parts like stem, leaves, etc. of woody vegetation, herbs and latter into the belowground plant parts, soil biota and carbon accumulation in different horizons of soil are divided into different sub-segments. In every part changes very much form the whole amount sequestered depending on a number of causes, counting the region, the type of system, *e.g.,* the nature of components and age of perennials such as trees, site quality, and earlier land-use. The total carbon accumulates in tree-based land-use systems on average the soil, and aboveground components are estimated to hold main portions roughly 60% and 30%, respectively (Lal, 2005, 2008). Trees are planted in croplands and pastures would affect more net carbon storage in above-ground and below-ground (Haile et al., 2008; Palm et al., 2004), agroforestry systems are supposed to have an advanced possible to sequester carbon as compared to pastures or field crops (Kirby and Potvin, 2007; Roshetko et al., 2002) (Table 4.4).

TABLE 4.4 Carbon Sequestering Potential of Different Agroforestry System in the Indian Subcontinent

Different agroforestry system	Carbon sequestering potential	Zone	Citation
Silvo-pastoral practices of 5 years old	9.5–19.7 tons Carbon per hectare	Moderately arid	Rai et al., (2001)
6 years old Silvo-pastoral practices	1.5–18.5 tons Carbon per hectare	Northwest part of India	Kaur et al., (2002)
6 years old Block plantation	24.1–31.1 tons Carbon per hectare	Madhya Pradesh and Chhattisgarh	Swamy et al., (2003)
8 years old Agrisilviculture practice	4.7–13.0 tons Carbon per hectare	Dry region	Singh (2005)
Agrisilviculture system (aged 11 years)			
Eucalyptus bund plantation			
Poplar block plantation	330,510 t		
Populus deltoides 'G-48' + wheat	18.53 tC/ha	26.0 tC/ha	Semiarid region NRCAF (2005) Gera et al., (2006)
P. deltoides + wheat boundary plantation	4.66 tC/ha	59,361 t	Punjab (Rupnagar district)
Silvopasture	31.71 tC/ha	Himachal Pradesh	Verma et al., (2008)
Natural grassland	19.2 tC/ha		
Agrihorti silviculture	18.81 tC/ha		
Hortipastoral	17.16 tC/ha		
Agrisilviculture	13.37 tC/ha		
Agri-horticulture	12.28 tC/ha		

4.9   ABOVEGROUND (VEGETATION) CARBON SEQUESTRATION

Aboveground carbon storage is the integration of carbon into plant material either in the harvested produce or in the parts residual on site in a living form. The quantity of biomass and subsequently carbon that accumulate depend to a big deal separately from the character of plant itself on the properties of the soil on which it produces with a top concentration of organic matter, nutrients and fine soil structure primary to larger biomass production. The aboveground biomass that is not separated from the site is finally reincorporated into the soil as plant remains and organic matter. For an approximation of carbon sequestration in the shoot part, it is generally assumed that 50% of carbon is stored in the branches and 30% of carbon (Shepherd and Montagnini, 2001; Schroth et al., 2002). In common agroforestry systems on the arid, semiarid and degraded sites have a lesser carbon sequestration potential than as compared on fertile humid sites, and the temperate agroforestry systems have comparatively lower carbon sequestration potential evaluate with tropical systems. In view of that, aboveground carbon sequestration approximation is a direct demonstration of above-ground biomass production, the fundamental mechanism of the two functions (carbon sequestration and above-ground biomass production) is the equal- uptake of atmospheric CO_2 through photosynthesis and move of fixed carbon into vegetation. A great number of ecological and management reasons control the rate at which these basic processes proceed.

4.10   BELOWGROUND (SOIL) CARBON SEQUESTRATION

Soils contribute a very important function in the total carbon cycle. Loss of organic carbon from tropical soils not only enhances the atmospheric CO_2 content but also decreases the fertility of those soils that are usually nutrient-poor. Soil organic matter contains extra reactive organic carbon than some other single terrestrial pool. Subsequently, soil organic matter plays a main role in influential carbon storage in ecosystems and in adaptable atmospheric CO_2 concentrations. As reported by Lal (2001) 1 picogram decrease in soil carbon level leads to increment of 0.47 ppm of CO_2 in the ambient atmosphere (Lal, 2001). Thus soil carbon that conventionally has been a sustainability sign of agricultural systems has now obtained the extra role as a display of environmental health. A common trend of increasing soil carbon sequestration under agroforestry evaluated to other land-use practices with the exclusion of forests generally the land-use systems was graded in terms of their soil organic carbon substance in the order forests > agroforestry systems > tree plantations > arable crops plantations. Agroforestry is difficult multi-strata systems like to home-gardens in structural complication but larger in size (Nair et al., 2009). There is a significant level of difference of soil carbon sequestering potential under different agroforestry systems which are regulated by the different biophysical and socio-economic condition of the agroforestry practice. The input to the soil carbon pool by the different agroforestry systems largely depends upon the biomass input by agroforestry species in terms of both quality and quantity along with non-tree components and associated soil structure and properties. Under silvopasture scheme of agroforestry tree plantation under grass-land community gives beneficial effect. Such effect is observed due to changes in the productivity rate of above and belowground plant parts which include rooting depth modification and alteration in the litter input both qualitatively and quantitatively (Connin et al., 1997; Jackson et al., 2000; Jobbagy and Jackson, 2000). Such modification in the strata of litter and soil component may also change the dynamics and storage of carbon in the ecosystem component (Ojima et al., 1991; Schlesinger et al., 1990). This, therefore, indicates that alternate land use practices such as agroforestry are governed by numerous factors such as agroecological conditions and different management practices.

4.11   MANAGEMENT CONSIDERATIONS FOR CARBON SEQUESTRATION

Decomposition of plant deposit and other organic resources in the soil is a resource of carbon and nutrients for fresh development of microbial areas and plants. A lot of this carbon is unrestricted return back into the atmosphere as CO_2 through respiration or is included in living biomass. On the other hand about one-third of soil organic substances break down a large amount more slowly and could still is there in the soil after 1 year (Angers and Chenu, 1997). This soil organic material symbolizes an important carbon store and can stay in the soil for complete periods as a part of soil collectives. The division of soil organic substance that is so confined from more quick decay is very significant from the point of examination of soil carbon sequestration. Agroforestry in relation to climate change is grounded on the principle that an agroforestry system has higher potential to sequester carbon as compared to pastures or field crops. The effectiveness of carbon sequestration depends on the type of agroforestry system implemented, climatic regime, soil type and fertility, use of wood and biomass, stage of agroforestry development, plant species composition, etc.

Carbon sequestration also highlights the significance of management exercise on Carbon sequestration particularly in soils (Lal, 2004). It is not obviously understood that how the management of included tree and crop production systems like agroforestry would modify the rate and magnitude of carbon sequestration. Tree part management impacts the carbon sequestration potential above ground and below ground in agroforestry systems. Carbon sequestration in the woody vegetation is directly related with the amount of dry weight present in the shoot part. Preparation of site through proper stand management from silvicultural system make changes in tree growth attributes and further adds material for both above and below ground carbon sequestration on a different time period which may lead to carbon emissions. Such exercises fall under two groups those positive to carbon sequestration, e.g., tree and stand management, choice of species, site quality, rotation length, etc. and adverse to carbon sequestration, e.g., harvesting, burning, etc. Depending on the magnitude and intensity the issue that further carbon sequestration also may have positive (+), negative (–) and neutral (0) effects. Exercise like weed control and fertilization may enhance stand growth and support carbon sequestration (Johnsen et al., 2001) but may also emit CO_2. For efficient utilization of agroforestry systems, best use of carbon gain needs to be designed. Carbon sequestration potentials and growth rate of woody vegetation are governed by species composition, site quality, age or rotation length, and tree management principles.

Phenology of leaf is the key criteria for selecting suitable species for agroforestry. Four categories of phonological species are (Williams et al., 1997): (1) Evergreen–presence of full canopy structure throughout the whole year by continuous production; (2) Brevi-deciduous–canopy structure decreases up to 50% and not occurs every year; (3) Semi-deciduous–reduction in canopy density every year up to 50%; and (4) Deciduous–remains leafless for a single month every year (Eamus and Prichard, 1998).

Uncertainty remains with respect to wood class variation and their carbon accumulation rate. Fast-growing species accumulates more carbon up to 10-years-old in comparison to slower-growing species who accumulates more carbon on a long-term basis (Redondo-Brenes, 2007). The woody part of slower-growing species due to their higher specific gravity improves the carbon sequestering potential on a long-term basis (Balvanera et al., 2005, Baker et al., 2004; Bunker et al., 2005; Redondo-Brenes and Montagnini, 2006). Higher specific gravity wood like construction timber, furniture, wood crafts acts as a sink for fixed carbon in comparison to low specific gravity wood used for short-lived reasons like packaging cases and poles, etc. Nitrogen (N) fixing trees comprise yet a different group of significant trees in agroforestry. Kumar et al., (1998a) reported that no significant differences in terms of carbon sequestering potential were observed between N fixing trees and other non-N fixers. However, N fixers are a valuable resource for agroforestry systems to promote soil fertility through N fixation and simultaneously promote the growth and output from secondary associated species. N fixing based mixed plantation yields more above ground dry weight or volume production in comparison to their monoculture stands (Bauhus et al., 2004; Forrester et al., 2006; Kumar et al., 1998b). The effects on plants and soil organic carbon accumulation may be positive, negative and neutral and it is probably to control biomass and soil carbon sequestration by choosing suitable tree species. Different agroforestry systems include multipurpose tree species whose growth rate are high leading to higher biomass and productivity along with N fixing property which improves soil health through efficient nutrient cycling and elevated level of soil carbon (Nair et al., 1999; Oelbermann et al., 2006) and therefore has higher carbon sequestration potential. Some profits are credited to a combination of species. Hartley (2002) reported that such approaches promote multifaceted of benefits which includes efficient resource utilization, development of site quality, catastrophic damage reduction such as mechanical damage and pest outbreak, wide produce, avoiding the risk of market failure along with ecological integrity. As per Stanley and Montagnini (1999) mixed type of plantation stands provides a better risk management opportunity through compensatory biomass and nutrient production along with diversified product and preventing soil erosion and biological nitrogen fixation leading towards the improvement of soil fertility (Montagnini et al., 2005).

The three major anthropogenic reasons that influence soil carbon sequestration are: (1) disturbance of the soil like tilling; (2) improvement like fertilization or irrigation; and (3) incorporation of organic matter (Jarecki and Lal, 2003). Agroforestry is measured as one of the management substitutes in the situation of transfer of degraded lands for cultivation of biofuel crops (Nair et al., 2011) as well as the new reduced emission from deforestation and forest degradation programme.

4.12   CONCLUSION

Multipurpose tree plantation in non-forest land provides dual support in terms of biodiversity and sequestering carbon. During crop failure, trees serve as an alternate source of income. More diversity and more species diversity are connected with top biomass of ecosystems leads to larger carbon sequestration. Conservation of existing tree cover, encouragement of agroforestry, land-use management methods have positive forces on biodiversity and also encourage the utilization of biomass fuels as substitution of the fossil fuels; leading to a reduction in emission of CO_2. Tree cultivation of trees on farming land promotes efficient utilization of nutrients from soil layers from variable depth. Avoiding burning of fuelwood and soil conservation through agroforestry systems help in reduction in the emission of CO_2. Capturing and sequestering carbon through different agroforestry systems may be a good technological advancement to overcome the problems in agriculture output and simultaneously combat climate change. Therefore agroforestry practice is one of the fruitful options towards adaptation and mitigation of climate change. Carbon sequestration potential of different agroforestry systems needs to be further explored. Tree integration within croplands and pastureland would promote above and below ground carbon sequestration. Methodological problems in estimating carbon stock of biomass particularly below ground and the area of soil carbon storage in varying circumstances are serious restrictions in exploiting this less cost environmental profit of agroforestry. The enhance in carbon storage in agroforestry systems evaluated with the non-agroforestry system under related ecological situations explain the degree of this potential particularly in soils and the significance of the nature and characters of soils in the scale of their carbon sequestration potential. In general, this is a clear signal of the function of the agroforestry system in climate-change mitigation.

KEYWORDS

- agroforestry
- agropastoral
- agrisilviculture
- agricultural crops
- biodiversity
- carbon sequestration
- carbon stock
- carbon storage
- crop production
- deforestation
- ecological sustainability
- environmental service
- food security
- multipurpose tree
- monocropping
- plantation
- reforestation
- shifting cultivation
- soil conservation
- soil organic carbon
- vegetation

REFERENCES

1. Alavalapati, J. R. R., & Nair, P. K. R. (2001). Socioeconomic and institutional perspectives of agroforestry. In M. Palao and J. Uusivuori (eds.) *World Forest Society and Environment-Markets and Policies*. Kluwer Academic Publishers, Dordrecht, pp. 71–81.
2. Angers, D. A., & Chenu, C. (1997). Dynamics of soil aggregation and C sequestration. In *Soil Processes and the Carbon Cycle* (R. Lal, J. M. Kimble, R. F. Follett, and B. A. Stewart, Eds.), pp. 199–206. *CRC Press*, Boca Raton.
3. Baes, C. F., Goeller, H. E., Olson, J. S., & Rotty, R. M. (1977). Carbon dioxide and climate: the uncontrolled experiment. *American Scientist, 65,* 310–320.
4. Baker, T. R., Phillips, O. L., Malhi, Y., Almeida, S., Arroyo, L., Di Fiore, A., Erwin, T., Killeen, T. J., Laurance, S. G., Laurance, W. F., Lewis, S. L., Lloyd, J., et al., (2004). Variation in wood density determines spatial patterns in Amazonian forest biomass. *Global Change Biol., 10,* 545–562.
5. Balvanera, P., Kremen, C., & Martinez-Ramos, M. (2005). Applying community structure analysis to ecosystem function: Examples from pollination and carbon storage. *Ecol. Appl., 15,* 360–375.
6. Bauhus, J., Van Winden, A. P., & Nicotra, A. B. (2004). Aboveground interactions and productivity in mixed-species plantations of *Acacia mearnsii* and *Eucalyptus globulus. Can. J. For. Res., 34,* 686–694.
7. Berner, R. A. (2003). The long-term carbon cycle, fossil fuels, and atmospheric composition. *Nature, 426,* 323–326.

8. Bunker, D. E., DeClerck, F., Bradford, J. C., Colwell, R. K., Perfecto, I., Phillips, O. L., Sankaran, M., & Naeem, S. (2005). Species loss and aboveground carbon storage in a tropical forest. *Science*, *310*, 1029–1031.

9. Chundawat, B. S., & Gautam, S. K. (2006). A textbook of agroforestry. *Oxford & IBH publishing Co. Pvt. Ltd.*, pp. 57–59.

10. Connin, S. L., Virginia, R. A., & Chamberlain, C. P. (1997). Carbon isotopes reveal soil organic matter dynamics following arid land shrub expansion. *Oecologia, 110,* 374–386.

11. Department of Energy (DOE) (1999). Carbon Sequestration: State of the Science. US DOE, Washington, DC.

12. Eamus, D., & Prichard, H. (1998). A cost-benefit analysis of leaves of four Australian savanna species. *Tree Physiol.*, *18,* 537–545.

13. Forrester, D. I., Bauhus, J., & Cowie, A. L. (2006). Carbon allocation in a mixed-species plantation of *Eucalyptus globulus* and *Acacia mearnsii*. *For. Ecol. Manage.*, *233,* 275–284.

14. FSI, India State of Forest Report. (2013). Forest Survey of India, (Ministry of Environment & Forests), Dehradun, India, 2013.

15. Gera, M., Mohan, G., Bisht, N. S., & Gera, N. (2006). Carbon Sequestration potential under agroforestry in Rupnagar district of Punjab. *Indian Forester*, *132*(5): 543–555.

16. Gordon, A. M., Thevathasan, N., & Nair, P. K. R. (2009). An agroecological foundation for temperate agroforestry. In "Temperate Agroforestry: Science and Practice" (H. E. Garrett and R. F. Fisher, Eds.), pp. 25–44. American Society of Agronomy, Madison, WI.

17. Haile, S. G., Nair, P. K. R., & Nair, V. D. (2008). Carbon storage of different soil-size fractions in Florida Silvo-pastoral systems. *J. Environ. Qual.*, *37,* 1789–1797.

18. Hartley, M. J. (2002). Rationale and methods for conserving biodiversity in plantation forests. *For. Ecol. Manage.*, *155,* 81–95.

19. ICFRE, Country report on poplars and willows period: 2008 to 2011. (2012). National Poplar Commission of India. Indian Council of Forestry Research and Education, Dehradun.

20. Jackson, N. A., Wallace, J. S., & Ong, C. K. (2000). Tree pruning as a means of controlling water use in an agroforestry system in Kenya. *For. Ecol. Manage.*, *126,* 133–148.

21. Jarecki, M. K., & Lal, R. (2003). Crop Management for Soil Carbon Sequestration. *Crit. Rev. Plant Sci.*, *22,* 471–502.

22. Jobbagy, E. G., & Jackson, R. B. (2000). The vertical distribution of soil organic carbon and its relation to climate and vegetation. *Ecol. Appl.*, *10,* 423–436.

23. Johnsen, K. H., Wear, D., Oren, R., Teskey, R. O., Sanchez, F., Will, R., Butnor, J., Markewitz, D., Richter, D., Rials, T., Allen, H. L., Seiler, J., et al., (2001). Carbon sequestration and southern pine forests. *J. Forest.*, *99,* 14–21.

24. Jose, S., Holzmueller, E. J., & Gillespie, A. R. (2009). Tree–crop interactions in temperate agroforestry. In "North American Agroforestry: An Integrated Science and Practice, 2nd edn." (H. E. Garrett, Ed.), pp. 57–74. American Society of Agronomy, Madison, WI.

25. Kaur, B., Gupta, S. R., & Singh, G. (2002). Carbon storage and nitrogen cycling in Silvo-pastoral systems on a sodic soil in northwestern India. *Agroforestry Systems,* 54, 21–29.

26. Kirby, K. R., & Potvin, C. (2007). Variation in carbon storage among tree species: Implications for the management of a small-scale carbon sink project. *For. Ecol. Manage., 246,* 208–221.

27. Kulkarni, H. D. (2013). Pulp and paper industry raw material scenario–ITC plantation a case study. *IPPTA, 25,* 79–89.

28. Kumar, A., Natarajan, S., Biradar, N. B., & Trivedi, B. K. (2011). Evolution of sedentary pastoralism in south India: case study of the Kangayam grassland. *Pastoralism Res. Policy Practice, 1,* 1–18.

29. Kumar, B. M. (2006). Carbon sequestration potential of tropical home gardens. In Tropical Homegardens: A Time-Tested Example of Sustainable Agroforestry (eds Kumar, B. M., & Nair, P. K. R.), Springer Science, Dordrecht, pp. 185–204.

30. Kumar, B. M., George, S. J., Jamaludheen, V., & Suresh, T. K. (1998a). Comparison of biomass production, tree allometry and nutrient use efficiency of multipurpose trees grown in woodlot and Silvo-pastoral experiments in Kerala, India. *For. Ecol. Manage., 112,* 145–163.

31. Kumar, B. M., Kumar, S. S., & Fisher, R. F. (1998b). Intercropping teak with Leucaena increases tree growth and modifies soil characteristics. *Agroforest. Syst., 42,* 81–89.

32. Lal, R. (2001). Soils and the greenhouse effect. In "Soil Carbon Sequestration and the Greenhouse Effect" (R. Lal, Ed.), pp. 1 26. Soil Science Society of America, Madison, WI.

33. Lal, R. (2004). Soil carbon sequestration impacts on global climate change and food security. *Science, 304,* 1623–1627.

34. Lal, R. (2005). Soil carbon sequestration in natural and managed tropical forest ecosystems. Environmental Services of Agroforestry Systems. First World Congress on Agroforestry, Orlando, Florida, USA, 27 June–2 July 2004, Vol. *21,* pp. 1–30, Food Products Press.

35. Lal, R. (2008). Soil carbon stocks under present and future climate with specific reference to European eco-regions. *Nutr. Cycl. Agroecosyst., 81,* 113–127.

36. Lorenz, K., & Lal, R. (2010). Carbon Sequestration in Forest Ecosystems. Springer, Dordrecht, The Netherlands.

37. Luukkanen, O., & Appiah, M. (2013). Good practices for smallholder teak plantations: keys to success. In Working Paper *173,* World Agroforestry Centre (ICRAF) Southeast Asia Regional Program, Bogor, Indonesia, 2013, p. 16, DOI:10.5716/WP13246.PDF.

38. Menalled, F. D., Kelty, M. J., & Ewel, J. J. (1998). Canopy development in tropical tree plantations: A comparison of species mixtures and monocultures. *For. Ecol. Manage., 104,* 249–263.

39. Montagnini, F., Cusack, D., Petit, B., & Kanninen, M. (2005). Environmental services of native tree plantations and agroforestry systems in Central *America. J. Sustain. Forest., 21,* 51–67.

40. Nair, P. K. R. (1993). An Introduction to Agroforestry. Kluwer Academic Publishers, Netherlands.

41. Nair, P. K. R. (2007). The coming of age of agroforestry. *J. Sci. Food Agric., 87,* 1613–1619.

42. Nair, P. K. R., Buresh, R. J., Mugendi, D. N., & Latt, C. R. (1999). Nutrient cycling in tropical agroforestry systems: Myths and science. In "Agroforestry in Sustainable

Agricultural Systems" (L. E. Buck, J. P. Lassoie, and E. C. M. Fernandes, Eds.), pp. 1–31. CRC Press, Boca Raton, FL.

43. Nair, P. K. R., Gordon, A. M., & Mosquera-Losada, M. R. (2008). Agroforestry.In "Encyclopedia of Ecology" (S. E. Jorgensen and B. D. Faith, Eds.), *Vol. 1*, pp. 101–110. Elsevier, *Oxford*, UK.

44. Nair, P. K. R., Kumar, B. M., & Nair, V. D. (2009). Agroforestry as a strategy for carbon sequestration. *J. Plant Nutr. Soil Sci., 172*, 10–23.

45. Nair, P. K. R., Saha, S. K., Nair, V. D., & Haile, S. G. (2011). Potential for greenhouse gas emissions from soil carbon stock following biofuel cultivation on degraded land. *Land Degrad. Develop*, *22*(4), 395–409.

46. NRCAF (2005). *Annual Report*. National Research Centre for Agroforestry, Jhansi.

47. Oelbermann, M., Voroney, R. P., Thevathasan, N. V., Gordon, A. M., Kass, D. C. L., & Schlonvoigt, A. M. (2006). Soil carbon dynamics and residue stabilization in a Costa Rican and southern Canadian alley cropping system. *Agroforest. Syst., 68*, 27–36.

48. Ojima, D. S., Kittel, T. G. F., & Rosswall, T. (1991). Critical issues for understanding global change effects on terrestrial ecosystems. *Ecol. Appl., 1*, 316–325.

49. Palm, C., Tomich, T., Noordwijk, M., Vosti, S., Gockowski, J., Alegre, J., & Verchot, L. (2004). Mitigating GHG emissions in the humid tropics: Case studies from the alternatives to slash-and-burn program (ASB). *Environ. Dev. Sustain., 6*, 145–162.

50. Rai, P., Yadav, R. S., Solanki, K. R., Rao, G. R., & Singh, R. (2001). Growth and pruned biomass production of multipurpose tree species in silvi-pastoral system on degraded lands in semi-arid region of Uttar Pradesh, India. *Forest Tree and Livelihood, 11*, 347–364.

51. Rao, M. R., Nair, P. K. R., & Ong, C. K. (1998). Biophysical interactions in tropical agroforestry systems. *Agroforest. Syst., 38*, 3–50.

52. Redondo-Brenes, A. (2007). Growth, carbon sequestration, and management of native tree plantations in humid regions of Costa Rica. *New For., 34*, 253–268.

53. Redondo-Brenes, A., & Montagnini, F. (2006). Growth, productivity, aboveground biomass, and carbon sequestration of pure and mixed native tree plantations in the Caribbean lowlands of Costa Rica. *For. Ecol. Manage., 232*, 168–178.

54. Roshetko, J. M., Delaney, M., Hairiah, K., & Purnomosidhi, P. (2002). Carbon stocks in Indonesian home garden systems: Can smallholder systems be targeted for increased carbon storage? *Am. J. Altern. Agric., 17*, 138–148.

55. Schlesinger, W. H., Reynolds, J. F., Cunningham, G. L., Huenneke, L. F., Jarrell, W. M., Virginia, R. A., & Whitford, W. G. (1990). Biological feedbacks in global desertification. *Science, 247*, 1043–1048.

56. Schroth, G., D'Angelo, S. A., Teixeira, W. G., Haag, D., & Lieberei, R. (2002). Conversion of secondary forest into agroforestry and monoculture plantations in Amazonia: Consequences for biomass, litter and soil carbon stocks after 7 years. *For. Ecol. Manage., 163*, 131–150.

57. Shepherd, D., & Montagnini, F. (2001). Above ground carbon sequestration potential in mixed and pure tree plantations in the humid tropics. *J. Trop. For. Sci., 13*, 450–459.

58. Singh, G. (2005). Carbon sequestration under an agrisilvicultural system in the arid region. *Indian Forester, 131*(4), 543–552.

59. Srinivasa, H. S. (2006). Large cardamom cultivation in India. Report of the Spices Board, Regional Office, Gangtok, Sikkim.

60. SSSA (2001). Carbon Sequestration: Position of the Soil Science Society of America (SSSA). Available from: www.soils.org/pdf/pos_paper_carb_seq.pdf.

61. Stanley, W. G., & Montagnini, F. (1999). Biomass and nutrient accumulation in pure and mixed plantations of indigenous tree species grown on poor soils in the humid tropics of Costa Rica. *For. Ecol. Manage.*, *113*, 91–103.

62. Sudha, P., Ramprasad, V., Nagendra, M. D. V., Kulkarni, H. D., & Ravindranath, N. H. (2007). Development of an agroforestry carbon sequestration project in Khammam district, India. *Mitigation and Adaptation Strategies for Climate Change*, *12*, 1131–1152.

63. Swamy, S. L., Puri, S., & Singh, A. K. (2003). Growth, biomass, carbon storage and nutrient distribution in Gmelina arborea Roxb. stands on red lateritic soils in central India. *Bioresource Technology*, *90*, 109–126.

64. Tewari, J. C., Sharma, A. K., Naraian, P., & Singh, R. (2007). Restorative forestry and agroforestry in hot region of India: a review. *J. Trop. For.*, *23*, 1–16.

65. UNFCCC (2007). Report of the conference of parties on its thirteenth session, Bali, Indonesia. In "United Nations Framework Convention on Climate Change" Geneva, Switzerland, UN.

66. Verma, K. S., Kumar, S., & Bhardwaj, D. R. (2008). Soil organic carbon stocks and carbon sequestration potential of agroforestry systems in H.P. Himalaya Region of India. *Journal of Tree Sciences*, *27*(1), 14–27.

67. Williams, R. J., Myers, B. A., Muller, W. J., Duff, D. A., & Eamus, D. (1997). Leaf phenology of woody species in a North Australian tropical savannah. *Ecology, 78*, 2542–2558.

68. Yadava, A. K. (2010). Biomass production and carbon sequestration in different agroforestry systems in Tarai region of central Himalaya. *Indian Forester*, *136*(2), 234–242.

CHAPTER 5

Conservation Issues, Challenges, and Management of Medicinal Plant Resources: A New Dimension Toward Sustainable Natural Resource Management

ABHISHEK RAJ,[1] M. K. JHARIYA,[2] D. K. YADAV,[2] and A. BANERJEE[2]

[1]PhD Scholar, Department of Forestry, College of Agriculture, I.G.K.V., Raipur–492012 (C.G.), India, Mobile: +00-91-8269718066, E-mail: ranger0392@gmail.com

[2]Assistant Professor, University Teaching Department, Department of Farm Forestry, Sarguja Vishwavidyalaya, Ambikapur–497001 (C.G.), India, Mobile: +00-91-9407004814 (M. K. Jhariya), +00-91-9926615061 (D. K. Yadav), +00-91-9926470656 (A. Banerjee), E-mail: manu9589@gmail.com (M. K. Jhariya), dheeraj_forestry@yahoo.com (D. K. Yadav), arnabenvsc@yahoo.co.in (A. Banerjee)

ABSTRACT

From ancient times, plants appeared to be very much supportive for feeding purpose along with medical treatment. Medicinal plants continue to be an important therapeutic aid for combating various ailments of mankind. Traditional medicines are still under the practice in Indian villages and have developed through the experience of traditional knowledge generation after generation. Environmental deformations including climate change are severely affecting these medicinal resources constantly. Therefore, it is urgent need of applying some strategies and legislative policy for conserving this global wealth. Sustainable utilization of medicinal plants may play an important role in conservation. Likewise, *in-situ* and *ex-situ* conservation methods are very important for maintaining their status and are excellent strategies to harness the economic power of these plants. Therefore, the purpose of this chapter is to review about the status of the medicinal plant of wild and cultivated, diversity, utilization pattern, threatened status, strategies for conservation and their inevitable role in socio-economic upliftments of rural peoples and tribal's.

5.1   INTRODUCTION

The world forests are the pivotal source of medicinal plants, for nurturing humankind and maintain rich biodiversity. In the forest, medicinal herbs, shrubs, climber, trees are the prominent resource which plays a major role in the healthcare of people and livestock. As per Mehta et al. (2015), approximately 90% of herbal raw drugs have been used in the manufacture of Ayurveda, Siddha, Unani and Homoeopathy systems of medicine and these herbal raw drugs are contributed by 20% of a forest of the country. Similarly, Baishya (2013) has reported about the various numbers of plants, which have been utilized in Indian system of medicine (Figure 5.1). In India, the medicinally recognized plant is used dates back to 5000 years. Around 2500 plant species have been recognized as medicinal properties, and over 6000 plants are estimated to be explored in traditional, folk and herbal medicine (Huxley, 1984). Prakasha et al. (2010) reported three thousand plants species with medicinal importance in India and globally forty percent people incorporates them in their primary health care system (WHO, 2003).

Although, traditional medicines are excellence source of health care for 65% of India's population and these medicines are based on medicinal plants and other plant resources (WHO, 2002). Western and Eastern Ghats in India represents two-third of 8000 medicinal plants under tropical condition (Kumar and Janagam, 2011) and 90% are the representatives from the forest. Their degradation and destructive harvesting cause loss of medicinal plant resources (Camm et al., 2002). However, due to the increasing demand of medicinal plants, the pressure for exploitation of forest resource is increasing in nature as time goes on, leading to unsustainable extraction of medicinal resources and creating an imbalance in the ecosystem in terms of biodiversity loss. Malcolm et al. (2006) and Bhardwaj et al. (2007) reported species such as *Aconitum heterophyllum,*

Coscinium fenestratum, Decalepis hamiltonii, Picrorhiza kurroa, Saraca asoca, and *Taxus wallichiana* are of higher concerned from conservation perspectives due to their higher utility in Indian herbal industry. It is also important to conserve the medicinal plant resources of our country to sustain 4635 ethnic communities to maintain their socioeconomic living condition along with health and hygiene (UNU-IAS, 2012).

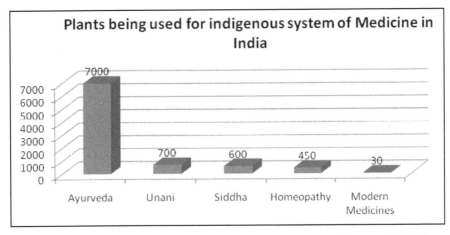

FIGURE 5.1 Numbers of utilized plants in Indian system of medicine (Modified from Baishya, 2013).

5.2 GLOBAL SCENARIO OF MEDICINAL PLANT RESOURCES

Globally 72000 (17.1% of world flora) plants are utilized for the medicinal purpose (Table 5.1), which is very indispensable for human health and disease preclusion. Majority (34.5%) of medicinal plants is utilized in Republic of Korea followed by Pakistan (30.3%), Bulgaria (21.0%), France (19.4%), Jordan (17.3%), Vietnam (17.1%), Sri Lanka (16.6%), India (16.1%), and so on. Malaysia comprises 7.7% of the medicinal plant out of total reported plant species (Schippmann et al., 2006). Globally, 50–70% of medicinal plant biomass is sourced from the wild in comparison to India having 77% of the medicinal plant from their natural source. In earlier times (1981), approx. About 121 prescribed drugs were derived from 95 species were used worldwide. Similarly, tropical forest contributed 328 plant drugs worth US$ 147 billion in 1995 (Rao and Rajput, 2010).

TABLE 5.1 Number of Plants Are Used Medicinally in the World (Modified from Schippmann et al., 2006)

Country	Plant species	MP species	%
China	32,200	4,941	15.3
USA	21,641	2,564	11.8
India	18,664	3,000	16.1
Malaysia	15,500	1,200	7.7
Thailand	11,625	1,800	15.5
Vietnam	10,500	1,800	17.1
Philippines	8,931	850	9.5
Nepal	6,973	900	12.9
Pakistan	4,950	1,500	30.3
France	4,630	900	19.4
Bulgaria	3,567	750	21.0
Sri Lanka	3,314	550	16.6
Republic of Korea	2,898	1,000	34.5
Hungary	2,214	270	12.2
Jordan	2,100	363	17.3
Average			17.1
World	422,000	72,000	

5.3 MEDICINAL PLANT BIODIVERSITY IN INDIA

World Health Organization (WHO) reported 80% dependency of global population on plant-based traditional medicines (Akerele, 1992) to fulfill their primary health care needs. India is richest, and remarkably diverse country comprises medicinal and aromatic plants called as botanical garden of the world (Painkra et al., 2015) and contributes one of the 12-mega bio-diverse countries of the world. As per Maheswari (2011), India is the home of 45,000 diverse floral species, out of which 15,000 are those of medicinal plants and their distribution pattern is shown in Table 5.2. Therefore, India is the storehouse of medicinal plants and contributed in export and import (Table 5.3) of these valuable medicinal resources of either cultivated or wild on the global economy basis (Ramawat and Goyal, 2008). Moreover, the biodiversity of medicinal flora in the forest is not confined to only one region, which depends upon forest type and varies from one region to others. Central India represents a large area of tropical dry and moist deciduous forests which supports a rich heritage of biodiversity (Raj and Toppo, 2014; Toppo et al., 2014). Chhattisgarh as a tribal-state (35% of the population tribal) has a rich heritage of medicinal plants often designating the state as an herbal state and medicinal hotspots. All the medicinal resources are intricately related with the socio-cultural development and primary health care of the local community stakeholders.

TABLE 5.2   Distribution of Medicinal Plants (Modified from Maheswari, 2011)

Country or Region	Total number of native species in flora	No of medicinal plant species reported	% of medicinal plants
World	297,000	52,885	10
India	17,000	7,500	44
Himalayas	8,000	1,748	22

TABLE 5.3   Exported and Imported Medicinal Plants of India (Modified from Ramawat and Goyal, 2008)

Export from India Botanical name	Import into India		
	Part of the plant	Botanical name	Native name
Aconitum species	Root	*Cuscuta epithymum*	Aftimumvilaiyti
Acorus calamus	Rhizome	*Glycyrrhiza glabra*	Mullathi
Adhatoda vasica	Whole plant	*Lavandula stoechas*	Ustukhudus
Berberis aristata	Root	*Operculina turpethum*	Turbud
Cassia angustifolia	Root	*Pimpinella anisum*	Anise fruit
Colchicum luteum	Rhizome and seed	*Smilax ornata*	Ushba
Hedychium spicatum	Rhizome	*Smilax china*	Chobchini
Juglans regia	Bark	*Thymus vulgaris*	Hasha
Juniperus macropoda	Fruit		
Plantago ovata	Seed and husk		
Punica granatum	Flower, root, and bark		
Rauvolfia serpentina	Root		
Saussurea lappa	Rhizome		
Valeriana jatamansi	Rhizome		
Zingiber officinale	Rhizome		

The dense forest of Chhattisgarh state has unique biodiversity and is a storehouse of several precious medicinal and aromatic plants which is an important source of food, income, and health to local and tribal peoples. Therefore, forest, tribal's and medicinal plants make synergistically links among themselves.

5.4 MEDICINAL PLANTS IN INDIAN HIMALAYA REGION (IHR)

Indian Himalaya region is mega hotspot of biodiversity (Myers et al., 2000) and higher level of species diversity representing 8,000 species and 40% endemic angiospermic plant and 44 species and 16% endemic gymnospermic plants, 600 species and 25% endemic pteridophytic plants, 1737 species and 33% endemic bryophytic plant, 1,159 species and 11% endemic lichen and 6,900 number of species with 27% endemic fungal species (Singh and Hajra, 1996; Samant et al., 1998). Aconitum, Dactylorhiza, Ephedra, Fritillaria, Nardostachys, Picrorrhiza, Podophyllum, Polygonatum, Taxus are the representative of Himalayan region of India and therefore are considered to have high medicinal value and have imminent potential in primary healthcare system and maintain trade values with huge market potential (Samant et al., 1998; Kala, 2008). Among the 675 wild edible plant species of Indian Himalayan region 118 were represented by oil yielding medicinal plants, 279 species fodder yielding plants, 155 scared plants and 121 species representing rare endangered plants species (Nayar and Sastry, 1987, 1988, 1990; Samant and Dhar, 1997; Samant and Pant, 2003).

Therefore, distribution and existence of medicinal plants are not confining to the only plain area but also exist at high altitude on Himalayas, and the Indian Himalaya region is a harbor of rich flora of medicinally important plant species with their unique features of adaptation at high altitude and characterized by genuine healing properties of any disease. Jain (1991), Samant et al. (1998) and Rai et al. (2000) have conducted several works on medicinal plants in the Himalayan region of India. They thoroughly studied the ethnobotanical perspective, floristic survey and biodiversity assessment (Collett, 1902; Aswal and Mehrotra, 1994; Sharma and Singh, 1996; Chauhan, 1989, 1990, 1999; Dhaliwal and Sharma, 1997, 1999; Sharma and Dhaliwal, 1997a, 1997b; Singh and Rawat, 2000; Kaur and Sharma, 2004). Topography and altitude seem to be the key factor in the distribution of diversity and the abundance of medicinal plant resources. Samant et al. (2007) reported that altitudinal gradients promote the diversity of medicinal plants and as per earlier reports in case of Himachal Pradesh at an altitude of < 1800 m to > 3800 m under tropical and subtropical nature of climate the population of medicinal plants varied considerably.

Acer spicatum, Aconitum falconeri, Aconitum heterophyllum, Aconitum violaceum, Acorus calamus, Ainsliaea aptera, Ajuga parviflora, Alangium rotundifolium, Allium humile, Aloe barbadensis, Angelica glauca, Arch-angelica himalaica, Arnebia benthamii, Arnebia euchroma, Artemisia absinthium, Asparagus adscendens, Asparagus filicinus, Asparagus racemosus, Azardirachta indica, A. japonica, Berberis aristata, Betula alnoides, Bunium persicum, Bupleurum falcatum, Calotropis gigantean, Carum carvii, Cassia fistula, Cinnamomum tamala, Corydalis govaniana, Corydalis meifolia, Corylus jacquemontii, Curculigo orchioides, Cyperus rotundus, Delphinium cashmerianum, Dioscorea bulbifera, Emblica officinalis, Eritrichium canum, Fritillaria roylei, Gentiana kurroo, Geranium nepalense, Geranium wallichianum, Hypericum perforatum, Impatiens gladulifera, Iris kumaonensis, Malaxis muscifera, Malaxisa cuminata, Mentha arvensis, Mentha piperata, Nardostachys grandiflora , Ocimum canum, Ocimum sanctum, Pediculari spectinata, Phytolacca acinosa, Polygala sibirica, Polygonum affine, Prinsepia utilis, Rauwolfia serpentina, Rheum australe, Rheum moorcroftianum, Rheum webbianum, Rhododendron anthopogon, Rhododendron campanulatum, Rosularia rosulata, Rubia cordifolia, Sagina saginoides, Salix lanata, Salvia plebeia, Saussurea auriculata, Saussurea costus, Saussurea gossypiphora, Saussurea graminifolia, Saussurea obvallata, Saussurea simsoniana, Skimmia laureola, Swertia angustifolia, Syzygium cuminii, T. tenuifolium, Tanacetum gracile, Taxus baccata subsp. wallichiana, Terminalia arjuna, Terminalia alata, Terminalia chebula, Terminalia tomentosum, Tinospora cordifolia, Woodfordia fruticosa, Zanthoxylum armatum, and Ziziphus mauritiana were the representative of the respective study area. Therefore, the process of speciation is regulated by some factors such as altitudinal range, topography and climatic conditions of those regions. Unscientific excavation leads to a gross fall back of medicinal plant resources of the Himalayan region (Samant et al., 1998).

5.5   WILD VS. CULTIVATED MEDICINAL PLANTS

There is a two school of thought for adoption/practices of wild and cultivated medicinal plants which depend upon the welfare of species and ecosystem, market demands and people needs. Adoption of cultivation practices of medicinal plants reduces harvesting pressure of rare and slow growing species in the context of the welfare of species and ecosystem. But sometimes cultivated species become invasive and do bad effect on the ecosystem. Further such incidences will lead to genetic erosion of wild relative of plants in nature. In the wild, the uncontrolled and unsustainable harvest may lead to an extinction of important medicinal resources. From the perspective of people, wild medicinal plants provide healthcare and cash income without any investment unless overexploitation of these resources (Anonymous, 1997). It has been reported that one-fourth of the medicinal plant resources belongs to wild plants (Hamilton, 2004). In order for a steady supply of herbal medicine, cultivation of medicinal plant resources is an essential prerequisite while the expenditure incurred in such a system would be high for small hold farmers. Each system has its own positive and negative consequences for example harvesting at the wild is a cost-effective approach as it does not require any sort of infrastructure development, but simultaneously it is under the curse of over-harvesting and overexploitation. The problem of unscientific exploitation can be regulated under cultivated conditions. Other benefits include standardization and improvement of genotype is possible, maintaining of quality standard and possibility of consistent botanical identification.

5.6   UTILIZATION TRENDS OF MEDICINAL PLANTS

Plants have been using as an ingredient of a medicine for a long time. There is no exact figure of the utility of a number of plants as medicine and these figures varies region wise or country wise (Table 5.4). The number of medicinally important plants in the world, China, India, Mexico, North America includes 35,000–70,000 or 53,000 (Farnsworth and Soejarto, 1991; Schippmann et al., 2002); 10,000–11,250 (He and Gu, 1997; Xiao and Yong, 1998; Pei, 2002a), 7500 (Shiva, 1996), 2237 (Toledo, 1995) and 2572 (Moerman, 1998), respectively. As per the projection of Maheswari (2011) the hike for medicinal plant resources would increase from the US $14 billion per year to the US $5 trillion in 2050. Among the forest produce, neem is perhaps the only tree that has been potentially exported (NABARD, 2011). Neem has been used in *Ayurvedic* medicine for more than 4,000 years, due to its medicinal and healing properties and regarded as "The Village Pharmacy" (Jhariya et al., 2013). From NTFPs perspective gum from *Acacia nilotica* shows significant promise through its multipurpose use in the food, industry and pharmaceutical bodies of India (Das et al., 2014; Raj, 2015a & 2015b; Raj and Singh, 2017). For the low economy, people such as the poor farming community under the rural system is solely dependent upon nature for their daily livelihood; therefore, poverty eradication through socio-economic upliftment along with biodiversity conservation is the combined issues of global forefront (Jhariya and Raj, 2014; Raj et al., 2015). Moreover, the utilization trends of medicinal plants are not confining to only human care, also useful for treating disease in domesticated animals.

TABLE 5.4 Numbers and Percentages of Medicinal Plant Species Recorded for Different Countries and Regions

Country or region	Number of species of medicinal plants	Total number of native species in flora	% of flora which is medicinal	Authors
China	11,146	27,100	41	Pei (2002a)
India	7,500	17,000	44	Shiva (1996)
North America	2,572	20,000	13	Moerman (1998)
Mexico	2,237	30,000	7	Toledo (1995)
World	52,885	297,000–510,000	10–18	Schippmann et al. (2002)

5.7 LIVELIHOOD UPLIFTMENTS

Conservation provides an essential foundation for sustainable rural life and livelihoods. The tribes primarily depend on locally available forest plants for their healthcare and treatment of various diseases. Beside medicinal value, these plants can provide a significant source of income for rural people and help in the improvement of livelihood. Various researches have been done by authors in this context. A total of five medicinal herbs *viz.* Rosemary (*Rosmarinus officinalis*), Satawar (*Asparagus racemosus*), Ban-tulsi (*Ocimum basilicum*), Tagar (*Valeriana jatamansi*) and Chamomile (*Matricaria chamomilla*) were choose for cultivation at farmer's field of selected village clusters, i.e., Dharaunj, Mudiyani, and Gumod through participatory action research approaches in Champawat district of Uttarakhand state in India. Community intervention in terms of meetings between the farming community and trading community along with other stakeholders were arranged in which Memorandum of Understanding (MoU) was undertaken between the various components. Such approaches promote skill development and technological progress among the farming community to go for MAPs cultivation which would meet the need of pharmaceutical industries along with improving the livelihood conditions of farming community (Phondaniet al., 2011).

5.8 TRADE OF MEDICINAL PLANTS IN INDIA

In India, about 960 medicinal plant taxa traded belong to 575 genera of 169 plant families which includes 152 families belong to Angiosperm (130 dicots and 22 monocots), 5 to Gymnosperm, 9 to Pteridophytes and 3 to Fungi and Lichen of Thallophytic group comprises 3 plant families (Fig. 5.2) and one-third of plant species belong to the top 10 families of total 169 plant families. Therefore, family wise break-up of highly traded medicinal plant species are as decreasing order as follow: Fabaceae (50), Asteraceae (42), Lamiaceae (33), Euphorbiaceae (32), Apiaceae (31), Rubiaceae (27), Cucurbitaceae (26), Caesalpiniaceae (25), Rutaceae (22), Zingiberaceae (22), respectively. Among 960 traded medicinal plant taxa comprise various life form of herbs (398 species), trees (251 species), shrubs (168 species) and climber including lianas (143 species). Similarly, amongst 960 traded species, the highest proportion of plant species is used under Ayurvedic system of medicine (688), followed by *Siddha* (501), Unani (328), Tibetan system (197) and Homeopathy (146). The traded data of different parts of medicinal plants are shown in decreasing order as follow: roots (338), fruits (333), whole plants (168), stem (162), leaves (140), flowers (84), exudates (37), wood (20), galls (4), oil (3), etc., respectively (Ved and Goraya, 2008). Similarly, source wise availability of highly traded medicinal plants is depicted in Table 5.5. During 2005–2006, the total trade of medicinal plants was 319500 mt of Rs.1068 crores of which herbal industry contributes 627.9 crores, Rural HHs contributes 86.0 crores and export value was 354.8 crores (NMPB, 2006).

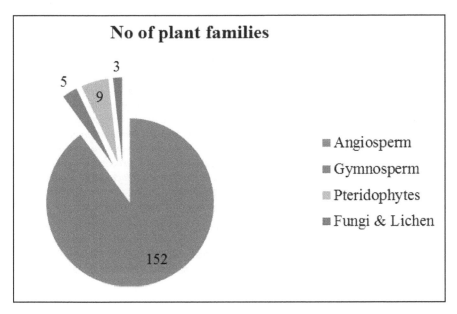

FIGURE 5.2   (See color insert.) Distribution of medicinal plants among various classes of plant kingdom (Source: Ved and Goraya, 2008).

5.9   IMPORT AND EXPORT VALUE OF MEDICINAL PLANTS

Imports and exports of medicinal plants were worth US\$ 1.59 and 1.44 billion/year with >40% growth rate/ annum in between 2004–2008 in the world (Table 5.6) (Rao and Rajput, 2010). From the global economic perspective, it is very difficult to determine the marketing and trading of medicinally plant resources due to lacking of an adequate database. In this connection analytical report of UNCTAD for the period of 1981–1998 has been reported in Table 5.7 (Lange, 2002) which addresses the problems and issues associated with the marketing and trade system of the plant resources. According to this report, India is the 2nd most leading country to export up to 36,750 tons of worth 57,400 US\$ after China (shows 1st rank) in export value of medicinal and aromatic plants. Therefore, India is the world second largest exporter of medicinal plants after China. Similarly, within 3000 medicinal plants traded internationally, only 900 are under cultivation and majority of the exported biomass were harvested from the wild.

TABLE 5.5   List of Highly Traded Plants from Various Sources (Source: SOSSMBC, 2010)

Highly traded medicinal plants from	
Tropical forest	*Acacia catechu, Acacia nilotica, Albizia amara, Alstonia scholaris, Anogeissus latifolia, Asparagus recemosa, Bombax ceiba, Boswellia serrata, Buchnania lanzan, Cassia fistula, Chlorophytum tuberosum, Cinnamomum sulphuratum, Emblia officinalis, Gmelina arborea, Gymnema sylvestre, Holoptelea integrifolia, Ixora coccinea, Madhuca indica, Messua ferrea, Mimusops elengi, Mucuna puriens, Oroxylum idicum, Pterocarpus marsupium, Pterocarpus santalinus, Rubia cordifolia, Saraca asoca, Semicarpus anacardium, Shorea robusta, Soymida febrifuga, Terminalia arjuna, Terminalia bellirica, Terminalia chebula, Vateria indicum, Ziziphus xylocarpus*
Temperate forest	*Abeis spectabils, Aconitum ferox, Aconitum heterophyllum, Cedrus deodara, Cinnamomum tamala, Juniperus communis, Jurinea macrocephala, Nardostachys grandiflora, Parmilia perlata, Reheum austral, Rhododendron anthopogon, Taxus wallichiana, Veleriana jatmansi, Viola pilosa*
Cultivated source	*Abelmoschus moscnatus, Acorus calamus, Adhatoda zeylanica, Aloe barbadensis, Azadirachta indica, Caesalpinia sappan, Carcuma angustifolia, Ficus benghalansis, Ficus religiosa, Indigofera tinctoria, Jatropha curcas, Lawsonia inermis, Ocimum basilicum, Piper longum, Pongamia pinnata, Prunus armeniaca, Saussurea costus, Withania somnifera, Ziziphus jujube*
Imported	*Aquilaria agallocha, Commiphora wightii, Quercus infectoria, Piper Chaba, Glycyrrhiza glabra*

TABLE 5.6   Import and Export (in billion US$) of Medicinal Plants in the World (Modified from Rao and Rajput, 2010)

Year	Import	Export
2004	1.36	1.21
2005	1.39	1.24
2006	1.52	1.35
2007	1.77	1.60
2008	1.94	1.78
Total	7.99	7.20
Average	1.59	1.44
Growth rate	47.2%	42.1%

TABLE 5.7 Import and Export of Medicinal and Aromatic Plant Material, 1991–1998 (*Source*: Lange, 2002)

Country of import	Volume (tons)	Value (1000 US$)	Country of export	Volume (tons)	Value (1000 US$)
Hong Kong	73,650	314,000	China	139,750	298,650
Japan	56,750	146,650	India	36,750	57,400
USA	56,000	133,350	Germany	15,050	72,400
Germany	45,850	113,900	USA	11,950	114,450
Rep. Korea	31,400	52,550	Chile	11,850	29,100
France	20,800	50,400	Egypt	11,350	13,700
China	12,400	41,750	Singapore	11,250	59,850
Italy	11,450	42,250	Mexico	10,600	10 050
Pakistan	11,350	11,850	Bulgaria	10,150	14,850
Spain	8,600	27,450	Pakistan	8,100	5,300
UK	7,600	25,550	Albania	7,350	14,050
Singapore	6,550	55,500	Morocco	7,250	13,200
Total	342,550	1,015,200	Total	281,550	643,200

5.10 MARKETING AND ECONOMIC CONSTRAINTS FOR MEDICINAL PLANT RESOURCES

From a marketing and economic perspective, the major problem of medicinal plant resources includes lack of a proper market to generate a considerable level of revenue from these products. This problem further aggravates with improper planning during collection, transportation up to the market and various other marketing function. Lack of scientific knowledge within the community stakeholders who are actively engaged in the collection process is a way of unsustainable harvesting of these valuable resources. Under cultivated condition lack of appropriate knowledge for plant species propagation along with lack of adequate resources and mismanagement of fund are some of the constraining factors to promote the market economy of medicinal plant resources. To overcome these constraints, it is necessary to aware and involves the people to adopt and cultivate these medicinal plants. Likewise, development of good agro-practice with managed harvest techniques, linked with the market with adequate infrastructure for processing and value addition of medicinal plants backed up by efficient administrative policies seems to be appropriate steps towards sustainability and production of medicinal resources.

5.11   THREATENED MEDICINAL PLANT RESOURCES

Globally, 34,000 species or 8% of the world's flora are threatened with extinction (Walter and Gillett, 1998). IUCN has reported that about 70% of plants species and 20.8% of the medicinal plant are threatened worldwide (Schippmann et al., 2006). In India, medicinally important species such as *Acquilaria malaccensis, Dioscorea deltoidea, Podophyllum hexandrum, Pterocarpus santalinus, Rauwolfia serpentina, Saussurea lappa,* and *Taxus wallichiana* have become endangered and need immediate attention for conservation (Rao, 2006). As we know, India is one of the richest reservoirs of medicinal biodiversity in the world and home of the great diversity of medicinally important plants. These resources are either wild or cultivated and do work for the welfare of people in term of both health and wealth. But there are several factors which threatened these medicinally important species.

Due to the increasing demand of herbal raw materials and products that results proportionally increasing demand of medicinal plant (some are depicted in Table 5.8) that leads overexploitation and unsustainable harvesting of these medicinal resources. Likewise, there are severally important world recognized and high demanded medicinal plants exist namely *Abies balsamea* (Balsam Fir), *Allium sativum* (Garlic), *Aloe vera* (Ghritkumari), *Althaea officinalis* (Marsh Mallow), *Asparagus racemosus* (Shatavari), *Citrus limon* (Lemon), *Crataegus monogyna* (Hawthorn), *Dioscorea batatas* (Chinese Yam), *Echinacea pallid* (Cone Flower), *Hippophaerham noides* (Sea Buckthorn), *Matricaria recutita* (German Camomile), *Panax ginseng* (Ginseng) and *Tanacetum parthenium* (Feverfew), etc. In the European, Asian and North American countries the rising demand of medicinal plant resources varied between 8–15% (Ross, 2005; Bentley, 2010). The World knows, the forests are the source of invaluable medicinal plants, but degradation of forest and deforestation are adversely affecting the resource base of medicinal plants and biodiversity. Likewise, other factors include market-demand driven harvesting without any concern for conservation,

TABLE 5.8 List of Important Medicinal Plants in Demand and Their Medicinal Use

Common name	Botanical name	Family	Distribution/habitat	Plant part use	Medicinal use
Aonla	*Emblica officinalis*	Euphorbiaceae	UP, Gujarat, Rajasthan, Maharashtra	–	Useful in anemia, jaundice, dyspepcia, hemorrhagic disorders, bilionsness, diabetes, asthma, bronchitis.
Guggal	*Commiphora wightii*	Burseraceae	–	Tree yield an oliogum-resin-guggulipid	Gum resin is commonly used for the treatment of rheumatoid arthritis.
Senna	*Cassia angustifolia*	Caesalpinaceae	Tamilnadu, Maharashtra, Gujarat, Rajasthan and Delhi	Leasves and pod.	Used in constipation.
Safedmusli	*Chlorophytum borivillanum*	Liliaceae	Southern Rajasthan, Western M P and North Gujarat	Tuberous roots	Used in Ayurvedic medicines
Satavari	*Asparagus racemosus*	Liliaceae	found almost all over India	Tuberous roots	Used in consumption, epilepsy, diarrhea, blood dysentery, hemophilic disorders, swellings
Long pepper	*Piper longum*	Piperaceae	–	Root	Used in cough, cold, chronic bronchitis, epilepsy, asthma, amorexia, piles, dyspepsia, leucoderma.
Brahmi	*Bacopa monnieri*	Scrophulariaceae	Throughout India	Whole plant especially leaves	Used as nervine tonic/ memory enhancer
Ashoka	*Saraca asoca*	Leguminosae	Himalayas, Bengal and Western Peninsula.	Bark	Used in menorrhagia and uterine affections, internal bleeding, and hemorrhagic dysentery

TABLE 5.8 *(Continued)*

Common name	Botanical name	Family	Distribution/habitat	Plant part use	Medicinal use
Ashwagandha	*Withania somnifera*	Solanaceae	Madhya Pradesh and Rajasthan	Root	Used in rheumatism, consumption, debility from old age
Indian barbery	*Berberis aristata*	Berberidaceae	Assam, Bihar and Himalayan Region	Root bark, stem, and wood fruit	Used in the treatment of diarrhea, jaundice, malaria, fever and skin diseases.
Isabgol	*Plantago ovata*	Plantaginaceae	Punjab plains, Sindh and Baluchistan.	Seeds and seek husk	Used as a single drug for the cure of constipation and dysentery.
Kokum	*Garcinia indica*	Clusiaceae	Maharashtra, Goa, Karnataka, Kerala, South Gujarat, Assam and West Bengal.	Fruit	Used for making Kokum syrup, and dried kokum rind, etc.
Kalmegh	*Andrographis paniculata*	Acanthaceae	Throughout India	Leaves	Used in chronic malaria, anemia, jaundice, and loss of appetite.

indiscriminate collection of essential regenerative components of a plant like roots, tubers, fruits, seeds, flowers, bark, etc. results to degradation, depletion and finally extinction of particular species. Due to destructive collection and overexploitation these medicinally important plants species becoming rare and some of them are critically endangered. As per the reports of IUCN (International Union for Conservation of Nature) and WWF (World Wildlife Fund) unsustainable way of harvesting has caused extinction of 15,000 species among 50,000 to 80,000 angiosperms which has wide global application (Bentley, 2010) and simultaneously 20% of the plant species were significantly affected by consumptive trends and biotic interference (Ross, 2005). Further habitat fragmentation and changing land use pattern have proved to be the significant factor in destructing medicinal plant resource in China (Nalawade, 2003; Heywood and Iriondo, 2003), India (Heywood and Iriondo, 2003; Hamilton, 2008), Kenya (Hamilton, 2008), Nepal (Hamilton, 2008), Tanzania (Zerabruk and Yirga, 2012), and Uganda (Zerabruk and Yirga, 2012).

5.12 IMPACT OF CLIMATE CHANGE ON MEDICINAL PLANT RESOURCES

Climate change is the most important global environmental challenge which is facing by all living organism, disturb natural ecosystems, agriculture, and health. It is a well-known fact that climatic irregularities in the form of changing pattern of temperature regime, precipitation pattern, inadequate humidity, and other associated phenomenon has promoted global climate change and therefore impacted floral and faunal biodiversity. Therefore, climate change disturbs both distribution and life cycle of vegetation including medicinally important plants. The impact on the medicinal plant is studied in several ways such as the effect on productivity, quality, secondary metabolites production, phonological changes, etc. Ziska (2005) has compiled several medicinal plants and their chemicals which is useful in several actions (Table 5.9), and their pharmaceutical value has a number of preparations (Table 5.10; Ved, 2001) that can be altered due to change in climate regime and weather pattern. Moreover, environmental parameters such as temperature, humidity, light intensity, minerals, nutrients and availability of water are affecting not only growing plant species of medicinal value but also disturb the production of secondary metabolites which is a very important source for food additives, flavors, and pharmaceutical purpose. Several studies have been conducted in this context, such as elevated temperature increased the secondary metabolites (Litvak et al., 2002) and volatile organic compounds (Loreto et al., 2006) but decreased that has been seen by Snow et al. (2003). Similarly, under the stressed condition growth of plants is inhibited more than photosynthesis results the fixation of carbon is not allocated to growth but this allocation is seen in secondary metabolites that lead to increased production of secondary metabolite (Mooney et al., 1991). Likewise, the availability of menthol crystals in field mint (*Mentha arvensis*) crops is affected by uneven and heavy rainfall by damaging this crop in Northern India (Cavaliere, 2009). Sometimes, the rise in CO_2 level enhances the production of beneficial substance or products in medicinal plants such as by 75% increase in CO_2 concentration leads towards increased spider lily bulb biomass by 56% which is used for treating cancer (Idso et al., 2000). This type of result has been seen in a ginger plant as increased in CO_2 resulted in high photosynthesis rate, increased plant biomass, increased water use efficiency by reducing stomatal conductance, increased in total flavonoids, total phenolics, total soluble carbohydrates and starch production (Ghasemzadeh and Jaafar, 2011). As per Salick et al. (2009), some cold adopted alpine species are migrating upward faced with extinction due to a continuous rise in temperature.

TABLE 5.9 Chemical Derivatives from Different Plants and Their Action (Compiled from Ziska, 2005)

Common name	Botanical name	Derive drugs	Action
Belladonna	*Atropa belladonna*	Atropine	Anticholinergic
Common barberry	*Berberis vulgaris*	Berberine	Bacillary dysentery
opium poppy	*Papaver somniferum*	Codeine	Analgesic, antitussive
Indian-Laburnum	*Cassia spp.*	Danthron	Laxative
Kiwach	*Mucuna spp.*	L-Dopa	Anti-Parkinson's
White willow	*Salix alba*	Salicin	Analgesic
Himalayan Yew	*Taxus baccata*	Taxol	Antitumor
Periwinkle or Sadabahar	*Catharanthus roseus*	Vincristine	Antileukemic agent
Digitalis purpurea	*Digitalis purpurea*	Digitoxin	Cardiotonic
Climbing Oleander	*Strophanthus gratus*	Ouabain	Cardiotonic

TABLE 5.10 Some Important Medicinal Plants Used in Pharmaceutical Preparations.

Common name	Botanical name	Family	No. of Pharmaceutical preparation
Vacha	*Acorus calamus*	Acoraceae	51
Bael	*Aegle marmelos*	Rutaceae	60
Daruharidra	*Berberis aristata*	Berberidaceae	65
Punarnava	*Boerhaavia diffusa*	Nyctaginaceae	52
Mustaka	*Cyperus rotundus*	Cyperacea	102
Yashtimadhu	*Glycyrrhiza glabra*	Fabaceae	141
Amla	*Emblica officinalis*	Phyllanthaceae	219
Bahera	*Terminalia bellirica*	Combretaceae	219
Hararrh	*Terminalia chebula*	Combretaceae	219
Gulancha	*Tinospora cordifolia*	Menispermaceae	88
Ashwagandha	*Withania somnifera*	Solanaceae	109

(Source: Ved, 2001).

5.13 CONSERVATION TECHNIQUES AND MANAGEMENT STRATEGIES TOWARDS SUSTAINABILITY

Medicinal plants products have the huge potentiality of curing various types of diseases based on indigenous knowledge of the tribal people. Their conservation and management are a very important step towards sustainability and a better future. This will start from the documentation and assessment of different medicinal plants in various regions of the world (Toppo et al., 2016). Diverse type of medicinal plant varies from region to region plays an important role in healing the diverse nature of the disease. Therefore, assessment and documentation of these medicinal resources are the prerequisites for the conservation of these biological resources (Raj and Toppo, 2014) on which tribals are much more dependent and essential for their survival.

As we know, medicinal resources are disappearing from their natural habitat at an alarming rate due to careless, poor quality, indiscriminate harvesting, and mismanagement. Loss and deterioration of forest (deforestation) due to the conversion of forested land to farmland, industrial establishment, mining practices, urbanization, etc. leads to depletion of medicinal plant resource base and primary health care system upon which various forests dwelling community depends upon. In order to promote conservation of medicinal plant resources, a bimodal approach is the need of the hour. Under this bimodal approach at one hand one should go for sustainable utilization of wild medicinal plant resources and on the other hand, go for cultivating this medicinal plant resource for germplasm protection. Such approaches under *ex-situ* and *in-situ* condition (Fig. 5.3) along with natural resource management in the form of sustainable farming and optimum use should be promoted for conservation perspective. Under *in-situ* strategies, protected area development through afforestation and reforestation techniques helps to maintain the germplasm reserves under natural condition, and the establishment of the botanical garden and seed bank serves as the future germplasm bank for propagation (Sheikh et al., 2002; Coley et al., 2003). Moreover, some biotechnological approaches such as tissue culture, micropropagation, molecular marker, and synthetic seed technology could improve growth and yield potential of medicinal plants resources (Chen et al., 2016).

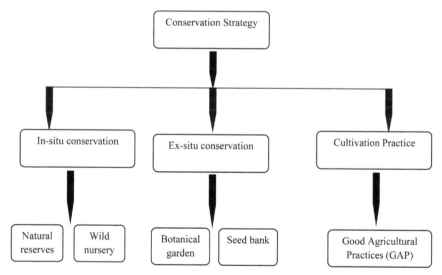

FIGURE 5.3   Methodological for the conservation of medicinal plants.

5.13.1   IN-SITU CONSERVATION

Medicinal plants make great importance for life exist on the earth. They form the backbone of the earth ecosystem and make a viable solution for any type of diseases. *In-situ* conservation of economically important medicinal plants is very important techniques by which valuable species are conserve and protect in their natural habitat or environment. A plant having medicinal properties often produces secondary metabolites under the stimuli of the natural environment which would be lacking under artificial conditions (Coley, 2003; Figueiredo and Grelle, 2009). *In-situ* conservation approach seems to be always beneficial as it promotes conservation of wild population and diversity (Gepts, 2006; Forest et al., 2007).

Therefore, an effective rule and regulation with proper compliance of medicinal plants are valuable measures for the success of *in-situ* conservation (Soule et al., 2005; Volis and Blecher, 2010). Although natural reserve under *in-situ* conservation is one of the commonest types of the protected area which helps to preserve and restore biodiversity (Chiarucci et al., 2001; Rodriguez et al., 2007). Around 13.2 million km² areas of 12,700 protected areas (approximately 9% of global land area) should be aimed to develop throughout the world (Huang, 2002). Wild population domestication under *in-situ* approach can be developed through establishing wild nurseries (Hamilton, 2004; Schippmann et al., 2005; Strandby and Olsen, 2008). Therefore, it is a valuable approach for *in-situ* conservation of endemic, endangered and in-demand medicinal plants (Li and Chen, 2007; Liu et al., 2011).

5.13.2 EX-SITU CONSERVATION

Ex-situ conservation is a good approach for conservation and protection of medicinal plants having slow growth with low abundances and high susceptibility to diseases and is under over-exploitation (Hamilton, 2004; Havens et al., 2006; Yu et al., 2010). Swarts and Dixon (2009) and Pulliam (2000) reported *ex-situ* conservation technique supports large resource base and helps to wild extinction of threatened species. A similar approach can be mediated through botanical garden under which rare and endangered plant species can be conserved for future use (Huang et al., 2002). Botanical gardens have been the representative of diverse taxonomic and ecological flora grown under similar conditions (Primack and Miller-Rushing, 2009). Further, seed bank reflects considerable promise in terms of preservation of genetic diversity of medicinal plants along with their wild relatives (Schoen and Brown, 2001; Li and Pritchard, 2009).

5.14 CULTIVATION PRACTICE

Cultivation practice has opportunities to adopt new improved techniques for solving several problems such as the presence of toxic components, pesticide contamination and low contents of active ingredients, etc. in medicinal plants (Raina et al., 2011). Under captivity, it has been reported that medicinally important compound level increases in plants under the appropriate level of a nutrient, water and environmental conditions (Liu et al., 2011; Wong et al., 2014).

5.15 SCIENTIFIC INTERVENTION FOR PROMOTING CONSERVATION OF MEDICINAL PLANT RESOURCES

From a conservation perspective, medicinal plants can be conserved under *in-situ* and *ex-situ* conservation schemes. *In-situ* conservation comprises national parks, biosphere reserves, sacred groves, medicinal plant conservation areas, joint forest management and some effective legislation and rules related to banning of exports. Likewise, *ex-situ* conservation includes herbal gardens, gene banks, seed banks, cryopreservation, tissue culture repositories, and other cultivation practices. Similarly, for the conservation of biodiversity and protection of endangered species, there is a need to adopt systematic cultivation of medicinal plants. Cultivation of this type of plants could only be promoted if there is a continuous demand for raw materials (Rao and Das, 2001). For scientific propagation screening of high yielding species needs to be done for effective cultivation of medicinal plant resources (Joshi and Joshi, 2014).

5.16 SUSTAINABLE HARVESTING OF MEDICINAL PLANT RESOURCES

Sustainable harvesting and use of medicinal plant are of high importance, play a major role in the conservation of these biological resources. The timing of harvesting, material to be harvested, harvesting techniques and harvesting equipment are the major parameters for achieving sustainable production and management of these resources. Therefore, a sustainable collection of good quality medicinal plants should be promoted which helps in maintaining the ecosystem and environment. In this context during collection sufficient experience is required for non-destructive sampling of medicinal plant species is an essential prerequisite. On the other hand, due consideration should be given for optimum use of these resources so that anthropogenic pressure does not lead to over-harvesting. Under natural condition, proper care should be given to natural regeneration possibility of these species through the building of natural seed bank.

5.17 AWARENESS AND ATTITUDE DEVELOPMENT AMONG COMMUNITY STAKEHOLDERS TO PROMOTE CONSERVATION OF MEDICINAL PLANT RESOURCES

An effective policy for conservation of natural resource relies on community participation. For effective community participation, awareness generation is a mandatory thing. Extension programmes, workshops, seminars, and other social networking systems can be utilized for awareness generation and developing conservative attitude among the stakeholders who are directly engaged with collecting, harvesting, and utilization of medicinal plant resources in the wild.

Proper education and development of technical skill for sustainable harvesting of medicinal plant resource need to be inculcated on local community stakeholders. In this regard, hands-on training, skill development programmes, conservation approach are the essential prerequisite. Over-harvesting needs to be avoided for the conservation of these valuable bio-resources. Local people should be stringent upon the optimum use of these resources.

5.18 FUTURE CHALLENGES FOR MEDICINAL PLANT RESOURCE INDUSTRY

Under the era of globalization, various pharmaceutical industries are actively exploring the medicinal plant resources for their own economic benefit. Such a condition has generated a competitive attitude among the various multinational companies leading to unsustainable harvesting of medicinal plant resources. Biopiracy is itself the biggest challenge to India from foreign multinational companies. Lack of adequate infrastructure, technical expertise, technological supports, and economic backup lead to a challenging condition of all countries in the global market. While there is a definite need to examine some of the complex issues relating to conservation, cultivation and the sustainable utilization of medicinal plants, emerging challenges, and opportunities in the global market require a critical look at the needs of commercialization in this industry. Although India's medicinal plants industry has been involved in the export of raw materials for a long time now, recent developments in the global market place have created new opportunities and challenges for Indian industry (Karki and Johari, 1999).

To conserve indigenous medicinal plant resources bio-piracy by the foreign multinational companies needs to be stopped immediately. Therefore, proper monitoring, the establishment of security set-up along with law enforcement should be strictly implemented. Under the medical treatment system of allopathy use of these medicinal resources has a blink future however they are indirectly over-exploited by the allelopathy system. Therefore, proper marketing and benefits should be shared to the gross population of the country for a total medical remedy.

5.19   CONCLUSION

Medicinal plant not only improves the health status of people but such type of herbal remedy also improves the socioeconomic status of the poor village dwellers. Unscientific exploration and abuse of medicinal plants affect the resource base and genetic diversity which becomes the ultimate cause of their extinction. Therefore, conservation of forest-based wild medicinal plants is very important because it supports livelihood security of tribal's and rural peoples along with maintaining the ecological integrity and biodiversity conservation. In this context, the adoption of in-situ and *ex-situ* conservation techniques could maintain these genetic resources and protect them from negative externalities. Likewise, adoption of some new improved techniques under good cultivation practices has played a great role for improving the contents of active ingredients in medicinal plants and checks the presence of toxic components, and reduces the chances of pesticide contamination. Prior necessity for developing effective policies which work in the field of research and development of medicinal plant resource base and therefore promote sustainability in the harvesting process. Also, people and community participation could be a viable option for betterment of medicinal resources for long-term utilization. Therefore, systematic conservation and large-scale cultivation of the concerned medicinal plants are thus of great importance.

KEYWORDS

- aromatic plants
- biodiversity
- community participation
- conservation areas
- conservation strategy
  cryopreservation
- extinction

- flora
- gene banks
- herbal gardens
- indigenous plants
- medicinal plant
- sustainable harvesting
- wild plant

REFERENCES

1. Akerele, O. (1992). WHO guideline for assessment of herbal medicines, *Fitoterapia*, *63*, 99–118.
2. Anonymous (1997). *Conservation Assessment and Management Plan Workshop Process*. WWF, India.
3. Aswal, B. S., & Mehrotra, B. N. (1994). *Flora of Lahaul-Spiti. (A Cold Desert in Northwest Himalayas)*. Dehradun: Bishen Singh Mahendra Pal Singh, Uttarakhand, India.
4. Baishya, R. A., Sarma, J., & Begum, A. (2013). Forest-based medicinal plants rendering their services to the rural community of Assam, India. *International Journal of Applied Biology and Pharmaceutical Technology, 4*(4), 10–20.
5. Bentley, R. (2010). *Medicinal Plants*. London: Domville-Fife Press; pp. 23–46.
6. Bhardwaj, J., Singh, S., & Singh, D. (2007). Hailstorm induced crop losses in India: some case studies. Abstract for presentation at 4th European Conference on Severe Storms in Trieste, Italy, 10–14, September 2007.
7. Camm, J., Norman, S., Polasky, S., & Solow, A. (2002). Nature reserve site selection to maximize expected species covered. *Oper Res, 50*, 946–55.
8. Cavaliere, C. (2009). The effects of climate change on medicinal and aromatic plants. *Herbal Gram, 81*, 44–57.
9. Chauhan, N. S. (1989). Potential of Aromatic Plants Flora in Himachal Pradesh. *Indian Perfumes, 33*(2), 118–122.
10. Chauhan, N. S. (1990). Medicinal Orchids of Himachal Pradesh. *Journal of Orchids Society of India, 4*, 99–105.
11. Chauhan, N. S. (1999). *Medicinal Aromatic Plants of Himachal Pradesh*. New Delhi: Indus Publishing Co, ISBN 81-7387-098-5, http://hillagricrepository.co.in/id/eprint/1373.

12. Chen, S., Yu, H., Luo, H., Wu, Q., Li, C., & Steinmetz, A. (2016). Conservation and sustainable use of medicinal plants: problems, progress, and prospects. *Chinese Medicine*, pp. 1–10.

13. Chiarucci, A., Maccherini, S., & De Dominicis, V. (2001). Evaluation and monitoring of the flora in a nature reserve by estimation methods. *Biol Conserv.*, *101*, 305–314.

14. Coley, P. D., Heller, M. V., Aizprua, R., Arauz, B., Flores, N., Correa, M., Gupta, M., Solis, P. N., Ortega-Barría, E., Romero, L. I., Gómez, B., Ramos, M., Cubilla-Rios, L., Capson, T. L., & Kursar, T. A. (2003). Using ecological criteria to design plant collection strategies for drug discovery. *Front Ecol Environ, 1,* 421–428.

15. Collett, H. (1902). *Flora Simlensis*. Thacker Spink. & Co Calcutta and Shimla, Reprinted 1971. Dehradun: Bishen Singh Mahendra Pal Singh, Uttarakhand, India.

16. Das, I., Katiyar, P. and, Raj, A. (2014). Effects of temperature and relative humidity on ethephon induced gum exudation in *Acacia nilotica. Asian Journal of Multidisciplinary Studies, 2*(10), 114–116.

17. Dhaliwal, D. S., & Sharma, M. (1997). Phytogeographic comments on the flora of Kullu district. In Sharma, T. A., Saini, S. S., Trivedi, M. L., & Sharma, M. (eds.), *Current researches in plant sciences* Vols. I–II (1997). Dehradun, Bishen Singh Mahendra Pal Singh, *1,* pp. 169–175.

18. Dhaliwal, D. S., & Sharma, M. (1999). *Flora of Kullu District (Himachal Pradesh)*. Dehradun: Bishen Singh Mahendra Pal Singh, Uttarakhand, India.

19. Farnsworth, N. R., & Soejarto, D. D. (1991). *Global Importance of Medicinal Plants.* In: Akerele O., Heywood V., & Synge H. (Eds.). The Conservation of Medicinal Plants. Cambridge University Press, Cambridge, UK, pp. 25–51.

20. Figueiredo, M. S. L., & Grelle, C. E. V. (2009). Predicting global abundance of a threatened species from its occurrence: implications for conservation planning. *Divers Distri, 15,* 117–121.

21. Forest, F., Grenyer, R., Rouget, M., Davies, T. J., Cowling, R. M., Faith, D. P., Balmford, A., Manning, J. C., Proche, S., Bank, M., Reeves, G., Terry, A. J., & Savolainen, V. (2007). Preserving the evolutionary potential of floras in biodiversity hotspots. *Nature, 445,* 757–760.

22. Gepts, P. (2006). Plant genetic resources conservation and utilization: the accomplishments and future of a societal insurance policy. *Crop Sci., 46,* 2278–2292.

23. Ghasemzadeh, A., & Jaafar, H. Z. E. (2011). Effect of CO_2 Enrichment on Synthesis of Some Primary and Secondary Metabolites in Ginger (*Zingiber officinale* Roscoe). *Int. J. Mol. Sci., 12,* 1101–1114.

24. Hamilton, A. C. (2004). Medicinal plants, conservation, and livelihoods. *Biodivers Conserv., 13,* 1477–1517.

25. Hamilton, A. C. (2008). Medicinal plants in conservation and development: case studies and lessons learned. In: Kala CP, editor. *Medicinal Plants in Conservation and Development.* Salisbury: Plantlife International Publisher; p. 1–43.

26. Havens, K., Vitt, P., Maunder, M., Guerrant, E. O., & Dixon, K. (2006). Ex situ plant conservation and beyond. *Bioscience, 56,* 525–31.

27. He, S. A., & Gu, Y. (1997). The challenge for the 21st Century for Chinese botanic gardens. In: Touchell D.H., & Dixon K.W. (eds), Conservation into the 21st Century. Proceedings of the 4th International Botanic Gardens Conservation Congress (Perth, 1995). Kings Park and Botanic Garden, Perth, Australia, pp. 21–27.

28. Heywood, V. H., & Iriondo, J. M. (2003). Plant conservation: old problems, new perspectives. *Biol Conserv., 113,* 321–35.
29. Huang, H., Han, X., Kang, L., Raven, P., Jackson, P. W., & Chen, Y. (2002). Conserving native plants in China. *Science, 297,* 935.
30. Huxley, A. (1984). *Green Inheritance: The World Wildlife Fund Book of India,* Collins/ Harvel, London, pp. 193, ISBN-13: 978-0385196031, ISBN-10: 0385196032.
31. Idso, S. B., Kimball, B. A., Petit III G. R., Graner, L. C and Backhaus, R. A. (2000). Effects of atmospheric CO_2 enrichment on the growth and development of *Hymnocallis littoralis* and the concentrations of several antineoplastic and antiviral constituents of its bulbs. *American Journal of Botany, 87,* 769–773.
32. Jain, S. K. (1991). *Dictionary of Indian Folk Medicine and Ethnobotany.* New Delhi: Deep Publications.
33. Jhariya, M. K., & Raj, A. (2014). Human welfare from biodiversity. *Agrobios Newsletter, 12*(9), 89–91.
34. Jhariya, M. K., Raj, A., Sahu, K. P., & Painkra, P. R. (2013). Neem-A Tree for Solving Global Problem. *Indian Journal of Applied Research, 3*(10), 66–68.
35. Joshi, B. C., & Joshi, R. K. (2014). The Role of Medicinal Plants in Livelihood Improvement in Uttarakhand. *International Journal of Herbal Medicine, 1*(6), 55–58.
36. Kala, C. P. (2008). High altitude medicinal plants: A promising resource for developing herbal sector. *Hima-Paryavaran, 20*(2), 7 9.
37. Karki, M., & Johari, R. (1999). Medicinal Plants Industry. In: Fostering Biodiversity Conservation and Rural Development. A Combined Report On The Proceedings of a National Colloquium Held on December 16–17, 1997, in New Delhi, India and A Workshop on Medicinal Plants as the Basis for the Relationship Between Industries and Rural Communities Held on February 17, 1998, in Bangalore, India. pp. 120.
38. Kaur, H., & Sharma, M. (2004). *Flora of Sirmaur* (Himachal Pradesh). Dehradun: Bishen Singh Mahendra Pal Singh, Uttarakhand, India.
39. Kumar, M. R., & Janagam, D. (2011). Export and import pattern of medicinal plants in India. *Indian Journal of Science and Technology, 4*(3), 245–248.
40. Lange, D. (2002). The role of east and southeast Europe in the medicinal and aromatic plants' trade. *Medicinal Plant Conservation, 8,* 14–18.
41. Li, D. Z., & Pritchard, H. W. (2009). The science and economics of ex situ plant conservation. *Trends Plant Sci., 14,* 614–21.
42. Li, X. W., & Chen, S. L. (2007). Conspectus of eco-physiological study on medicinal plant in wild nursery. *Zhong Guo Zhong Yao ZaZhi., 32,* 1388–92.
43. Litvak, M. E., Constable, J. V. H., & Monson, R. K. (2002). Supply and demand processes as controls over needle mononterpene synthesis and concentration in Douglas fir [*Pseudotsuga menziesii*(Mirb.) Franco]. *Oecologia, 132,* 382–391.
44. Liu, C., Yu, H., & Chen, S. L. (2011). Framework for sustainable use of medicinal plants in China. *Zhi Wu Fen Lei Yu Zi Yuan XueBao, 33,* 65–68.
45. Loreto, F., Barta, C., Brilli, F., & Nogues, I. (2006). On the induction of volatile organic compound emissions by plants as consequence of wounding or fluctuations of light and temperature. *Plant Cell Environ., 29,* 1820–1828.

46. Maheswari, J. (2011). Patenting Indian medicinal plants and products. *Indian Journal of Science and Technology*, 4(3), 298–301.
47. Malcolm, J. R., Liu, C., Neilson, R. P., Hansen, L., & Hannah, L. (2006). Global warming and extinctions of endemic species from biodiversity hotspots. *Conservation Biology, 20*(2), 538–548.
48. Mehta, R., Singh, P., & Singh, A. K. (2015). Medicinal plant diversity, its traditional use, and phytochemistry at Dugli forest area of Chhattisgarh, India. *Global Journal of Multidisciplinary Studies*, 4(7), 184–189.
49. Moerman, D. E. (1998). *Native North American food and medicinal plants: epistemological considerations.* In: Prendergast, H. D. V., Etkin, N. L., Harris, D. R., & Houghton, P. J. (eds) Plants for Food and Medicine. Proceedings from a Joint Conference of the Society for Economic Botany and the International Society for Ethnopharmacology, London, 1–6 July 1996. Royal Botanic Gardens, Kew, UK, pp. 69–74.
50. Mooney, H. A., Winner, W. E., & Pell, E. J. (1991). Response of plants to multiple stresses. Academic Press, San Diego, California, USA.
51. Myers, N., Muttermeier, R. A., Muttermeier, C. A., da Fonseca, A. B. G., & Kent, J. (2000). Biodiversity hotspots for conservation priorities. *Nature, 403,* 853–858.
52. NABARD (2011). Neem- A Versatile Tree. Model Bankable Projects. http://www.nabard.org/modelbankprojects/forestrywasteland.asp.
53. Nalawade, S. M., Sagare, A. P., Lee, C. Y., Kao, C. L., & Tsay, H. S. (2003). Studies on tissue culture of Chinese medicinal plant resources in Taiwan and their sustainable utilization. *Bot Bull Acad Sin., 44,* 79–98.
54. National Medicinal Plant Board (NMPB) (2006). Demand and Supply of Medicinal Plants in India. NMPB & FRLHT, Bangalore, National Medicinal Plants Board, New Delhi, India.
55. Nayar, M. P., & Sastry, A. R. K. (1987, 1988, 1990). *Red Data Book of Indian Plants*, Vol. I–III. Calcutta: Botanical Survey of India.
56. Painkra, V. K., Jhariya, M. K., & Raj, A. (2015). Assessment of knowledge of medicinal plants and their use in tribal region of Jashpur district of Chhattisgarh, India. *Journal of Applied and Natural Science*, 7(1), 434–442.
57. Pei, S. (2002a). A brief review of ethnobotany and its curriculum development in China. In: Shinwari, Z. K., Hamilton, A., & Khan, A. A. (eds) Proceedings of a Workshop on Curriculum Development in Applied Ethnobotany, Nathiagali, 2–4 May 2002. WWF-Pakistan, Lahore, Pakistan.
58. Phondani, P. C., Negi, V. S., Bhatt, I. D., Maikhuri, R. K., & Kothyari, B. P. (2011). Promotion of Medicinal and Aromatic Plants Cultivation for Improving Livelihood Security: A Case Study from West Himalaya, India. *Int. J. Med. Arom. Plants, 1*(3), 245–252.
59. Prakasha, H. M., Krishnappa, M., Krishnamurthy, Y. L., & Poornima, S. V. (2010). Folk medicine of NR PuraTaluk in Chikamaglur district of Karnataka. *Indian Journal of Traditional Knowledge*, 9(1), 55–60.
60. Primack, R. B., & Miller-Rushing, A. J. (2009). The role of botanical gardens in climate change research. *New Phytol., 182,* 303–313.
61. Pulliam, H. R. (2000). On the relationship between niche and distribution. *Ecol Lett., 3,* 349–361.

62. Rai, L. K., Pankaj, P., & Sharma, E. (2000). Conservation threats to some important plants of the Sikkim Himalaya. *Biodiversity and Conservation, 93,* 27–33.

63. Raina, R., Chand, R., & Sharma, Y. P. (2011). Conservation strategies of some important medicinal plants. *Int J Med Aromat Plant, 1,* 342–347.

64. Raj A. (2015a). *Evaluation of Gummosis Potential Using Various Concentration of Ethephon.* M.Sc. Thesis, I.G.K.V., Raipur (C.G.), pp. 89.

65. Raj A. (2015b). Gum exudation in Acacia nilotica: effects of temperature and relative humidity. In Proceedings of the National Expo on Assemblage of Innovative ideas/ work of postgraduate agricultural research scholars, Agricultural College and Research Institute, Madurai (Tamil Nadu), pp. 151.

66. Raj, A., & Singh, L. (2017). Effects of girth class, injury and seasons on Ethephon induced gum exudation in *Acacia nilotica* in Chhattisgarh. *Indian Journal of Agroforestry, 19*(1), 36–41.

67. Raj, A., & Toppo, P. (2014). Assessment of floral diversity in Dhamtari district of Chhattisgarh. *Journal of Plant Development Sciences, 6*(4), 631–635.

68. Raj, A., Haokip, V., & Chandrawanshi, S. (2015). *Acacia nilotica:* a multipurpose tree and source of Indian gum Arabic. *South Indian Journal of Biological Sciences, 1*(2), 66–69.

69. Ramawat, K. G., & Goyal, S. (2008). *The Indian Herbal Drugs Scenario in Global Perspectives.* In: Bioactive Molecules and Medicinal Plants. Ramawat, K. G., & Merillon, J. M. (eds.) Springer, pp. 323–345.

70. Rao, B. R. R., & Rajput, D. K. (2010). Global Scenario of Medicinal Plants. CMP-HPU-2010, Conference Proceedings, pp. 17–20.

71. Rao, N. S., & Das, S. K. (2001). Herbal gardens of India: A statistical analysis report. *African J Biotechnol, 10*(31), 5861–5868.

72. Rao, R. R. (2006). Proceedings of the 29th All India Botany Conference, Oct 9–11, Department of Botany, ML Sukhadia University, Udaipur, India, Abstract 1.

73. Rodriguez, J. P., Brotons, L., Bustamante, J., & Seoane, J. (2007). The application of predictive modeling of species distribution to biodiversity conservation. *Divers Distrib., 13,* 243–51.

74. Ross, I. A. (2005). Medicinal plants of the world (volume 3): chemical constituents, traditional and modern medicinal uses. New Jersey: Humana Press Inc; p. 110–132.

75. Salick, J., Fangb, Z., & Byg, A. (2009). Eastern Himalayan alpine plant ecology, Tibetan ethnobotany, and climate change. *Global. Environ. Chang., 19*(2), 147–155.

76. Samant, S. S., & Dhar, U. (1997). Diversity, endemism and economic potential of wild edible plants of Indian Himalaya. *International Journal of Sustainable Development and World Ecology, 4,* 179–191.

77. Samant, S. S., & Pant, S. (2003). Diversity, distribution pattern and traditional knowledge of Sacred Plants in Indian Himalayan Region. *Indian Journal of Forestry, 26*(3), 201–213.

78. Samant, S. S., Dhar, U., & Palni, L. M. S. (1998). *Medicinal Plants of Indian Himalaya: Diversity Distribution Potential Values,* Nainital: *Gyanodaya Prakashan,* pp. 455–457.

79. Samant, S. S., Pant, S., Singh, M., Lal, M., Singh, A., Sharma, A., & Bhandari, S. (2007). Medicinal plants in Himachal Pradesh, northwestern Himalaya, India. *International Journal of Biodiversity Science and Management, 3,* 234–251.

80. Schippmann, U., Leaman, D. J., & Cunningham, A. B. (2002). Impact of Cultivation and Gathering of Medicinal Plants on Biodiversity: Global Trends and Issues. Inter-Department Working Group on Biology Diversity for Food and Agriculture, FAO, Rome, Italy.

81. Schippmann, U., Leaman, D., & Cunningham, A. B. (2006). A comparison of cultivation and wild collection of medicinal and aromatic plants under sustainability aspects. In: R. J. Bogers, L. E. Craker & D. Lange (eds.). *Medicinal and Aromatic Plants*, pp. 75–95. Springer, The Netherlands.

82. Schippmann, U., Leaman, D. J., Cunningham, A. B., & Walter, S. (2005). Impact of cultivation and collection on the conservation of medicinal plants: global trends and issues. III WOCMAP Congress on medicinal and aromatic plants: conservation, cultivation and sustainable use of medicinal and aromatic plants, Chiang Mai.

83. Schoen, D. J., & Brown, A. H. D. (2001). The conservation of wild plant species in seed banks. *Bioscience, 51,* 960–966.

84. Sharma, M., & Dhaliwal, D. S. (1997a). Additions to the Flora of Himachal Pradesh from Kullu District. *Journal of Bombay Natural History and Society, 94*(2), 447–50.

85. Sharma, M., & Dhaliwal, D. S. (1997b). Biological Spectrum of the flora of Kullu district (Himachal Pradesh). *J Ind Bot Soc, 76,* 283–294.

86. Sharma, M., & Singh, H. (1996). Phytogeographic observations on the Flora of Chamba District (Himachal Pradesh). Part-II. *Neo Botanica, 23,* 103–112.

87. Sheikh, K., Ahmad, T., & Khan, M. A. (2002). Use, exploitation and prospects for conservation: people and plant biodiversity of Naltar Valley, northwestern Karakorum, Pakistan. *Biodivers Conserv, 11,* 715–742.

88. Shiva, V. (1996). Protecting our Biological and Intellectual Heritage in the Age of Biopiracy. The Research Foundation for Science, Technology and Natural Resources Policy, New Delhi, India.

89. Singh, D. K., & Hajra, P. K. (1996). Floristic diversity. In: Gujral, G. S., Sharma, V. (Eds.) *Biodiversity Status in the Himalaya*. New Delhi: British Council; 23–38.

90. Singh, S. K., & Rawat, G. S. (2000). *Flora of Great Himalayan National Park; Himachal Pradesh*. Dehradun: Bishen Singh Mahendra Pal Singh, Uttarakhand, India.

91. Snow, M. D., Bard, R. R., Olszyk, D. M., Minster, L. M., Hager, A. N., & Tingey, D. T. (2003). Monoterpene levels in needles of Douglas fir exposed to elevated CO_2 and temperature. *Physiol. Plantarum., 117,* 352–358.

92. SOSSMBC (Society for Social Services Madhya Bharat Chapter) 2010. Assessment of export potential of medicinal plants and its derivatives from Chhattisgarh and its Policy Implications. Chhattisgarh State Medicinal Plants Board (CGSMPB), Raipur Chhattisgarh, pp. 171.

93. Soule, M. E., Estes, J. A., Miller, B., & Honnold, D. L. (2005). Strongly interacting species: conservation policy, management, and ethics. *Bioscience, 55,* 168–176.

94. Strandby, U., & Olsen, C. S. (2008). The importance of understanding trade when designing effective conservation policy: the case of the vulnerable *Abies guatemalensis Rehder. Biol Conserv., 141,* 2959–68.

95. Swarts, N. D., & Dixon, K. W. (2009). Terrestrial orchid conservation in the age of extinction. *Ann Bot., 104,* 543–56.

96. Toledo, V. M. (1995). New paradigms for a new ethnobotany: reflections on the case of Mexico. In: Schultes, R. E., & von Reis, S. (Eds) *Ethnobotany: Evolution of a Discipline*. Chapman & Hall, London, pp. 75–88.

97. Toppo, P., Raj, A., & Harshlata (2014). Biodiversity of woody perennial flora in Badal Khole sanctuary of Jashpur district in Chhattisgarh. *Journal of Environment and Bio-Sciences, 28*(2), 217–221.

98. Toppo, P., Raj, A., & Jhariya, M. K. (2016). Agroforestry systems practiced in Dhamtari district of Chhattisgarh, India. *Journal of Applied and Natural Science, 8*(4), 1850–1854.

99. UNU-IAS. (2012). Biodiversity, Traditional Knowledge and Community Health: Strengthening Linkages, Yokohama 220-8502, Japan.

100. Ved, D. K., & Goraya, G. S. (2008). Demand and Supply of Medicinal Plants in India. Bishen Singh Mahendra Pal Singh, Dehra Dun, and FRLHT, Bangalore, National Medicinal Plants Board, New Delhi, India.

101. Ved Prakash (2001). Indian Medicinal Plants: Current Status. In: Samant, S. S., Dhar, U., Palni, L. M. S. (Eds), *Himalayan Medicinal Plants: Potential and Prospects*. Nainital, Gyanodaya Prakashan, pp. 45–65.

102. Volis, S., & Blecher, M. (2010). Quasi in situ: a bridge between ex-situ and in situ conservation of plants. *Biodivers Conserv., 19,* 2441–2454.

103. Walter, K. S., & Gillett (1998). 1997 IUCN Red List of Threatened Plants. Gland, Switzerland, and Cambridge, UK.

104. WHO (2002). Traditional Medicine Strategy 2002–2005.

105. WHO (2003). Guidelines for the Assessment of Herbal Medicine Programme on Traditional Medicine. Doc. World Health Organization WHO/TRM/91.4.WHO, Geneva.

106. Wong, K. L., Wong, R. N., Zhang, L., Liu, W. K., Ng, T. B., Shaw, P. C., Kwok, P. C. L., Lai, Y. M., Zhang, Z. J., Zhang, Y. B., Tong, Y., Cheung, H., Lu, J., & Wing, S. C. (2014). Bioactive proteins and peptides isolated from Chinese medicines with pharmaceutical potential. *Chin Med., 9,* 19.

107. Xiao P. G., & Yong P. (1998). Ethnopharmacology and research on medicinal plants in China. In: Prendergast, H. D. V., Etkin, N. L., Harris, D. R., & Houghton, P. J. (Eds) Plants for Food and Medicine. Proceedings from a Joint Conference of the Society for Economic Botany and the International Society for Ethnopharmacology, London, 1–6 July 1996. Royal Botanic Gardens, Kew, UK, pp. 31–39.

108. Yu, H., Xie, C. X., Song, J. Y., Zhou, Y. Q., & Chen, S. L. (2010). TCMGIS-II based prediction of medicinal plant distribution for conservation planning: a case study of *Rheum tanguticum. Chin Med., 5*: 31.

109. Zerabruk, S., & Yirga, G. (2012). Traditional knowledge of medicinal plants in Gindeberet district, Western Ethiopia. *S Afr J Bot, 78,* 165–169.

110. Ziska, L. H. (2005). The impact of recent increases in atmospheric CO_2 on biomass production and vegetative retention of Cheatgrass (Bromus tectorum): Implications for fire disturbance. *Global Change Biology, 11,* 1325–1332.

CHAPTER 6

Silvo-Pasture System: A Way Ahead for Sustainable Development in India

S. SARVADE[1] and V. B. UPADHYAY[2]

[1]Assistant Professor, College of Agriculture, Balaghat, Jawaharlal Nehru Krishi Vishwa Vidyalaya, Jabalpur–481 331 (MP), India, Mobile: +00-91-8989851720, E-mail: somanath553@gmail.com

[2]Dean, College of Agriculture, Balaghat, Jawaharlal Nehru Krishi Vishwa Vidyalaya, Jabalpur–481 331 (MP), India, Mobile: +00-91-9179085067, E-mail: deanbalaghat@gmail.com

ABSTRACT

Demands of green and dry fodder are steadily increasing from the last two-three decades in India due to the increasing trend of livestock population. India leads in milk production in the world. Rather than focusing on fodder production on agriculture field, we have to focus more on the forest (69.41 million ha.) and resources other than the forests which were used by common peoples (54.1 million ha.), which are the main grazing resources. Whereas, the country's 146.82 million hectare area comes under different land degradation categories. An establishment of silvo-pasture systems on such areas is a big opportunity to bridge the gap between production and supply of fodder. At an environmental, economic and social level, such production practices help in the country's sustainable development. This system has lots of direct and indirect benefits such as yields fodder, fuelwood, biomass production, carbon sequestration, carbon storage and other environment ameliorating benefits. The 5.58 M ha area is under the silvo-pasture system in India, is the potential area for fodder production. The perennial nature of woody tree species and perennial grass species, the system has advantages over other systems for biomass production and carbon storage potentials. Biomass production and carbon storage in vegetation and in the soil help to improve fodder production and environmental conditions in coming years, when the country's whole degraded area brought under the silvo-pasture system.

6.1 INTRODUCTION

The 16.7% of the world's human population (Sarvade, 2014a) and 18% livestock population (Sarvade et al., 2017) is supported by the 329 million ha. area of the country which accounts for 2.4% of the world's area. The country's population showed an increasing trend which realized many problems relating to the daily needs of the people. For feeding such huge population, country farmers go for intensive agriculture practices, which leads to the degradation of soil and environmental health conditions (Sarvade et al., 2014).

Livestock production and agriculture are linked with each other and are most important for food security in India. Milk production is most important an integral part of food at the global level, where India is the leading milk producing country in the world (Singh and Tulachan, 2002; Gangasagare and Karanjkar, 2009). The high milk production creates 'White Revolution' in India, and millions of smallholders are the pillars of revolution. Most of the required forage for livestock production comes from the agriculture fields, grasslands, forest, and agroforests. The intensive agriculture practices of the country for food grain production (Kushwaha, 2008; Sarvade et al., 2014; Anonymous, 2016) gives less attention towards forage production; whereas, the condition of grasslands is not so good to produce high quantity of green and dry fodders (Dabadghao and Shankarnayan, 1973; Pathak and Roy, 1994).

In the present situation, forest ecosystems are degrading and exploited more for developmental activities. Agroforests are the best options (resources) for fodder production on a sustainable basis. Their components have the potential for the conservation of natural resources. Such appliance helps in sustainable developmental activities. Without solving the problem to global socioeconomic and environmental degradation, we cannot go for developmental activities. Sustainable development involves the meeting of people's present needs with confirming the ability of future generations to meet out their needs (Harris, 2000). Sustainable development (SD) is aimed at the long-term stability of the economy and environment which will be achieved through the integration of economic, environmental, and social concerns. The concept of sustainability describes a healthy and dynamic condition of the different systems, which creates a balance between productive and protective functions for the well-being of human (Heintz, 2004).

In the case of agroforests, silvo-pasture system is best for fodder production on a sustainable basis. This system has lots of direct and indirect benefits such as yields fodder, fuelwood, biomass production, carbon sequestration, carbon storage and other environment ameliorating benefits (Feldhake, 2001, 2002; Garrett et al., 2004; Anonymous, 2013). This system produces high biomass and stores carbon in aboveground vegetation biomass and in belowground vegetation biomass with carbon stored in the soil. At an environmental, economic and social level, silvo-pasture is sustainable which helps in sustainable development (Mosquera-Losada et al., 2005).

Complex nature of silvo-pasture creates habitat diversity in above and below ground area through their different functions mentioned above for stability of an environment. Perennial fodder tree and shrub species have more potential for nutrient cycling and conservation of natural resources. Many researchers concluded that the silvo-pasture system has the potential for combating climate change through high carbon sequestration and storage in vegetation and soil (Gordon et al., 2005). As the multifunctional nature of the system, it can provide a range of economic benefits and ecosystem services. The diversity of products such as fodder, fuelwood, and other NTFPs were obtained which improves the economic output of the system. Based on such outputs, silvo-pasture is more profitable than grassland (McAdam et al., 1999; Doyle and Thomas, 2000; Ram, 2009).

6.2   MATERIALS AND METHODS

The study is completely based on a literature review. The various forms of published and unpublished journal articles, books and book chapters, reports, conference proceedings, management, development plans were collected relating to the biomass production and carbon storage in different components of the silvo-pasture systems. For this, we searched documents online Research Gate, Google Scholar, Springer, etc. (Anand and Radhakrishna, 2017; Sarvade et al., 2017).

The keywords such as livestock population, gazing resources, deficit in green and dry fodder production, silvo-pasture system, system components, study zones for agroforestry, biomass production, carbon storage, Soil carbon storage, etc. used for the compilation of the information to the formulation of this manuscript.

6.3   RESULTS AND DISCUSSION

6.3.1   *LIVESTOCK STATUS*

Livestock population of the country was highest in 2007 and showing an increasing trend from the year 1992. The livestock census of 2012 reported 512.10 population of livestock (Figure 6.1), in which cattle contributes highest (37.28%) followed by goat, buffalo, sheep, pigs and other (Figure 6.2). Livestock population has increased significantly in Gujarat (15.36%), Uttar Pradesh (14.01%), Assam (10.77%), Punjab (9.57%), Bihar (8.56%), Sikkim (7.96%), Meghalaya (7.41%), and Chhattisgarh (4.34%). Cow and buffalo population has increased (6.75%) from 111.09 to 118.59 million. It contributes 4.11% in total GDP of the country during 2012–13 (Anonymous 2012), and provides an employment and the livelihood for 70% peoples of rural areas (Roy and Singh, 2013).

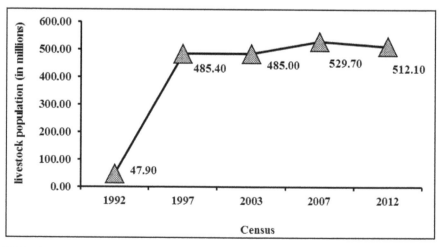

FIGURE 6.1   Trend in livestock population from 1992 to 2012.

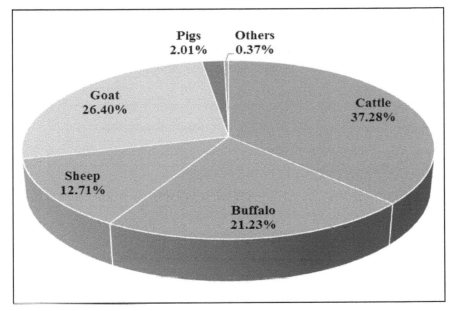

FIGURE 6.2 (See color insert.) Share of different livestock categories in the total population (2012).

6.3.2 GRAZING RESOURCES OF INDIA

Forest (69.41 million ha) and common property resources other than forests (54.1 million ha.) contribute higher in available gazing resources in India (Figure 6.3). They contribute 22.70% and 17.70%, respectively (Anonymous, 2013). Primary productivity of such gazing resources of India is very low as compared to grazing lands of the world (Dabadghao and Shankarnayan, 1973; Shankar and Gupta, 1992; Pathak and Roy, 1994). They reported 0.20 to 1.47 adult cattle units (ACU)/ha, carrying capacities of these areas but the present stocking rates are much higher. The 1 to 51 ACU/ha against the carrying capacity of 1 ACU/ha is the present stocking rates in semi-arid areas, whereas 1 to 4 ACU/ha against the carrying capacity of 0.2–0.5 ACU/ha in arid areas. Therefore, their restoration through silvo-pasture systems is needed. National Bureau of Soil Survey & Land Use Planning (NBSS & LUP) in 2004 reported 146.82 million ha. area as a degraded land under different land degradation categories (Anonymous, 2010). This is a potential area for development of grasslands and other fodder producing systems, which helps in bridging the gap between fodder (green and dry) demand and supply. In the year 2010, 10.95% and 35.66% dry and green fodder was a deficit, which is predicted as 13.20% and 18.43% in the year 2050 (Table 6.1).

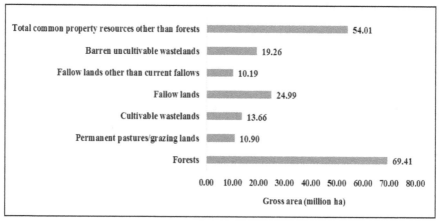

FIGURE 6.3   Grazing resources available in India.

TABLE 6.1   Net (million tons) and Percent (%) Deficit in Dry and Green Forages

Year	Net deficit		% Deficit	
	Dry	**Green**	**Dry**	**Green**
2010	55.72	291.3	10.95	35.66
2020	62.85	260.9	11.85	30.65
2030	68.07	224.2	11.98	24.59
2040	70.57	193.0	11.86	20.22
2050	83.27	186.6	13.20	18.43

Source: Ghosh et al. (2016).

6.3.3   WHAT IS SILVO-PASTURE?

Silvo-pasture system is an ecologically viable land use system in which growth of quality fodder yielding crops, grasses, legumes, and trees in combination on the same land unit. The term 'silvo' means 'tree,' and 'pasture' means 'grasses' or 'grass + legume' mixtures. Trees on pasture lands (Silvo-pastoral systems) – trees + pastures/grasses + animals.

Silvo-pasture combines fodder yielding tree species with forage yielding grasses/crops and livestock production. Integration of trees and ruminants is very complex due to the diverse system components and dynamic nature of biophysical factors and the natural production resources. There are several terminologies, but the term Silvo-pastoral systems are being widely accepted and repeatedly used for defining such systems (Devendra, 2012, 2014). According to him, the Silvo-pastoral system involves trees (e.g., coconuts, oil palm, and rubber) and animals. Silvo-pasture can be developed through using four different approaches viz. Scattered trees and shrubs on pasture, Intensive Silvo-pastoral (fodder trees with agri-fodder crops), live fence of fodder trees or boundary systems and hedge, and protein bank (cut and carry), explained in Figure 6.4, Plates 6.1 and 6.2 (Nair, 1984; Dwivedi, 1992; Nair, 1993; Puri, 2007; Panda, 2013; Gurung and Temphel, 2015; Soni et al., 2016; Mathukia et al., 2016).

According to Klopfenstein et al. (2008) and Dwivedi (1992), silvo-pasture is an agroforestry practice, specifically designed and managed for the growing of trees, shrubs, grasses, bamboo, and crops to get tree products, forage, and livestock. The principles of silvo-pasture system are as follows (Nair, 1984; Dwivedi, 1992; Martin and Sherman, 1992; Nair, 1993).

6.3.3.1 PRINCIPLES OF SILVO-PASTURE SYSTEM

- System should involve two or more fodder yielding plant; at least one of them should be woody perennial.
- System has to produce two or more outputs, where the main output should be forage production.
- Structurally, functionally, ecologically and economically system is multifaceted than a mono-cropping system.
- Should increases–number of grasses, legumes, trees, and animals.
- Should assure maximum resource combinations.
- Should optimize the production of fodder from the same unit of land.
- Should improve environmental conditions and should have rich biodiversity.

The fodder yielding tree species should have multipurpose uses such as provide high-value timber, firewood, shade and shelter for livestock, forage production, stress reduction and ameliorating micro-climate which helps in increasing forage production. Many scientists and researchers

(Huxley and Westley, 1989; Mishra et al., 2009; Chhetri, 2010; Devendra and Leng, 2011; Dawson et al., 2014; Gunasekaran et al., 2014; Soni et al., 2016) studied on different fodder tree species and identified some important traits which we have to take in consideration while selecting tree species for fodder production under silvo-pasture system.

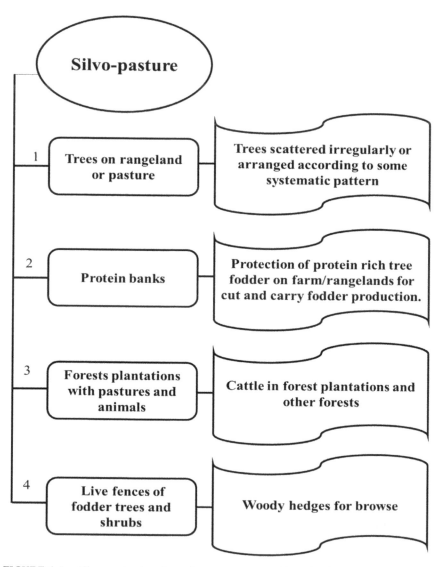

FIGURE 6.4   Silvo-pastoral systems (trees + pasture and/or animals).

6.3.3.2 SUITABLE TRAITS OF FODDER TREE SPECIES

- Easy to establish and adoptive under existing agro-ecological situations.
- The orientation of tree row should be in an east-west direction for minimizing competition between tree and grass/legumes.
- Tolerance to side shading.
- Sparse branching and optimum light penetration to the ground.
- Suitable for fodder with an ability to tolerate lopping, pruning and coppicing as well as having resistant to pest and diseases.
- MPTs (Multipurpose trees and shrubs) which are able to produce fodder, fuel, fruit, food, and fiber.
- Leaf litters having rapid decomposition ability, e.g., *Leucaena, Grevillea*, etc.
- Ability to fix atmospheric nitrogen: species should have the nitrogen-fixing ability. Most of the leguminous species are Nitrogen Fixing Tree species (NFTs).
- No allelopathic effect.
- Deciduous nature, leaf falling during crop season like *Melia azedarach*.
- Should not act as an alternative host for common insect pest and pathogen in the crop field.
- Should yield nutritive and palatable fodder.
- Absence of competition at root zone level for moisture and nutrients.

Likewise, the suitable characteristics of the fodder yielding tree are important for high productivity; the grasses and crop species also have more importance, because quality fodder is needed for high productivity of livestock in India.

6.3.3.3 SUITABLE TRAITS OF FODDER YIELDING GRASSES AND CROPS

- Ease to establish
- Resistance to drought condition
- Resistance to insect-pest attack
- Should have high palatability

- Should have high crude protein
- Should have high digestible crude protein
- Should have high total digestible nutrients
- Should have high crude fiber (%)
- Should have high Ca (%) and P (%)
- Should produce high aboveground and belowground biomass
- Should store high aboveground and belowground carbon
- Should store a high amount of soil carbon

Cubbage et al. (2012) reported that the silvo-pasture is a specific type of agroforestry and it has adequate conservation objectives and livelihoods security in small-scale cattle-farming based social-ecological systems. In general, the silvi-pastoral system benefited through diversify farm enterprise, enhanced the growth of trees, reduced stress and improved animal productivity, increased nutrient cycling and enhanced wildlife habitat.

6.3.4   STATUS OF SILVO-PASTURE SYSTEMS: GLOBAL SCENARIO

Forest or woodland and open forest areas are most important for traditional or natural silvo-pasture practices in many countries. In case of improved systems, quality fodder crops and grass species are incorporated with perennial fodder yielding tree and shrub species. The lack of definite procedures and the very complex nature of SPs becomes an obstacle for area estimation. Whereas, Beer et al. (2000) estimated 9.2 M ha area under silvo-pasture systems in Central America and Nair et al. (2009) was estimated 516 M ha globally under Silvo-pastoral systems and 307 M ha area under other agroforestry systems. Whereas, Reisner et al. (2007) estimated 65.2 M ha under European silvo-arable systems. Dixon (1995), reported 585–1215 M ha area covered by agro-Silvo-pastoral and agroforestry systems in Africa, Asia, and America. Tree components and understory components of silvo-pasture systems practicing in different countries of the world are given in Table 6.2.

6.3.5   STATUS OF SILVO-PASTURE SYSTEMS IN INDIA

India accounts 5.58 M ha area under silvo-pasture (trees on grazing/ rangelands) system (NRCAF, 2007). The components of the system have

TABLE 6.2 Status of Silvopasture Systems in the World

Region country	Main tree species component	Main understory component
Mesopotamia, Argentina	Mainly evergreen *Pinus taeda*, *P. elliotti*, and hybrid pines	*Brachiaria brizantha, Axonopus spp.*
Delta, Argentina	Deciduous *Populus spp., Salix spp.*	Natural grasslands, *Lolium multiflorum, Brommus catharticus*
Chaco, Argentina	*Schinopsis lorentzii, Aspidosperma uebrachoblanco Prosopis spp.*	Native grasslands, *Cenchrus ciliaris* cv. Texas (Buffel grass); *Panicum maximum* (Gatton panic)
Patagonia, Argentina	*Pinus ponderosa, Deciduous, Nothofagus antarctica*	Natural grasslands, *Festuca pallescens, Dactylis glomerata, Trifolium repens*
Arid and semiarid zones, Chile	*Acacia spp., Atriplex spp., Prosopis spp.*	Alfalfa
Temperate Mediterranean zones, Chile	*Pinus radiata, P. pinea,*	Natural grasslands,
	Nothofagus obliqua, Juglans regia, Prunus avium, Populus spp.	*Avena strigosa, A. sativa, Dactylis glomerata, Festuca spp., Lolium perenne L. multifl orum, Medicago sativa, Phalaris acuatica, Trifolium repens, T. subterraneum*
Patagonia, Chile	*Pinus contorta, P. ponderosa, Pseudotsuga menziesii, Nothofagus antarctica, N. pumilio*	*Dactylis glomerata, Festuca sp., Medicago sativa, Trifolium spp., Lolium perenne, Holcus lanatus, Poa pratensis*
Cool zone, Southern Brazil	*Eucalyptus spp., Pinus spp., Acacia mearnsii, Grevillea Robusta, Corymbia spp., Araucaria angustifolia* and various native trees	*Brachiaria spp., Paspalum spp.,* Ryegrass, other annual and perennial temperate and tropical forages
North America	*Pinus taeda,* black walnut, *Quercus douglasii, Quercus wislizenil* and *Pinus sabiniana*	*Bromus inermis, Poa pratensis, Trifolium repens, Kummerowia spp.*

TABLE 6.2 *(Continued)*

Region country	Main tree species component	Main understory component
Colombia	*Leucaena leucocephala, Tithonia diversifolia, Crescentia cujete, Erythrina fusca, Gliricidia sepium, Guazuma ulmifolia, Moringa Oleifera, Cratylia argentea, Acacia decurrens, and Sambucus peruviana*	Natural grasses
Australia	*Eucalyptus spp. Leucaena leucocephala, Acacia aneura, Prosopis spp. Faidherbia albida and Gleditsia triacanthos*	*Themeda spp., Bothriochloa spp., Heteropogon contortus and Sorghum spp. Stylosanhes spp.*
Queensland	*Leucaena leucocephala*	*Chloris gayana, Panicum maximum, Cenchrus ciliaris, Chloris gayana, Panicum maximum or Cenchrus ciliaris*
Africa	*Pinus radiata, Acacia melifiora,*	*Brachiaria humidicola B. brizantha, B. decumbens, B. mutica Cenchrus ciliaris L., Chloris gayana, Kunth, Cynodon plectostachyus, Dichanthium aristatum,*
	Acacia spp. and Eucalyptus	
	Parkia biglobosa, Balanites aegyptica, Adansonia digitata, Bauhinia rufescens, Faidherbia albida, and Sclerocarya birrea, Sesbania	*Digitaria decumbens, Panicum maximum, Paspalum Dilatatum, Pennisetum purpureum and P. landestinum*
	sesban, D. sissoo, Albizia spp. Eucalyptus, P. juliflora and L. leucocephala, Codariocalyx gyroides and Annona senegalensis	
Tanzania	*Albizia spp. and Dalbergia melanoxylon. Acacia tortilis, A. nilotica, and A. polyacantha, Adansonia digiata and Tamarindus indica*	Savanna grasses
Israel	*Acacia victoriae, Malva nicaeensis, and Salvia fruticose, Acacia nilotica, Acacia albida, Betula alnoides*	Natural grasses

TABLE 6.2 (Continued)

Region country	Main tree species component	Main understory component
India	*Acacia catechu, Acacia tortilis, Leucaena, Tectona grandis, Prosopis juliflora, Eucalyptus tereticornis, Populus deltoides, Terminalia arjuna, Acacia auriculiformis, Syzigium. cumini, Albizia lebbek, Dalbergia sissoo, Sesbania sesban, Ziziphus nummularia, and Pongamia pinnata*	*Pennisetum purpureum, Eulaliopsis binate, Leptochloa fusca (L.) Urochloa mutica, Cymbopogon, and Setaria grass*
Sri Lanka	*Coconuts*	Forage grasses
Indonesia	*Acacia vilosa, Ficus poacellie, Leucaena leucocephala, Gliricidia sepium, Hibiscus tillaeceus and Lannes corromadilica*	*Cenchrus ciliaris, Urochloa mosambiensis, Panicum maximum* and legumes (*Stylosanthes guyanensis, Centrosema pubescence, Stylosanthes hamata*
Philippines	*L. leucocephala, Artocarpus heterophylius, G. sepium, Trema orientalis, Sesbania grandijiora, Samanea saman, Pithecollobium duice, Passijiora edulis and Erythrina variegata,*	Natural grasses
Vietnam	*Sesbania grandijiora, Tephrosia candida, L. leucocephala, Acacia auriculiformis, Ficus auriculata, F. racemosa, Streblus asper, T.rems orientalis. Bauhinia spp., Broussonetia papyri/era. F. fulva, F.hespida. F. gibbosa. F. gallose. F. laccar, F. variegata. Garuga pinnata*	Natural grasses
Peninsular Malaysia	*Acacia mangium*	*Setaria anceps var: splendia*

Source: Gintings and Lai, 1994; Reeves, 2003; Garrett et al., 2004; Prasad, 2007; Dhillon and Dhillon, 2009; Mor-Mussery and Leu, 2013; Peri et al., 2016.

a variation with the different agroforestry zones identified for AICRP on Agroforestry. Central Agroforestry Research Institute, Jhansi identified five zones, i.e., Himalayan, Indo-Gangetic, Humid and Sub-humid, Arid and Semi-Arid and Tropical zone, for the carrying effective research in different agroforestry systems (CAFRI, 2016). Whereas, Dhyani et al. (2009) added one zone, i.e., all Islands, which has its own importance for the study of different components of agroforestry systems. Each fodder yielding species needs special environmental and edaphic conditions which are suitable for their growth and development. Zone wise different fodder yielding components of silvo-pasture systems are given in Table 6.3. Cultivation of fodder yielding trees species like *Albizia, Melia, Acacia, Leucaena* and understorey grasses such as *Pennisetum, Dichanthium, Cenchrus, Chrysopogon, Panicum* and legumes like *Stylosanthes hamata* and *Macroptilium* are very popular with the farmers in India (Misri, 2006).

6.3.6   BIOMASS PRODUCTION

Biomass is the total amount of organic matter existing in different parts of a plant species per unit area, described by weight on a dry basis. The unit of biomass is g/m^2 or kg/m^2 or tons/ha. In general, biomass includes aboveground (leaf, branches, stem, fruit, and flowers) and belowground (roots and dead mass of fine and coarse litter associated with the soil) living mass. During the productive season, CO_2 from the atmosphere is taken up by the vegetation and stored as plant biomass (Losi et al., 2003). Climatic, edaphic and biotic factors like temperature, moisture content, soil texture, photosynthesis, microbial activity, and available nutrients, which affects the biomass production capacity (Limbu and Koirala, 2011).

Country's' high production of total grass biomass (dry matter) was observed under the silvo-pasture systems (14.05 M tons) as compared to the natural (1.08 M tons) and an improved (6.33 M tons) grasslands (Figure 6.5). Such results showed the complimentary reactions between fodder tree species and the grasses grown in the understory of silvo-pasture systems. The tree species helps to improve soil health (physicochemical-biological parameters) which serves benefits for growing understory fodder grasses (Mahanta et al., 2017).

TABLE 6.3 The Silvopasture Systems Practiced in Different Regions of India

Region/Zone	Tree component (Fodder/Fuelwood)	Herb/crop component	Source
Himalayan Zone			
North Eastern Himalaya	*Alnus nepalensis, B. variegata* and *Ulmus* spp., *Grewia* spp, *Bauhinia* spp.	*Chrysopogon fulvus Themeda- Arundinella*	Dabadghao and Shankarnarayan (1973); Pathak et al. (2000); AFRI (2016); Murthy et al. (2016)
Western Himalayas	*Grewia optiva, Morus alba, Populus ciliate, Salix alba,* and *Celtis australis*	*Setaria* spp	Dhyani et al. (2009)
Part of Eastern Himalayas	Bamboo, *Artocarpus heteropiyllus, Litsea polyantha, Streblus asper, Albizia, Parkia roxburghii, B. variegate, Morus alba, Gmelina arborea, Ficus spp,* and *Symingtonia*	Napier	Dhyani et al. (2009)
Himachal Pradesh	*Acacia catechu, Albizia lebbeck, Albizia stipulata, Artocarpus chaplasha. Bauhinia variegata, Bombax ceiba, Cedrella toona, Celtis australis, Dalbergia siesoo, Dendrocalamus hamiltonii, Ficus sp., Grewia optiva, Morus alba, Quercus sp., Robinia pseudoacacia,* and *Terminalia alata*	*Andropogon vulmis, Chrysopogon gryllus, Chrysopogon montanus, Cymbopogor martiri, Dichanthium sp., Eragrostis* sp. Fescue grass, *Heteropogon contortus, Imperata cylindrica, Paspalum notatum, Pennisetum clandestinum, Saccharum spontaneum,* and *Sorghur. halepense. Lotus corniculatus* and *Trifolium sp.*	Dev et al. (2013)
Indo-Gangetic Zone			
Lower Gangetic Plains	*Morus alba, Albizia lebbeck*	*Dicanthium* and *Pennisetum*	Dhyani et al. (2009)
Middle Gangetic Plains	*Albizia lebbeck*	*Chrysopogaa, Dicanthium*	Dhyani et al. (2009)

TABLE 6.3 *(Continued)*

Region/Zone	Tree component (Fodder/Fuelwood)	Herb/crop component	Source
Trans Gangetic Plains	*Albizia lebbeck* and *Bauhinia variegata*	*Cenchrus, Pennisetum*	Dhyani et al. (2009)
Upper Gangetic Plains	*Albizia lebbeck* and *Bauhinia variegata*	*Chrysopogan, Dicanthium, Pennisetum*	Dhyani et al. (2009)
IGFRI, Jhansi, UP	*Sesbania sesban*	*Trifolium alexandrinum* and *Zea mays*	Shukla et al. (2002)
Hill region, Uttarakhand	*Bauhinia purpuria*	*Digitaria decumbence*	Dabadghao and Shankarnarayan (1973);
	Quercus incana	*Digitaria decumbence*	Pathak et al. (2000); CAFRI (2016)
	Grewia optiva	*Digitaria decumbence*	
	Celtius australis	*Cenchrus ciliaris*	
Doon Valley, Uttarakhand	*Grewia optiva*	Hybrid	Singh and Dadhwal (2013)
		Napier	
Humid and Sub-Humid Zone			
East Coast Plains & Hills	*Artocarpus* spp	*Chrysopogan, Cenchrus,* Napier	Dhyani et al. (2009)
West Bengal	Eucalyptus, *Acacia auriculiformis, Sesbania grandiflora, Leucaena leucocephala*	*Pennisetum pedicellatum* (Dinanath), *P. olystachyan* (thin napier), *Stylosanthus guinensis, S. scabra, S. humilis* and *S. hemata*	Lahiri (1992)
Arid and Semi-Arid Zone			
(Up to 500 mm annual rainfall), Rajasthan	*Tecomela undulata, Colophospermum mopane* and *Acacia senegal, Acacia amura, Acacia albida, Acacia tortilis, Azadirachta indica, Eucalyptus camaeldulensis, Hardwikia binate, Leucaena leucocephala*	*Lasiurus sindicus, Cenchrus ciliaris, C. setigerus, Panicum antidotale, P. turgidum, Dichanthium annulatum, Stylosanthes scabra, Desmodium triflorum, Macroptilium atropurpureum*	Soni et al. (2016)

TABLE 6.3 *(Continued)*

Region/Zone	Tree component (Fodder/Fuelwood)	Herb/crop component	Source
	Parkinsonia aculeate, Pithecelobium dulce, Prosopis cineraria, Albizia lebbeck Ziziphus mauritiana, Ziziphus nummularia	and *Sporobolus marginatus*	
(500–1150 mm annual rainfall), Rajasthan	*Acacia albida, Acacia nilotica* var. *cupressiformis Albizia lebbek, Azadirahta indica, Dalbergia sissoo, Gliricidia sepium, Grevillea robusta, Leucaena leucocephala*	*Cenchrus ciliaris, Cenchrus setigerus, Sehima nervosum, Chrysopogon fulvus, Heteropogon contortus, Stylosanthes hamata, Atyloxia scarabasoides, Clitoria ternatea, Macroptilium atropurpureum*	Soni et al. (2016); Mishra et al. (2010)
(> 1150 mm rainfall), Rajasthan	*Acacia nilotica, Azadirachta indica, Albizia amara, Bauhinia variegata, Dalbergia sissoo, Gmelina arborea, Grevia optiva, Grevillea robusta, Emblica officinalis*	*Chrysopogon fulvus, Dichanthium annulatum, Setaria ancep, Macroptilium Atropurpureum, Stylosanthes hamata, Clitoria ternatea*	Soni et al. (2016)
West Coast Plains & Hills	*Hardwickia binnata, Albizia lebbeck*	*Cenchrus*	Dhyani et al. (2009)
Gujrat, Coast Plains & Hills	*Leucaena leucocephala*	*Cenchrus* and *Setaria*	Dhyani et al. (2009)
Western Dry Region	*Hardwickia binnata, Albizia lebbeck*	*Cenchrus*	Dhyani et al. (2009)
Tropical Zone			
Karnataka	*A. auriculiformis*	IGFRI-7, Coimbatore-2, Guinea grass, IGFRI-3	Narendra et al. (2011)
Tamil Nadu	*Acacia leucophloea, A. nilotica, Albizia* sp., *Azadirachta Indica*	*Cenchrus ciliaris, C. setigerus, Chrysopogon fulvus, Dichanthium annulatum, Andropogon pumilus*	Anonymous (2004)

TABLE 6.3 *(Continued)*

Region/Zone	Tree component (Fodder/Fuelwood)	Herb/crop component	Source
		Aristida funiculate, Setaria glauca. Phaseolus trilobus, Dolichos biflorus, Atylosia scarabaeoides	
Kerala	*Leucaena leucocephala, Morus indica* and *Calliandra calothyrsus*	hybrid Napier (HN), *Stylosanthes hamata, Desmanthus virgatus* and *Vigna unguiculata*	Raj et al. (2016); Varsha et al. (2016); Raj et al. (2015)
Central Plateau and Hills	*Albizia amara, Leucaena leucocephala, Dichrostycus cineria*	*Chrysopogan, Stylosanthus hamata, S. scabra*	Dhyani et al. (2009)
Western Plateau and Hills	*Acacia mangium, Albizia amara*	*Cenchrus*	Dhyani et al. (2009)
All Islands			
Islands	*Bauhinia spp. Erythrina indica, Leucaena leucocephala*	*Cenchrus, Pennisetum*	Dhyani et al. (2009)

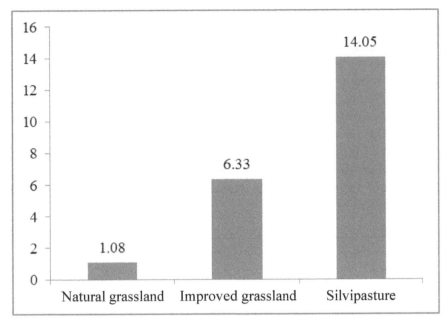

FIGURE 6.5   Total grass production (DM MT).

The total tree leaves the production of the country is 12.96 million tons (Figure 6.6), in which the tree-based systems contributed around 71% and remained by boundary plantations (Mahanta et al., 2017).

The comparative study at NRCAF, Jhansi revealed that on an average 4.55 t from pasture + 0.51 t from tree leaves and pods produced from a silvi-pastoral system which is just double the yield than natural grassland. These results indicate the possibilities to get more biomass by establishing the silvi-pasture, which is producing less than 2 t/ha/year forage through natural vegetation. The forage production from one tier (*Cenchrus* pasture alone), two-tier (*Cenchrus* + *Ailanthus excelsa*) and three-tier (*Cenchrus* + *A. excelsa* + *Dichrostachys cinerea*) systems were compared with natural pasture at CSWRI, Avikanagar, Rajasthan under semi-arid condition. Results revealed that the three-tier silvi-pasture system provided maximum forage production (2.78 t/ha dry forage from pasture + 0.95 t/ha green tree leaves).

Roy et al. (1987) studied 8–9 years old silvi-pasture system in semi-arid conditions and reported that the mean green and dry leaf fodder of 9.63 and 5.28 t/ha, respectively with annual lopping at 1/3 intensity in

Albizia procera, whereas *A. lebbeck* provide of 6.24 and 2.78 t/ha green and dry leaf fodder, respectively in case of through biannual lopping at 2/3 intensity. *Bahunia purpurea* yielded green leaf fodder of 10.02 kg/tree/year at Jhansi. Fodder production from top vegetation layer i.e. from fodder yielding tree species from six trees of Bundelkhand region yields high dry leaf fodder (11.38 kg/tree) in *A. procera* which was followed by *A. amara* (11.20 kg/tree), *A. lebbeck* (4.21 kg/tree), *Hardwickia binata* (3.67 kg/tree) and *D. cinerea* (2.76 kg/tree) and a minimum of 0.51 kg/tree in *A. tortilis*, respectively. In *S. grandiflora* and *S. sesban*, the dry leaf fodder yield of 0.3 kg/tree was recorded after 3.5 years of the establishment when grown at the density of 5000/ha (Gupta et al., 1983).

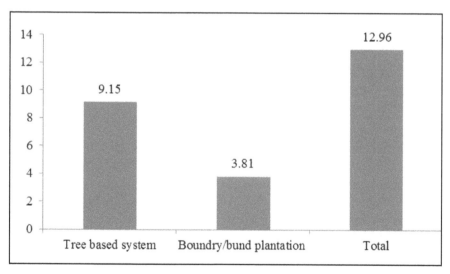

FIGURE 6.6   Estimate of tree leaves production (MT).

A three-tier silvi-pastoral system developed at NRCAF, Jhansi with woody components of *Dichrostachys cinerea* + *A. amara* + *L. leucocephala* showed top feed (dry leaf + pod) production of 0.04, 0.26, 0.48, 1.15, 0.53, 0.55 and 0.54 t/ha during 1 to 7 years, respectively, when pruned up to 50% height of the tree species from ground level. An average fodder yields from the leaf of young trees of *A. excelsa*, *P. cineraria*, *A. indica*, and *Dichrostachys cinerea* were 21.13, 25.70, 12.65 and 0.67 kg/tree, respectively. The corresponding yields of older trees were 46.14, 57.81, 46.45

and 1.12 kg/tree, respectively. Raj et al. (2016) conducted two year field study involving nine modules They observed that the highest dry matter (31.49Mg ha^{-1}) and the yield of crude protein (4.75Mg ha^{-1}) was obtained from two-tier Hybrid Napier + trees (mulberry + *Calliandra*) system, which is a better option for producing quality forage at smallholder stall-fed dairy farms under humid tropical climate.

Samra et al. (1999) studied the biomass production of trees and grasses in a silvi-pasture system on marginal lands of Doon valley in Uttarakhand. They observed that the total biomass production was highest in intercropping of *Bauhinia purpurea* with grass (2929 kg ha^{-1}) followed by *Grewia optiva* (2372 kg ha^{-1}) which appeared to be most suitable tree species along with *Eulaliopsis binata* for sustainable silvo-pasture development on the marginal land.

Significantly higher fresh and dry weight was recorded for guinea grass (4781 and 2732 kg/ha) among different forage grass species, which was followed by IGFRI-7 and Coimbatore-2 under *Acacia auriculiformis* based silvo-pastoral system during 2007–08 (Narendra et al., 2011). Kumar et al. (2001) reported possible reasons of the higher dry weight in guinea grass (63%) may be due to shade tolerant nature compared to fairly shade intolerant grasses like hybrid napier (42%) under multipurpose tree species (MPTs) compared to their sole crop.

On average of 2 yr, *C. ciliaris* had produced 12.78 t ha^{-1} green and 3.72 t ha^{-1} dry biomass under the tree canopies of *A. tortilis*. The dry matter yield reduced to 38% under the tree canopies over the open-grown grasses (Mishra et al., 2010). In the semi-arid state (UP), silvo-pasture system in a natural grassland where introduced by fodder yielding tree species of *Albizia procera*, *Albizia lebbeck*, *Eucalyptus tereticornis*, *Dalbergia sissoo*, and *Emblica officinalis* which accumulated 8.6 to 5.41 t/ha/yr of biomass (Rai et al., 2009).

6.3.7   CARBON STORAGE

The information on biomass production by any system is important for understanding their functioning in case of nutrient cycling, energy fixation, and atmospheric CO_2 sequestration. It is also important to define the carbon budget in terrestrial ecosystems. Tree-based systems are the natural storehouses of biomass and carbon. Carbon sequestration process involves the net amputation of CO_2 and storage in aboveground plant and belowground biomass of long-lived pools such as roots, soil microorganisms. The carbon accumulated in different parts of plants through photosynthesis and added to the topsoil after the breakdown of leaf litter and dead root system. So that the above ground litter and below ground organic matter is the main source of carbon in any kind of an ecosystem.

The human activities like fossil fuel burning, land use change, deforestation activities during the last few decades accelerate the increasing rate of an atmospheric CO_2 concentration. Disturbances due to natural and human influences in any kind of system lead to more carbon released into the atmosphere than the amount sequestered or stored by vegetation during photosynthesis (Brown, 2002). The health and functioning of any kind of an ecosystem are indicated by carbon sequestration and carbon stored in that system (Giri et al., 2014). The knowledge of biomass C densities of different tree based land use systems (such as silvo-pasture systems) is one of the important components for assessing their contribution to the global C cycle (Gairola et al., 2012; Singh et al., 2015). Carbon stored in vegetation and soil to act as significant sinks of atmospheric CO_2 and it is highly highlighted under the *Kyoto Protocol*. Dry matter production (biomass) from the naturally occurred grasslands and silvi-pastoral systems having *Albizia amara*, *Dichrostachys cinerea* and *Leucaena leucocephala* as woody perennials along with *Chrysopogon fulvus* as grass and *Stylosanthes hamata* and *S. scabra* as legume was studied at CAFRI, Jhansi (UP). The results revealed that rate of carbon accumulation in silvi-pastoral was 6.72 t C/ha/yr, which was two times higher than naturally occurred grassland (3.14 t C/ ha/yr) (NRCAF, 2007). An amalgamation of the silvi-pastoral system in a naturally occurred grasslands was studied by Rai et al. (2009) and reported that the 1.89–3.45 tC/ha stored in silvi-pasture and 3.94 tC/ha in pure pastures. Narain (2008) also reported that the growing *Prosopis juliflora* and *Dalbergia sissoo* with grasses in silvi-pastoral systems increase their carbon flux in net primary productivity. The carbon storage in the system was 1.89–3.45 t C/ha in silvi-pastoral and which is lower pure pasture (3.94 t C/ha). Kaul et al. (2010) and Santos et al. (2011) reported that the *Vetiveria zizanioides* grass sequester the high amount of carbon as compared to the other agricultural fodder crops (Table 6.4). Such results occurred may be due to growth habit and nature of fodder yielding species and also depends on several factors, including climate, soil type, type of crop or vegetation cover and different management practices.

TABLE 6.4   Carbon Sequestrations in Different Forage Species

Tree/crop/cropping system	C sequestration rate $(Mg\ ha^{-1}\ yr^{-1})$
Vetiveria zizanioides	15.24
Lemongrass	5.38
Palmarosa	6.14
Vetch (V)–maize (M)–oat(O)–soybean(S)–wheat (W)	7.26
O–M–W–S	8.56
V–M–W–S	7.58
Ryegrass–M–R–S	8.44

Source: Kaul et al. (2010) and Santos et al. (2011).

6.3.8   SOIL CARBON

Carbon in soils can be divided into two major pools, i.e. Soil organic carbon (SOC) and Soil inorganic carbon (SIC). Soil organic carbon (SOC) is derived from organic matter and is important in the maintenance and improvement of soil fertility. Soil inorganic carbon (SIC) includes carbonates derived from weathering of rocks (lithogenic) and carbonates derived from the direct absorption of CO_2 into the soils (pedogenic). The SIC pool is relatively stable and is thought neither to be a net sink nor to be strongly affected by land management and hence not of much significance in the climate change perspective (Lal, 2009). The soil organic carbon was increased by 1.7 to 2.3 times in a Silvo-pastoral system having *Leucaena leucocephala* as a tree component, and grass like *Cenchrus ciliaris* and legume as *Stylosanthes hamata* compared to a control (Kaur et al., 2002).

Hazra (1995) studied different tree based silvo-pasture systems for the dry forage yield, leaf litter production, soil carbon storage and relative increase in total soil organic carbon (Table 6.5). *Cenchrus ciliaris* was the common grass species for these different fodder tree based silvo-pasture systems. He reported that the higher total soil organic carbon content was observed under *Leucaena leucocephala* followed by *Albizia lebbeck, Dichrostachys mutants,* and *Albizzia proceura.* The relative increase in total soil organic carbon was higher under *the Leucaena leucocephala* based silvo-pasture system as compared to others. In similar studies, Venkatswaralu (2010) reported that the high levels of soil organic carbon accumulation in dry sub-humid and arid ecosystems where grass species were grown with annual crops in a silvi-pastoral system. In the case of the arid ecosystem, he also reported that the no change in organic carbon under tree based system incorporating grasses. Planting fodder yielding tree species and grass species in wastelands of arid areas may help to increase in soil carbon stock from 24.30Pg to 34.90 Pg (CAZRI, 2008). Kumar, (2010) studied such systems for soil carbon storage potential and reported that the total belowground carbon accumulated (1.60 t/ha) by a silvi-pastoral system having *Acacia tortilis* and *Cenchrus setegerus* was 23.40% of the total aboveground carbon stock.

TABLE 6.5   Influence of Tree Species on Forage Yield, Leaf Litter, and TOC Content of Soil in *Cenchrus ciliaris* Silvo-Pasture

Trees	Dry forage yield (Mg/ha)	Leaf litter (Mg/ha)	TOC (g/kg soil)	Relative increase
Albizia lebbeck	3.40	5.63	9.31	3.04
Albizia procera	2.30	4.21	8.37	2.73
Acacia tortilis	0.70	4.36	7.44	2.43
Acacia nilotica	2.20	4.47	7.98	2.61
Leucaena leucocephala	3.20	5.28	10.37	3.39
Dichrostachys nutan	2.00	5.46	9.04	2.96
Hardwikia binata	1.50	5.34	7.18	2.35
Eucalyptus sp.	2.00	3.87	6.91	2.26
Open (without tree)	-	5.95	6.91	2.26
Initial*	-	0.62	3.05	

Source: Hazra (1995).

6.4   CONCLUSION

Grazing resources of the country are limited, and their condition is very poor. Therefore the country faces 32.40% green fodder, 10.70% dry fodder/crop residue and 44% concentrate feed ingredients deficit. The country also faces land degradation problems, in 2004 India has 146.82 million ha. degraded land under different land degradation categories. The development of silvo-pasture systems on such areas is a big opportunity to combat such problems. Silvo-pasture can be developed through using four different approaches viz. Scattered trees and shrubs on pasture, Intensive silvi-pastoral (fodder trees with agri-fodder crops), live fence of fodder trees or boundary systems and hedge, and protein bank (cut and carry). An area under tree based fodder production systems was 585–1215 M ha in Africa, Asia, and America. Whereas, India accounts only 5.58 M ha area under silvo-pasture (trees on grazing/rangelands) system.

PLATE 6.1   (See color insert.) Improved silvo-pasture systems A. Leucaena based B. Ber Based and C. *Hardwickia* based (*Source*: Mahanta et al., 2017).

PLATE 6.2   (See color insert.) Traditional silvo-pasture systems.

Developing a silvo-pasture system on degraded lands of the country is one of the best options for sustainable farming practices and solving the problem of food insecurity. It also helps in strengthening the country's GDP through the production of diverse produces. It minimizes the risk that occurred in agriculture during adverse conditions and helps to improve the socio-economic condition and livelihood security of rural people. Other than direct benefits, the system helps indirectly to improve environmental health through carbon sequestration and carbon storage in aboveground and belowground vegetation parts, and carbon storage in soil. Due to the perennial nature of fodder yielding tree species and some grass species, the system has advantages over other systems for biomass production and carbon storage potentials. Biomass production and carbon storage in vegetation and in the soil help to improve environmental conditions in coming years if the area under silvo-pasture increases.

KEYWORDS

- adult cattle unit
- agroforestry systems
- biomass production
- biotic and abiotic factors
- carbon sequestration
- carbon storage
- carrying capacity
- dry matter production
- fodder deficit
- fodder tree species
- grasslands
- grazing resources
- gross domestic production
- lithogenic
- livestock population
- milk production
- pasturelands
- pedogenic
- silvo-pasture
- soil inorganic carbon
- soil organic carbon
- sustainable development
- system components
- white revolution

REFERENCES

1. Anand, S., & Radhakrishna, S. (2017). Investigating trends in human-wildlife conflict: is conflict escalation real or imagined? *Journal of Asia-Pacific Biodiversity, 10,* 154–161.

2. Anonymous (2004). *Annual Report: Grassland and Silvi-pasture management.* Indian Grassland and Fodder Research Institute, Jhansi (UP), India, pp. 47–60.
3. Anonymous (2010). *Degraded and Wastelands of India: Status and Spatial Distribution.* Indian Council of Agricultural Research, Krishi Anusandhan Bhavan I, Pusa, New Delhi. 155p.
4. Anonymous (2012). *19th Livestock Census-2012 All India Report.* Ministry of Agriculture Department of Animal Husbandry, Dairying and Fisheries Krishi Bhawan, New Delhi. 121p.
5. Anonymous (2013). *IGFRI: Vision 2050.* Indian Grassland and Fodder Research Institute (Indian Council of Agricultural Research). Jhansi (UP). 25p.
6. Anonymous (2016). State of Indian Agriculture 2015–16. Government of India, Ministry of Agriculture & Farmers Welfare, Department of Agriculture, Cooperation & Farmers Welfare, Directorate of Economics and Statistics, New Delhi. 252p.
7. Beer, J., Ibrahim, M., & Schlonvoigt, A. (2000). Timber production in tropical Agroforestry systems of Central America. *In:* Krishnapillay et al. (ed). *Forests and society: The role of research.* Vol. 1 Sub-plenary Sessions. XXI IUFRO World Congress, Kuala Lumpur. pp. 777–786.
8. Brown (2002). Measuring carbon in forests: Current status and future challenges. *Environmental Pollution, 116*(3), 363–372.
9. CARI (2009). *Trends in Arid Zone Research in India. In*: Amal K, Garg BK, Singh MP, Kathju S (eds.). Central Arid Zone Reassert Institute, Jodhpur. 481p.
10. Chhetri, R. B. (2010). Some fodder yielding trees of Meghalaya, Northeast India. *Indian Journal of Traditional Knowledge, 9*(4), 786–790.
11. Cubbage, F., Balmelli, G., Bussoni, A., Noellemeyer, E., Pachas, A. N., Fassola, H., Colcombet, L., Rossner, B., Frey, G., Dube, F., Silva, M. L., Stevenson, H., Hamilton, J., & Hubbard, W. (2012). Comparing silvo-pastoral systems and prospects in eight regions of the world. *Agroforestry Systems, 86,* 303–314.
12. Dabadghao, P. M., & Shankarnarayan, K. A. (1973). The grass cover of India. ICAR, New Delhi.
13. Dawson, I. K., Carsan, S., Franzel, S., Kindt, R., van Breugel, P., Graudal, L., Lilleso, J-P. B., Orwa, C., & Jamnadass, R. (2014). Agroforestry, livestock, fodder production and climate change adaptation and mitigation in East Africa: issues and options. ICRAF Working Paper No. 178. Nairobi. 22p.
14. Dev, I., Radotra, S., Misri, B., Sareen, S., Pathania, M. S., & Singh, J. P. (2013). Participatory appraisal of grazing resources and socioeconomic profile of livestock keepers of Himachal Pradesh. *Indian Journal of Agroforestry, 15*(1), 64–70.
15. Devendra, C. (2012). Agroforestry and silvo-pastoral systems potential to enhance food security and environmental sustainability in South East Asia. *Open Access Scientific Reports, 1*(2), 163–170.
16. Devendra, C. (2014). Perspectives on the potential of silvo-pastoral systems. *Agro-technology, 3*(1), 1–8.
17. Devendra, C., & Leng, R. A. (2011). Feed resources for animals in Asia: issues, strategies for use, intensification, and integration for increased productivity. *Asian-Aust. J. Anim. Sci, 24*(3), 303–321.
18. Dhillon, S. K., & Dhillon, K. S. (2009). Phytoremediation of selenium-contaminated soils: the efficiency of different cropping systems. *Soil Use Manage, 25,* 441–453.

19. Dhyani, S. K., Kareemulla, K., Ajit and Handa, A. K. (2009). Agroforestry potential and scope for development across agro-climatic zones in India. *Indian J For, 32,* 181–190.

20. Dixon, R. K. (1995). Agroforestry systems: sources or sinks for greenhouse gases? *Agroforestry Systems, 31,* 99–116.

21. Doyle, C., & Thomas, T. (2000). The social implications of agroforestry. *In:* Agroforestry in the UK. Bulletin 122. A.M. Hislop and J. Claridge (eds). Forestry Commission, Edinburgh.

22. Dwivedi, A. P. (1992). *Agroforestry: Principles and Practices.* Oxford & IBH Publishing Company. New Delhi. 365p.

23. Feldhake, C. (2001). Microclimate of a natural pasture under planted *Robinia pseudoacacia* in central Appalachia, West Virginia. *Agroforestry Systems, 53,* 297–303.

24. Feldhake, C. (2002). Forage frost protection potential of conifer silvo-pastures. *Agricultural and Forest Meteorology, 112*(2), 123–130.

25. Gairola, S., Sharma, C. M., Ghildiyal, S. K., & Suyal, S. (2012). Chemical properties of soils in relation to forest composition in moist temperate valley slopes of Garhwal Himalaya, India. DOI 10.1007/s10669-012-9420-7.

26. Gangasagare, P. T., & Karanjkar, L. M. (2009). Status of milk Production and economic profile of dairy farmers in the Marathwada region of Maharashtra. *Veterinary World, 2*(8), 317–320.

27. Garrett, H. E., Kerley, M. S., Ladyman, K. P., Walter, W. D., Godsey, L. D., Van Sambeek, J. W., & Brauer, D. K. (2004). Hardwood silvo-pasture management in North America. *Agroforestry Systems, 61,* 21–33.

28. Ghosh, P. K., Palsaniya, D. R., & Srinivasan, R. (2016). Forage research in India: issues and strategies. *Agric Res J., 53*(1), 1–12.

29. Giri, N., Rawat, L., & Kumar, P. (2014). Assessment of biomass carbon stock in a *Tectona grandis* Linn F. plantation ecosystem of Uttarakhand, India. *International Journal of Engineering Science Invention, 3*(5), 46–53.

30. Gintings, A. N., & Lait, C. K. (1994). Agroforestry in Asia and the Pacific: with special reference to silvo-pasture systems. *In:* Copland, J.W., Djajanegra. A., & Sabrani, M (eds). *Agroforestry and Animal Production for Human Welfare.* Proceedings of an international symposium. 7th AAAP Animal Science Congress, Bali. Indonesia. ACIAR Proceedings No. *55,* 125p.

31. Gordon, A. M., Naresh, R. P. F., & Thevathasn, V. (2005). How much carbon can be stored in Canadian agro-ecosystems using a Silvo-pastoral approach? *In:* Mosquera-Losada, M. R., McAdam, J. H., & Riguerio-Rodriguez, A. *Silvo-pastoralism and Sustainable Land Management.* CABI Publishing, Wallingford UK. pp. 210–218.

32. Gunasekaran, S., Viswanathan, K., & Bandeswaran, C. (2014). Selectivity and palatability of tree fodders in sheep and goat fed by cafeteria method. *International Journal of Science, Environment, 3*(5), 1767–1771.

33. Gupta, S. K., Pathak, P. S., & Roy, R. D. (1983). Biomass production of *Sesbania sesban* Pers. on different habitats. *My Forest, 19*(3), 145–155.

34. Gurung, T. R., & Temphel, K. J. (2015). *Technological advancement in agroforestry systems: Strategy for climate smart agricultural technologies in SAARC Region.* BARC Complex, Farmgate, Dhaka-1215, Bangladesh. 294p.

35. Harris, J. M. (2000). Basic Principles of Sustainable Development. Global Development and Environment Institute, working paper 00–04. Tufts University. 25p.

36. Hazra, C. R. (1995). Soil and water conservation in relation to land use and biomass production. *In*: R.P. Singh (ed.). Forage Production and Utilization. Indian Grassland and Fodder Research Institute, Jhansi.
37. Heintz, H. T. (2004). Applying the concept of sustainability to water resources management. *Water Resour Update, 127,* 6–10.
38. Huxley, P. A., & Westley, S. B. (1989). *Multipurpose Trees: Selection and Testing for Agroforestry.* International Council for Research in Agroforestry. English Press. Nairobi, Kenya. 109p.
39. Kaul, M., Mohren, G. M. J., & Dadhwal, V. L. (2010). Carbon storage and sequestration potential of selected tree species in India. *Mitigation and Adaption Strategies for Global Change, 15,* 489–510.
40. Kaur, B., Gupta, S. R., & Singh, G. (2002). Carbon storage and nitrogen cycling in silvi-pasture systems on sodic soil in Northwestern India. *Agroforestry Systems, 54,* 21–29.
41. Klopfenstein, N. B., Rietveld, W. J., Carman, R. C., Clason, T. R., Sharrow, S. H., Garret, G., & Anderson, B. E. (2008). Silvo-pasture: an agroforestry practice. Adapted from http://www.agroforestry.net/overstory/oversory36.html.
42. Kumar, A. K. (2010). Carbon sequestration: underexplored environmental benefits of Tarai agroforestry. *Indian Journal of soil Conservation, 38,* 125–131.
43. Kushwaha, N. (2008). Agriculture in India: Land use and sustainability. *International Journal of Rural Studies, 15*(1), 1–17.
44. Lahiri, A. K. (1992). Silvo-pasture practices on lateritic tract of southwest Bengal. *Indian Forester,* 887–892.
45. Lal, R. (2009). Carbon sequestration in saline soils. *Journal of soil salinity and water quality, 1,* 30–40.
46. Limbu, D. K., & Koirala, M. (2011). Above-ground and below-ground biomass situation of Milke-Jaljale rangeland at different altitudinal gradient. *Our Nature, 9,* 107–111.
47. Losi, C. J., Siccama, T. G., Condit, R., & Morales, J. E. (2003). Analysis of alternative methods for estimating carbon stock in young tropical plantations. *Forest Ecology and Management, 184*(1–3), 355–368.
48. Mahanta, S. K., Koli, P., Singh, K. K., Das, M. M., Misra, A. K., Rokde, S. N., Singh, S., Maity, S. B., Kushwah, B. P., & Nayak, S. (2017). Feeding strategies in relation to climate resilient forage and livestock production. ICAR-Indian Grassland and Fodder Research Institute, Jhansi (UP). 350p.
49. Martin, F. W., & Sherman, S. (1992). *Agroforestry Principles.* ECO Technical notes. ECHO, 17391 Durrance Rd., North Ft. Myers FL 33917, USA. 11p.
50. Mathukia, R. K., Sagarka, B. K., & Panara, D. M. (2016). Fodder production through agroforestry: a boon for profitable dairy farming. *Innovare Journal of Agri. Sci, 4*(2), 13–19.
51. McAdam, J. H., Crowe, R., & Sibbald, A. (1999). Agroforestry as a sustainable land use option. *In*: Burgess, P. J., Brierley, E. D. R., Morris, J., & Evans, J. (eds). *Farm Woodlands for the Future.* Bios Scientific Publishers, Oxford, United Kingdom, pp. 127–137.
52. Misri, B. (2006). Country Pasture/Forage Resource Profiles: India. Regional Research Centre, Indian Grassland and Fodder Research Institute, HPKV Campus, Palampur-176 062 (India). 12p.

53. Mishra, A., Nautiyal, S., & Nautiyal, D. P. (2009). Growth characteristics of some indigenous fuelwood and fodder tree species of sub-tropical Garhwal Himalayas. *Indian Forester*, *135*(3), 373–379.

54. Mishra, A. K., Tiwari, H. S., & Bhatt, R. K. (2010). Growth, biomass production and photosynthesis of *Cenchrus ciliaris* L. under *Acacia tortilis* (Forssk.) Hayne based Silvo-pastoral systems in semi-arid tropics. *Journal of Environmental Biology*, *31*(6), 987–993.

55. Mor-Mussery, A., Leu, S., & Budovsky, A. (2013). Modeling the optimal grazing regime of Acacia victoriae silvo-pasture in the Northern Negev, Israel. *Journal of Arid Environments*, *94*, 27–36.

56. Mosquera-Losada, M. R., McAdam and Rigueiro-Rodriquez, A. (2005). Declaration, summary, and conclusions. *In:* Mosquera-Losada, M. R., McAdam, J., & Riqueiro-Rodriquez, A. (eds), *Silvo-pastoralism and Sustainable Land Management*. CABI Publishing, Wallingford, England UK, pp. 418–421.

57. Murthy, I. K. Subhajit Dutta, Vinisha Varghese, Joshi, P. P., & Kumar, P. (2016). Impact of Agroforestry Systems on Ecological and Socioeconomic Systems: A Review. *Global Journal of Science Frontier Research: H Environment & Earth Science*, *16*(5), 14–28.

58. Nair, P. K. R. (1984). Tropical agroforestry systems and practices. *In*: Furtado, J.I., & Ruddle, K. (Eds). *Tropical Resource Ecology and Development*. John Wiley, Chichester, England. 39 p.

59. Nair, P. K. R. (1993). *An Introduction to Agroforestry*. Kluwer Academic Publishers. The Netherlands. 491p.

60. Nair, P. K. R., Kumar, B. M., & Nair, V. D. (2009). Agroforestry as a strategy for carbon sequestration. *J. Plant Nutr. Soil Sci. 172*, 10–23.

61. Narain, P. (2008). Dryland management in an arid ecosystem. *Journal of the Indian Society of Soil Science*, *58*, 337–347.

62. Narendra, A. G., Madiwalar, S. L., & Channabasappa, K. S. (2011). Studies on *Acacia auriculiformis* based Silvo-pastoral system. *Karnataka J. Agric. Sci.*, *24*(3), 350–353.

63. NRCAF (2007). Perspective Plan Vision 2025. National Research Centre for Agroforestry, Jhansi, Uttar Pradesh.

64. Panda, S. C. (2013). *Cropping and Farming Systems*. Agrobios (India), Jodhpur. 413p.

65. Pathak, P. S., Pateria, H. M., & Solanki, K. R. (2000). *Agroforestry Systems in India: A Diagnosis and Design Approach*. Indian Council of Agriculture Research, New Delhi.

66. Pathak, P. S., & Roy, M. M. (1994). Silvo-pastoral systems of production: a research bulletin. IGFRI, Jhansi. 55p.

67. Peri, P. L., Dube, F., & Varella A. C. (2016). Opportunities and challenges for Silvo-pastoral systems in the subtropical and temperate zones of South America. In: Peri, P. L., Dube, F., & Varella A. C (eds.). *Silvo-pastoral Systems in Southern South America*. Springer Cham Heidelberg New York Dordrecht London. pp. 257–270.

68. Prasad, M. N. V. (2007). Phytoremediation in India. *Method Biotechnol.*, *23*, 435–454.

69. Puri, S. (2007). *Agroforestry: Systems and Practices*. New India Publishing. New Delhi. 657p.

70. Rai, P., Ajit, Caturvedi, P. O., Singh, R., & Singh, U. P. (2009). Biomass production in multipurpose tree species in natural grasslands under semiarid conditions. *Journal of Tropical Forestry*, *25*, 11–18.

71. Raj, A. K., Kunhamu, T. K., & Kiroshima, S. (2015). Optimizing management practices in the mulberry for intensive fodder production in humid tropics of Kerala. *Indian Journal of Agroforestry*, *17*(2), 36–41.
72. Raj, A. K., Kunhamu, T. K., Jamaludheen, V., & Kiroshima, S. (2016). Forage yield and nutritive value of intensive silvo-pasture systems undercut and carry scheme in humid tropics of Kerala, India. *Indian Journal of Agroforestry*, *18*(1), 47–52.
73. Ram, S. N. (2009). Productivity, forage quality and economics of guinea grass Caribbean stylo intercropping system. *Annals of Arid Zone*, *48*(2), 159–163.
74. Reeves, R. D. (2003). Tropical hyper-accumulators of metals and their potential for phytoextraction. *Plant Soil*, *249*, 57–65.
75. Reisner, Y., de Filippi, R., Herzog, F., Palma, J. (2007). Target regions for silvo-arable agroforestry in Europe. *Ecol. Eng., 29*, 401–418.
76. Roy, A. K., & Singh, J. K. (2013). Grasslands in India: Problems and perspectives for sustaining livestock and rural livelihoods. Tr*opical Grasslands–Forrajes Tropicales, 1*, 240–243.
77. Roy, M. M., Deb Roy, R., & Pathak, P. S. (1987). *Dichrostachys cinerea* a potential fodder shrub for wasteland plantation, IGFRI, Jhansi Silver Jubilee Publication No. 6.
78. Samra, J. S., Vishwanatham, M. K., & Sharma, A. R. (1999). Biomass production of trees and grasses in a silvo-pasture system on marginal lands of Doon Valley of north-west India 2. Performance of grass species. *Agroforestry Systems*, *46*(2), 197–212.
79. Santosh, N. Z., Dieckow, J., Bayer, C., Molin, R., Favaretto, N., Pauletti, V., & Piva, J. T. (2011). Forages, cover crops and related shoots and root additions in no-till rotation to C sequestration in subtropical Ferrallisols. *Soil and Tillage Research, 111*, 208–218.
80. Sarvade, S. (2014a). Agroforestry: refuge for biodiversity conservation. *International Journal of Innovative Research in Science & Engineering*, *2*(5), 424–429.
81. Sarvade, S., Gautam, D. S., Kathal, D., & Tiwari, P. (2017). Waterlogged wasteland treatment through agroforestry: A review. *Journal of Applied and Natural Science*, *9*(1), 44–50.
82. Sarvade, S., Singh, R., Ghumare, V., Kachawaya, D. S., & Khachi, B. (2014). Agroforestry: an approach for food security. *Indian J. Ecol*, *41*(1), 95–98.
83. Shankar, V., & Gupta, J. N. (1992). Restoration of degraded rangelands. *In*: J. S. Singh (ed.). *Restoration of Degraded Lands-Concepts and Strategies*. Rastogi Publications, Meerut, India, pp. 115–155.
84. Shukla, N. P., Rekib, A., & Burman, D. (2002). Productivity of Sesbania alley cropping system with sequentially cropped berseem and maize at different moisture regimes and fertilizers. *Tropical Agricultural Research and Extension*, *5*(1 & 2), 35–42.
85. Singh, C., & Dadhwal, K. S. (2013). Hybrid Napier-*Grewia* optiva-based silvipastoral system for degraded bouldery riverbed lands in Doon Valley. *Indian Journal of Agroforestry*, *15*(1), 60–63.
86. Singh, M., Gupta, B., Sarvade, S., & Awasthe, R. K. (2015). Biomass and carbon sequestration potential in different agroforestry systems in Giri catchment of North Western Indian Himalaya. *Indian J. of Agroforestry*, *17*(2), 42–48.
87. Singh, V., & Tulachan, P. M. (2002). A dynamic scenario of livestock and dairy production in Uttaranchal Hills. In: ENVIS Bulletin- Himalayan Ecology. S. K. Rao (Eds.). G.B. Pant Institute of Himalayan Environment and Development, Kosi-Katarmal, Almora, Uttarakhand, India, 81p.

88. Soni, M. L., Subbulakshmi, V., Yadava, N. D., Tewari, J. C., & Dagar, J. C. (2016). Silvo-pastoral agroforestry systems: lifeline for dry regions. *In*: Agroforestry Research Developments. (Eds): J. C. Dagar and J. C. Tewari. Nova Science Publishers, Inc. New York. pp. 245–305.

89. \Varsha, K. M., Raj, A. K., Kurien, E. K., Bastin, B., Kunhamu, T. K., & Pradeep, K. P. (2017). High-density silvo-pasture systems for quality forage production and carbon sequestration in humid tropics of Southern India. *Agroforestry Systems*, 1–14. doi:10.1007/s10457–016–0059.

90. Vekatswaralu, J. (2010). Rainfed agriculture in India: research and developmental scenario. Indian Council of Agricultural Research, New Delhi, 508p.

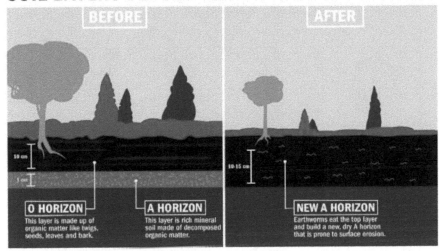

SOIL LAYERS BEFORE AND AFTER EARTHWORMS

BEFORE

AFTER

10 cm

1 cm

10-15 cm

O HORIZON
This layer is made up of organic matter like twigs, seeds, leaves and bark.

A HORIZON
This layer is rich mineral soil made of decomposed organic matter.

NEW A HORIZON
Earthworms eat the top layer and build a new, dry A horizon that is prone to surface erosion.

FIGURE 2.2 Diagram of soil layers before and after vermicompost application (*Source*: Great Lakes Worm Watch).

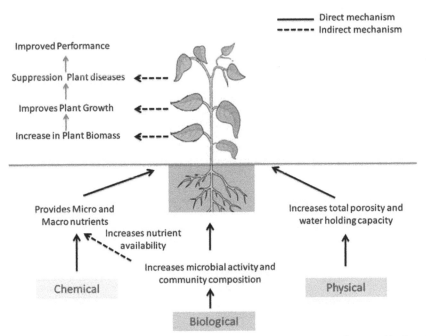

——— Direct mechanism
------- Indirect mechanism

Improved Performance

Suppression Plant diseases ◄----

Improves Plant Growth ◄----

Increase in Plant Biomass ◄----

Provides Micro and Macro nutrients

Increases nutrient availability

Increases total porosity and water holding capacity

Increases microbial activity and community composition

Chemical

Physical

Biological

FIGURE 2.5 Vermicompost and plant growth regulating mechanisms (direct and indirect) (*Source*: Modified from Lazcano and Domínguez, 2011).

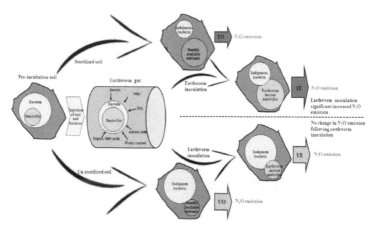

FIGURE 2.6 Hypothetical model demonstrating the stimulation of growth of denitrifiers in earthworm gut and evolution of nitrogen gas after vermicompost treatment (*Source*: Reproduced from Wu, H., Lu, M., Lu, X., Guan, Q., & He, X. (2015a). Interactions between earthworms and mesofauna has no significant effect on emissions of CO2 and N2O from soil. *Soil Biol. Biochem, 88*, 294–297. Copyright © 2015 Elsevier Masson SAS. All rights reserved.).

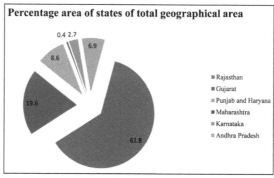

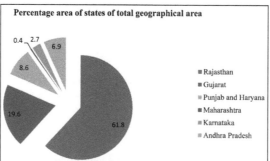

FIGURE 3.1 Area of arid region in different state of India (Bhandari et al., 2014).

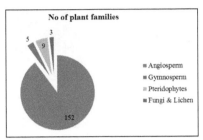

FIGURE 5.2 Distribution of medicinal plants among various classes of plant kingdom (Source: Ved and Goraya, 2008).

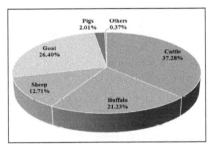

FIGURE 6.2 Share of different livestock categories in the total population (2012).

PLATE 6.1 Improved silvopasture systems A. Leucaena based B. Ber Based and C. *Hardwickia* based (*Source*: Mahanta et al., 2017).

PLATE 6.2   Traditional silvo-pasture systems.

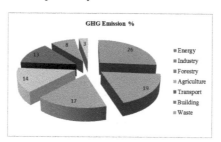

FIGURE 7.3   Contributions of different sectors to greenhouse gas emissions (Modified from IPCC, 2007; Smith et al., 2008).

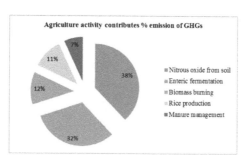

FIGURE 7.4   Contributions of different agricultural practices to greenhouse gas emissions (Modified from IPCC, 2007; Smith et al., 2008).

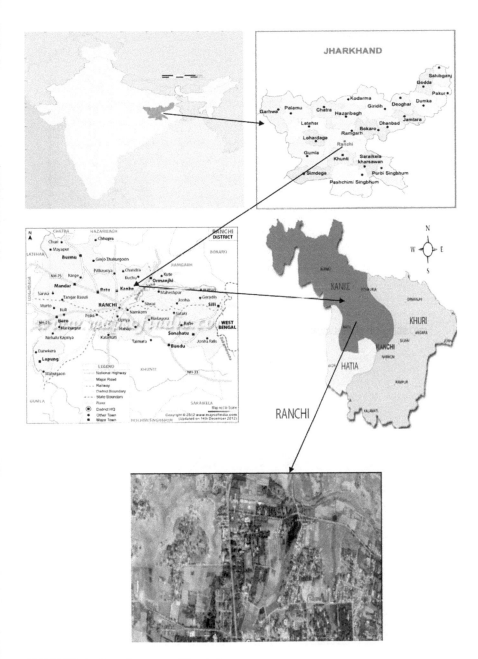

FIGURE 8.1 Location map of the study area.

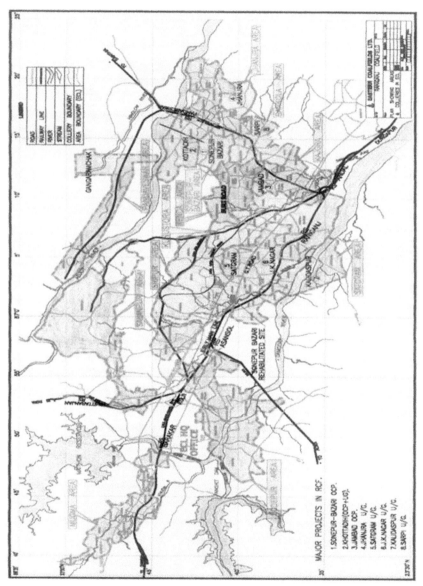

FIGURE 9.1   Location of study sites in Raniganj Coalfield areas.

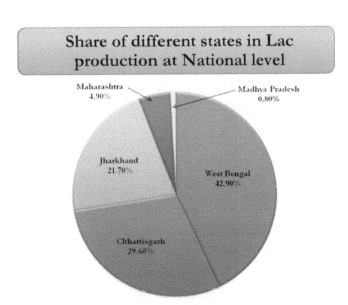

FIGURE 10.3   Percentage share of different states in lac production in India (IINRG, 2011).

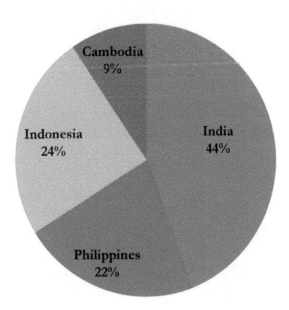

FIGURE 10.7   Percentage contribution of country in NTFPs trade in South East Asia (NTFPs Annual report, 2011).

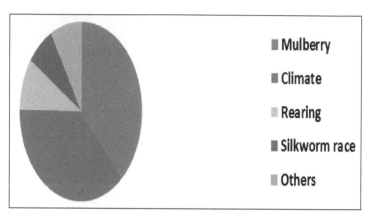

FIGURE 11.1   Regulating factors for efficient silkworm rearing.

FIGURE 11.2   Mulberry tree growing in Ladakh (Cold desert).

FIGURE 11.3   A woman rearer harvesting vegetables are grown under mulberry trees in Kashmir.

CHAPTER 7

Agroforestry for Climate Mitigation and Livelihood Security in India

ABHISHEK RAJ,[1] M. K. JHARIYA,[2] A. BANERJEE,[3] and D. K. YADAV[2]

[1]PhD Scholar, Department of Forestry, College of Agriculture, I.G.K.V., Raipur–492012 (C.G.), India, Mobile: +00-91-8269718066, E-mail: ranger0392@gmail.com

[2]Assistant Professor, University Teaching Department, Department of Farm Forestry, Sarguja Vishwavidyalaya, Ambikapur–497001 (C.G.), INDIA. Mobile: +00-91-9407004814 (M. K. Jhariya), +00-91-9926615061 (D. K. Yadav), E-mail: manu9589@gmail.com (M. K. Jhariya), dheeraj_forestry@yahoo.com (D. K. Yadav)

[3]Assistant Professor, University Teaching Department, Department of Environmental Science, Sarguja Vishwavidyalaya, Ambikapur–497001 (C.G.), INDIA. Mobile: +00-91-9926470656, E-mail: arnabenvsc@yahoo.co.in

ABSTRACT

Agroforestry known as sustainable land use farming system includes three elements *viz.*, trees as woody perennials, herbaceous crops and livestock's on the same piece of land. This system contributes to the high production of crops and various tree products such as timber and non-timber products that helps to generate income of farmers. It comprises different models in different region worldwide, which plays a vital role to combat the negative impact of climate change by increasing tree-crop diversification that leads to more carbon storage (carbon-sequestration) capacity than alone cultivation of sole agricultural crops. Therefore, agroforestry systems should be managed in such a way that can fulfill the both production (fuelwood, nutritive fruits, food, timber, NTFPs, etc.) and protection (shade, efficient nutrient cycling, watershed management, reducing soil erosion, improving soil health and fertility, etc.) functions with greater economic benefits to farmers. In short, agro-forestry has tremendous potential to enhance farm production, income, and employment opportunities and address the crisis of food and nutritional uncertainty in burgeoning population by providing multiple outputs to smallholder's farmers and which increases rural living standards. Thus, agroforestry plays both climate change mitigation and adaptation role for farmers.

7.1   INTRODUCTION

It is well understood about the term of agroforestry as sustainable land use management comprises the integration of trees, crops and livestock/pasture on a unit piece of land which is socially, economically and environmentally acceptable and practically feasible. Agroforestry plays a wider role in natural resource management, conservation, and their proper utilization and plays a potential role to mitigate global climate change issue due to several natural and anthropogenic factors like deforestation, industrialization and rapid land use transformation. Today, deforestation is the biggest crisis which can be reduced by practice of several agroforestry models in different agroclimatic zones which leads towards a reduction in dependency of peoples on natural forest for their needs like fuel, food, and timber and the activity of agroforestry helps in maintaining the forest cover/area as per recommended by national forest policy (Jhariya et al., 2015; Singh and Jhariya, 2016; Raj and Jhariya, 2017). Similarly, under the crisis of changing climate, the large farmers have initiated mitigation and adaptation activities by changing the land management and agronomic practices. But due to lack of knowledge and poor socioeconomic conditions of small and marginal farmers, they could not be able to shift from traditional agriculture. Under these situations, the agroforestry systems could help in adapting this greater climatic variability as well as mitigating greenhouse gases through storage and sequestration of atmospheric carbon (Verchot et al., 2007; Nair and Nair, 2014). Thus, a combination of woody perennials trees, herbaceous crops, and animals in agroforestry system stand for scientific and sustainable land use systems and signifies as socially and ecologically acceptable to the farmers (Nair, 1979). As per Fanish and Priya (2013), agroforestry could play a potential role in the enhancement of overall biomass production and carbon sequestration, improvement of soil fertility, nutrient cycling and microclimate of the systems and production of bioenergy and biofuel. Compatibility between tree and crops with the incorporation of multipurpose trees and N_2 fixing trees with agricultural crops can play a key role in overall biomass production and improve the soil health through improvement of biological and physical attributes which helps in efficient nutrient cycling and ecosystem maintenance (Raj et al., 2014a, 2014b).

Agroforestry contributes significantly towards mitigating climate change through carbon sequestration process. It is a process in which atmospheric carbon dioxide is trapped by green plants and incorporated into the woody parts in the form of biomass. In this connection, carbon stock/storage is the level of carbon present in the standing crop biomass and soil components. Although, agroforestry could potentially sequesters carbon both in vegetation and soil results climate change mitigation and produces woody, and non-woody materials contribute to farmer's socio-economic development (Alavalapati and Nair, 2001; Sudha et al., 2007; Parihaar, 2016). Higher biomass and productivity of agroforestry system range between 2–10 t ha^{-1} per annum under the rainfed condition as well as in the arid and semiarid regions (Dhyani et al., 2013). Thus, agroforestry has gaining a broad recognition for potential storage and sequestration of carbon to mitigating climate change and maintaining agricultural sustainability. In the view of above points, this chapter review about agroforestry systems, the role of different models in carbon storage and sequestration, enhancement in productivity and economic gain for farmer's livelihood security under changing the climate.

7.2 AGROFORESTRY CONCEPTS AND AREA

As we know, agroforestry is not new; it is an old set of farming practices that prevailed in the entire world. It is an ecologically sustainable and economically viable land use system that maintains total yield productivity through a suitable combination of crops with tree and/or livestock on the same unit of land. Moreover, the agroforestry system enhances carbon sequestration, maintains water tables and soil productivity and helps in maintaining greenery through diversification of crops, results in reducing pressure on natural forest. Nair et al. (2009) have reported 1,023 million ha area under agroforestry in the world which is comparatively much higher than India having 25.32 Mha (Dhyani et al., 2013). Moreover, the area coverage of Agroforestry systems in India is divisible into 20.0 M ha of cultivated land (includes both irrigated and rainfed areas of 7.0 and 13.0 Mha) and 5.32 Mha area of others that comprise both homestead agroforestry and shifting cultivation of 2.93 and 2.28 Mha respectively. Similarly, Nair and Nair (2003) have reported around 235.2 Mha area of alley-cropping, Silvo-pastoral, windbreaks and riparian buffers in the USA. Kumar (2006) estimated the area of the home garden in South and Southeast Asian home garden as 8.0 Mha.

7.3   SCOPE AND POTENTIAL

The scope of agroforestry is inevitable. Globally, a large hectare area is available for the adoption of these sustainable farming practices (Raj and Jhariya, 2017). Different models of agroforestry are adopted either on the plain or sloppy (hilly) area. Integration of some MPTs and N_2 fixing trees such as *Acacia nilotica* (babul), *Azadirachta indica* (neem), *Butea monosperma* (palas), *Terminalia arjuna* (arjun), etc. and commercial fruit-bearing trees such as *Carica papaya, Citrus spp, Mangifera indica, Psidium guajava*, etc. with some agricultural crops in fields can provides both tangible and intangible benefits such as fuelwood, fodder, fruit and provides shelter as well as improve socioeconomic status of poor farmers. Jhariya et al. (2015) reported *Acacia nilotica, Albizia procera, Azadirachta indica, Butea monosperma, Gmelina arborea, Terminalia tomentosa, Terminalia arjuna, Mangifera indica, Ziziphus mauritiana*, etc. are the multipurpose tree species grown in the paddy field beside rice. Beside it, the benefits of neem plantation in agroforestry systems, social forestry program and other afforestation works can reduce deforestation through minimizing pressure on natural forest, combat desertification and maintain soil health (Jhariya et al., 2013).

However, agroforestry deserves multifarious potentials such as supplement multiple products of either timber and non-timber products, ecological restoration and ecosystem maintenance, address negative impacts of climate due to global warming, enhance carbon sequestration, maintenance of soil fertility, improve soil health by reducing soil erosion and nutrient loss, combat desertification, provide rural employment opportunities and increases income of farmers by diversifying farming practices.

7.4   CLIMATE CHANGE OVERVIEW

Of course, climate change can't be denied and is the major issue over the globe. Greenhouse gas (GHGs) emission such as chlorofluorocarbons (CFC), methane, carbon dioxide (CO_2) has led to an excessive rise in temperature and therefore hampering agricultural productivity. It also promotes the origin of infectious diseases and alteration of nutrient cycling leading to huge loss of biodiversity. Among all these GHGs, CO_2 is foremost cause of the warming of the earth (Forster et al., 2007) and their emission from 1751 to 2010 in a global context is shown in Figure 7.1 (CDIAC, 2013). Therefore, emission results rise in global temperature which is graphically shown in Figure 7.2 (GISS, 2015). In 2007, Mauna Loa, Hawaii has measured the concentration of CO_2 in the atmosphere which increased from 280 to 383 ppm (Forster et al., 2007; Keeling et al., 2008). Similarly, the burning of fossil fuel and industrialization are accelerating the changing climate. The different sectors contribute emission of GHGs which is depicted in Figure 7.3. Among the contributors, energy sectors are the major emitter (26%) of GHGs than waste sector (3%). Moreover, different practices/activity under agriculture contributes in GHGs emission (Figure 7.4) (IPCC. 2007; Smith et al., 2008). Although, deforestation and fossil fuel combustion enhance climate change phenomenon, which is a becoming threat to lives and livelihoods in the world (Ackerman, 2009). Moreover, the issue of climate changes has a great impact on the worldwide loss of biodiversity (Corlett and Primack, 2008).

7.5   LIVELIHOOD UNDER CHANGING CLIMATE

From an economic perspective, forest plays a key role for human civilization. Tangible benefits of fuel, fodder are obtained, and NTFPs includes fruit, medicinal plants product which serves to 300 million rural people for maintaining their daily livelihood (Basu, 2014). But in this era, consequences of climate change and global warming are clearly visible causes high temperature, uncertain rainfall and increase in infectious disease results loss of overall productivity and crisis of farmer's livelihood security. Although, the forest is a potential source for both tangible and intangible benefits to poor's but due to unsustainable harvesting and illegal cutting cause deforestation resulting change in the pattern of weather in region and climate in the globe. Therefore, afforestation programme, social forestry, and agroforestry practices could potentially mitigate these problems through huge plantation either on field bund and other wastelands/degraded areas and meet outs farmer daily needs like food, fodder, NTFPs, and other forest produces result enhancement of socioeconomic condition and improve the livelihood security.

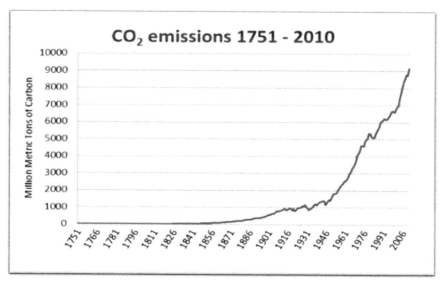

FIGURE 7.1   Global projection of CO_2 emission (CDIAC, 2013).

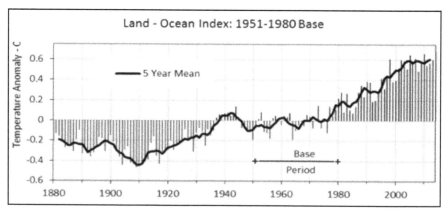

FIGURE 7.2   Rise in global temperature from 1880 to 2014 (GISS, 2015).

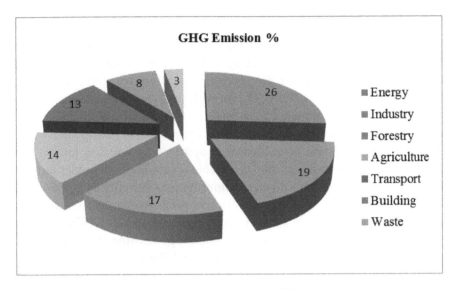

FIGURE 7.3 (See color insert.) Contributions of different sectors to greenhouse gas emissions (Modified from IPCC, 2007; Smith et al., 2008).

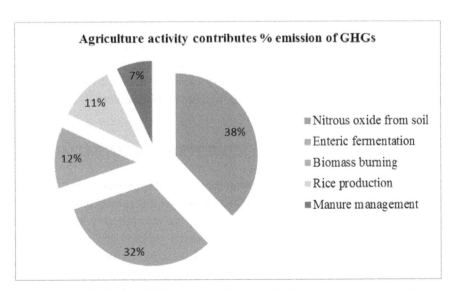

FIGURE 7.4 (See color insert.) Contributions of different agricultural practices to greenhouse gas emissions (Modified from IPCC, 2007; Smith et al., 2008).

7.6   MITIGATION STRATEGY

Climate change not only affects human and animal but this change in the pattern of weather may affect agricultural crops results unsustainable production. In this condition, the practice of agroforestry works as a viable solution by absorbing GHGs (CO_2) through the process of carbon-sequestration and mitigates climate change and reduces global warming (Singh et al., 2013). Therefore it is the only viable option to meet out people needs in the burgeoning population. Moreover, agroforestry which is known as a form of climate-smart agriculture is a viable option for smallholder farmers throughout the world, and diverse or multifarious benefits of agroforestry have been seen in developing countries in Asia, Africa, and Central and South America (Colin, 2013). As per an estimate strategies GHGs emission reduction, elevated carbon sequestration through agroforestry approaches, reducing the dependency towards conventional fuels are very much promising to account for 20–60% mitigation of climate change (Rose et al., 2012).

In 2009, the FAO coined a new term, "climate-smart agriculture" (CSA). CSA is not a specific tool or technology, but an approach to improving the productivity and sustainability of agriculture under climate change, for the purposes of advancing food security and strengthening the resilience of rural livelihoods (FAO, 2013b). Therefore, CSA is defined by three objectives: firstly, increasing agricultural productivity to support increased incomes, food security, and development; secondly, increasing adaptive capacity at multiple levels (from farm to nation); and thirdly, decreasing GHGs emissions and increasing carbon sinks (Campbell et al., 2014).

7.7   CARBON STORAGE AND SEQUESTRATION POTENTIAL

Agroforestry systems are a good option for sustainable production and protection of integrated trees and crops and work as a sequestration of atmospheric carbon in their tree components. Although, the storage and sequestration capacity depends upon the type of agroforestry models and their growing nature including management practices. Likewise, carbon sequestration and their allocation pattern are varies among different tree components. Prasad et al. (2010) reported the variability in carbon content among various species which includes the following sequences *Eucalyptus tereticornis* = *Azadirachta indica* = *Acacia nilotica* = *Butea monosperma>Albizia procera* = *Dalbergia sissoo>Emblica officinalis* = *Anogeissus pendula*. The further highest level of carbon in tree components was shown by branch = stem followed by root, foliage and least in stem bark = branch bark. As a result agroforestry system contributes towards carbon storage capacity and biomass accumulation over the sole crop or other tree plantation.

Izaurralde et al. (2001) reported higher degraded forest area (2 billion ha) globally, and under tropical conditions, it is 1.5 billion ha. Afforestation and agroforestry practices under tropical and temperate region reveal carbon sequestrations rate happens to be 8.7×10^9 and 4.9×10^9 Mg carbon per annum (Dixon and Turner, 1991). As per Yadav et al. (2015) aboveground biomass, carbon stock and carbon stock equivalent carbon dioxide (CO_2) varied from 10.8 to 37.8 Mg ha^{-1}, 4.8 to 17.0 Mg ha^{-1}, 17.6 to 62.3 Mg ha^{-1}, subsequently in the orchard based agri-horticulture practice in the Indian Himalayan region. The significantly (<0.05) higher biomass (37.8 Mg ha^{-1}), carbon stock (17 Mg ha^{-1}) and carbon stock equivalent CO_2 (62.3 Mg ha^{-1}) was recorded in the pear + wheat, and the lowest was observed in wheat mono-cropping. Under pear + wheat sequence biomass, carbon and CO_2 accumulation rate appeared as 12.0, 5.3, 19.6 Mg ha^{-1} year^{-1}, respectively and varied significantly in diverse fruit species. Fruit tree biomass showed a significant and positive relationship with total biomass, total carbon, and total CO_2 mitigation. Conversion of unproductive land into agroforestry systems may promote carbon sequestering potential value up to 391,000 Mg C y^{-1} by 2010 to 586,000 Mg C y^{-1} by 2040 (IPCC, 2000).

Moreover, agroforestry potentially represents 0.3 to 15.2 Mg C/ha/yr carbon storage capacity and the highest being in humid tropics (Nair et al., 2011). As per Sathaye and Ravindranath (1998), the Indian agroforestry contributes 25 t C ha^{-1} on average purpose for more than 95 million hectare land. Singh and Pandey (2011) reported that the annual carbon sequestration rate (Mg ha^{-1}yr^{-1}) for populus, eucalyptus, bamboo, and tropical home gardens happens to be 8, 6, 6.51–8.95 and 16–36, respectively. Table 7.1 depicts the carbon sequestering potential of various agroforestry systems under different agroclimatic zones in India (Maikhuri et al., 2000). Agrisilvicultural system reflected significant promise in the carbon storage capacity under humid tropical lands (12–228 MgC ha^{-1}) and dry lowlands regions of South East Asia (68–81 MgC ha^{-1}) (Murthy et al., 2013). Also, various researches have been conducted for the potential of storage and sequestration capacity of carbon in tree-crop components (Table 7.2) and in soil under different models of agroforestry systems (Table 7.3) which helps in mitigation climate change. Agroforestry practices, therefore, leads to soil conservation by building up soil carbon pool in soil and tree.

TABLE 7.1   Carbon Sequestration Potential Under Agroforestry in Different Agro-Climatic Regions in India (Maikhuri et al., 2000)

Agro-climatic regions	Carbon sequestration rate
Central Himalayas	3.9 t ha^{-1} yr^{-1}
Indo-Gangetic plains	8.5–15.2 t.ha^{-1} yr^{-1}
Agricultural soils of Indo-Gangetic plains	12.4–22.6 t.ha^{-1} yr^{-1}
Strip plantation aged 5.3 yr. in Haryana	15.5 t ha^{-1} during the first rotation
Degraded forest land	1.1 t.ha^{-1} yr^{-1}

TABLE 7.2   The Carbon Assimilation/Sink Potential of Different Agroforestry Models in Different Region

Region	Agroforestry models	Age	Carbon storage capacity	Author
Semiarid region	Agrisilviculture system	11 years old	26.0 tC/ha	NRCAF (2005)
Central India	Block plantation	6 years old	24.1–31.1 tC/ha	Swamy et al. (2003)
	Populus deltoids + wheat	-	18.53 tC/ha	
Himachal Pradesh	Silvopasture	-	31.71 tC/ha	Verma et al. (2008)
	Agrisilviculture	-	13.37 tC/ha	
	Agri-horticulture	-	12.28 tC/ha	
Kerala, India	Silvo-pastoralism	5 years old	6.55 Mg ha^{-1} y^{-1}	Kumar et al. (1998a)
Sumatra	Indonesian home gardens	13.4 years old	8.00 Mg ha^{-1} y^{-1}	Roshetko et al. (2002)
Semiarid	Silvo-pastoral system	5 years old	9.5–19.7 tC/ha	Rai et al. (2001)
Arid region	Agrisilviculture system	8 years old	4.7–13.0 tC/ha	Singh (2005)
Punjab (Rupnagar district)	Poplar block plantation	-	330,510 t	Gera et al. (2006)
	Eucalyptus bund plantation	-	59,361 t	
Puerto Rico	Agroforestry woodlots	4 years old	12.04 Mg ha^{-1} y^{-1}	Parrotta (1999)
Chattisgarh, Central India	Agrisilviculture	5 years old	1.26 Mg ha^{-1} y^{-1}	Swamy and Puri (2005)
S. Gou, Mali, W African Sahel	Fodder bank	7.5 years old	0.29 Mg ha^{-1} y^{-1}	Takimoto et al. (2008b)

TABLE 7.3 Carbon-Sequestration Potential of Soil Under Agroforestry Systems

Region	Agroforestry system/species	Age (years)	Soil depth (cm)	Soil C sink (Mg ha⁻¹)	Author
Puerto Rico	Mixed stands, Eucalyptus + Casuarina, Casuarina + Leucaena, and Eucalyptus + Leucaena	4	0 to 40	61.9, 56.6, and 61.7	Parrotta (1999)
Chhattisgarh (Central India)	Agrisilviculture model comprising *Gmelina arborea* and eight field crops)	5	0 to 60	27.4	Swamy and Puri (2005)
Ipet-Embera (Panama)	Home and outfield gardens		0 to 40	45.0 to 2.3	Kirby and Potvin (2007)
Florida (USA)	*Pinus Elliottii* and *Paspalum notatum* based tree based pasture model	8–40	0 to 125	6.9 to 24.2	Haile et al. (2008)
South Canada	Alley cropping system comprised hybrid poplar + wheat, soybeans, and maize rotation	13	0 to 40	1.25	Oelbermann et al. (2006)

7.8   AGROFORESTRY TOWARDS LIVELIHOOD SECURITY UNDER CHANGING CLIMATE

In any circumstance, the occurrence of tree species on the farms guarantees not only farm diversification but also ensure income generation through the delivery of additional resources like timber, fruits, nuts, vegetables, fodder, etc. and buffer the higher temperature under changing the climate. Moreover, agroforestry practices are the source of many products such as timber or non-timber products and which could play a valuable role in economic enhancement and livelihoods upliftments of poor farmers (WAC, 2010) and they depend on nature for these valuable products (Jhariya and Raj, 2014).

Several practices such as apiculture, lac culture, sericulture along with gum, and resin-producing tree in agroforestry system could enhance the socioeconomic status of farmers and maintain livelihood security (Dhyani, 2012). Although, gums and resins are a good source of NTFPs and help in augmenting the farmer's household income by 14 to 23% for small producers in Ethiopia (Abtew, 2014). Majorly, central region of Indian subcontinent (represent tropical dry and moist deciduous forest) encompasses gum-producing trees and tapping of babul tree for gum can enhance the socioeconomic condition of farmers and maintaining biodiversity with environment conservation (Das et al., 2014, Raj and Toppo, 2014; Toppo et al., 2014, Raj et al., 2015; Raj, 2015a; Raj, 2015b; Raj and Singh, 2017).

There are various research has been conducted on economic evaluation and livelihood upliftment in agroforestry systems. One estimates show that integration of *Psidium guajava* (guava) with agricultural crops under agri-horticultural based system gave an approximately three times higher economic benefit in comparison to sole crop system in Meghalaya (Bhattacharya and Mishra, 2003). Such output proves the immense potentiality of horticulture based agroforestry systems to play the key role in socioeconomic upliftment of the community stakeholders in any area. Agroforestry also creates an opportunity of employment generation up to 943 million person-days annually from 25.4 million ha (Figure 7.5) (NRCAF, 2007). In this context, Dhyani et al. (2005) reported agroforestry potential for rural development and income generation appears to be 5.763 million person-days yr^{-1} in the Himalaya region. Figure 7.6 reflects the scenario of net return in comparison to investment associated with different agroforestry systems (NRCAF, 2007). Based on net economic output highest output appears for rainfed agrisilviculture and irrigated agrisilviculture, subsequently followed by agri-horticulture and silviculture.

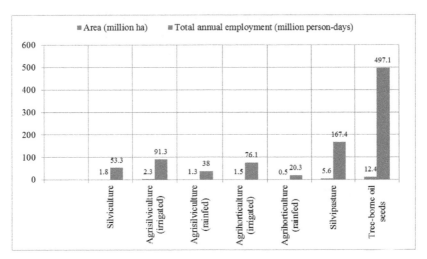

FIGURE 7.5 Employment generation potential of agroforestry in India (Adapted and modified from NRCAF, 2007).

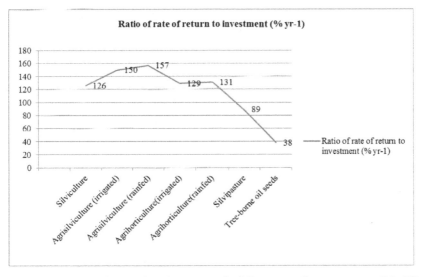

FIGURE 7.6 Rates of return from investment in different agroforestry systems (Modified from NRCAF, 2007).

In Chhattisgarh state, integration of aonla as horticultural tree with groundnut and gram as field crops under agri-horticulture model resulted analysis of several type of economic parameters such as total expense of tree and crops (86,494 Rs. per ha), total benefits (93,903 Rs. per ha), net benefit (7,410 Rs. per ha.) and B:C ratio (1.09). Likewise, integration of *Gmelina arborea* tree with paddy and linseed under agrisilviculture system resulted in a total expense of tree and crops (69139 Rs. per ha), total benefits (119,997 Rs. per ha), net benefit (50,858 Rs. per ha.) and B:C ratio (1.74). Therefore, the economic analysis of these agroforestry models is able to give an idea about the acceptability of these model under given agroclimatic zones (GoI, 2001).

7.9   CONCLUSION

Varied agroforestry models under varied agroclimatic conditions of the world play a key role to reduce the impact of climate change through diversion of traditional agriculture system to agroforestry system which promotes more carbon storage and sequestration and sustainability in comparison to sole cropping system. This leads to both adaptability and mitigatory effect on climate change. People must be aware about the multifarious benefits and potential of agroforestry systems, and they should practice this farming technique efficiently. Policy framing should be done keeping in mind the view related to extension and research and development activities in the field of agroforestry which causes diversification in the agriculture outputs and maintaining ecological services in the form of climate change mitigation, decontamination of environment and biodiversity conservation.

KEYWORDS

- afforestation
- agricultural productivity
- agroforestry
- carbon sequestration
- carbon storage
- climate-resilient
- climate-smart agriculture
- climatic change
- ecosystem
- food security
- livelihood security
- NTFPs
- soil health
- sustainable farming
- sustainable land use

REFERENCES

1. Abtew, A. A., Pretzsch, J., Secco, L., & Mohamod T. E. (2014). Contribution of Small-Scale Gum and Resin Commercialization to Local Livelihood and Rural Economic Development in the Dry-lands of Eastern Africa. *Forests, 5,* 952–977.
2. Ackerman, F. (2009). Financing the Climate Mitigation and Adaptation Measures in Developing Countries, Stockholm Environment Institute, Working Paper WP-US-0910, pp. 1–17.
3. Alavalapati, J. R. R., & Nair, P. K. R. (2001). Socioeconomic and institutional perspectives of agroforestry. In M. Palao and J. Uusivuori (eds.) *World Forest Society and Environment-Markets and Policies.* Kluwer Academic Publishers, Dordrecht, pp. 71–81.
4. Bashu, J. P. (2014). Agroforestry, climate change mitigation and livelihood security in India. *New Zealand Journal of Forestry Science, 44,* 2–10.
5. Bhattacharya, B. P., & Misra, L. K. (2003). Production potential and cost-benefit analysis of agri-horticulture agroforestry systems in Northeast India. *Journal of Sustainable Agriculture, 22,* 99–108.
6. Campbell, B. M., Thornton, P., Zougmore, R., Asten, P. V., & Lipper, L. (2014). Sustainable intensification: What is its role in climate-smart agriculture? *Current Opinion in Environmental Sustainability, 8,* 39–43.
7. CDIAC (2013). Carbon dioxide information analysis center. http://cdiac.ornl.gov/trends/emis/overview_2010.html
8. Colin, M. (2013). "Agroforestry and Smallholder Farmers: Climate Change Adaptation through Sustainable Land Use." Capstone Collection Paper 2612.
9. Corlett, R. T., & Primack, R. B. (2008). Tropical rainforest conservation: a global perspective. In: Carson WP, Schnitzer SA (Eds.) Tropical Forest Community Ecology, Blackwell Science, UK, pp. 442–457.
10. Das, I., Katiyar, P., & Raj, A. (2014). Effects of temperature and relative humidity on ethephon induced gum exudation in *Acacia nilotica. Asian Journal of Multidisciplinary Studies, 2*(10), 114–116.
11. Dhyani, S. K. (2012). Agroforestry interventions in India: Focus on environmental services and livelihood security. *Indian Journal of Agroforestry, 13*(2), 1–9.
12. Dhyani, S. K., Handa, A. K., & Uma (2013). Area under agroforestry in India: an assessment for present status and future perspective. *Indian Journal of Agroforestry, 15*(1), 1–11.
13. Dhyani, S. K., Sharda, V. N., & Sharma, A. R. (2005). Agroforestry for sustainable management of soil, water, and environmental quality: Looking back to think ahead. *Range Management and Agrofores*try, *26*(1), 71–83.
14. Dixon, R. K., & Turner, D. P. (1991). The global carbon cycle and climate change: responses and feedback from below-ground systems. *Environmental Pollution, 73,* 245–261.
15. Fanish, S. A., & Priya, R. S. (2013). Review on benefits of agroforestry system. *International Journal of Education and Research, 1*(1), 1–12.
16. FAO (2013b). Climate-smart agriculture sourcebook. Retrieved from: http://www.fao.org/climatechange/climatesmart/en/.

17. Forster, P., Ramaswamy, V., Artaxo, P., Berntsen, T., Betts, R., Fahey, D. W., Haywood, J., Lean, J. D., Lowe, C., Myhre, G., Nganga, J., Prinn, R., Raga, G., Schulz, M., & Van Dorland, R. (2007). Changes in Atmospheric Constituents and in Radiative Forcing. In: *Climate Change 2007: The Physical Science Basis. Contribution of Working Group I to the Fourth Assessment Report of the Intergovernmental Panel on Climate Change* [Solomon, S., D. Qin, M. Manning, Z. Chen, M. Marquis, K.B. Averyt, M.Tignor and H.L. Miller (eds.)]. Cambridge University Press, Cambridge, United Kingdom and New York, NY, USA.

18. Gera, M., Mohan, G., Bisht, N. S., & Gera, N. (2006). Carbon Sequestration potential under agroforestry in Rupnagar district of Punjab. *Indian Forester, 132*(5), 543–555.

19. GISS (2015). Surface Temperature Analysis, NASA, accessed January *25, 2015*; Global temperature, 1800- 2006, ProcessTrends.com, accessed October *27, 2009*.

20. GoI (2001). Planning commission report of the task force on growing India for livelihood security and sustainable development, 231p.

21. Haile, S. G., Nair, P. K. R., & Nair, V. D. (2008). Carbon storage of different soil-size fractions in Florida Silvo-pastoral systems. *J. Environ. Qual., 37,* 1789–1797.

22. IPCC (2000). Land use, Land-use Change, and Forestry. A Special Report of the IPCC. Cambridge University Press Cambridge, UK, p. 375.

23. IPCC (2007). Climate Change 2007: Synthesis Report. Contribution of Working Groups I, II and III to the Fourth Assessment Report of the Intergovernmental Panel on Climate Change. Geneva.

24. Izaurralde, R. C., Rosenberg, N., & Lal, R. (2001). Mitigation of climatic change by soil carbon sequestration: issues of science, monitoring, and degraded lands. *Advances in Agronomy, 70,* 1–75.

25. Jhariya, M. K., & Raj, A. (2014). Human welfare from biodiversity. *Agrobios Newsletter, 12*(9), 89–91.

26. Jhariya, M. K., Bargali, S. S., & Raj, A. (2015). Possibilities and Perspectives of Agroforestry in Chhattisgarh. In: Miodrag Zlatic, editors. Precious Forests–Precious Earth. InTech E- Publishing Inc; pp. 238–257.

27. Jhariya, M. K., Raj, A., Sahu, K. P and Paikra, P. R. (2013). Neem–a tree for solving global problem. *Indian Journal of Applied Research, 3*(10), 66–68.

28. Keeling, R. F., Piper, S. C., Bollenbacher, A. F., & Walker, J. S. (2008). Atmospheric CO_2 records from sites in the SIO air sampling network. In Trends: A Compendium of Data on Global Change. Carbon Dioxide Information Analysis Center, Oak Ridge National Laboratory, U.S. Department of Energy, Oak Ridge, Tenn., U.S.A.

29. Kirby, K. R., & Potvin, C. (2007). Variation in carbon storage among tree species: Implications for the management of a small-scale carbon sink project. *For. Ecol. Manage., 246,* 208–221.

30. Kumar, B. M. (2006). Carbon sequestration potential of Tropical homegardens. In: Tropical Homegardens: A time tested example of sustainable agroforestry. Advance in Agroforestry 3. Edited by Kumar, B.M., & Nair, P.K.R. (eds.), Springer, Dordrecht, the Netherlands, pp. 185–204.

31. Kumar, B. M., George, S. J., Jamaludheen, V., & Suresh, T. K. (1998a). Comparison of biomass production, tree allometry and nutrient use efficiency of multipurpose

trees grown in wood lot and Silvo-pastoral experiments in Kerala, India. *For. Ecol. Manage.*, *112*, 145–163.

32. Maikhuri, R. K., Semwa, R. L., Rao, K. S., Singh, K., & Saxena, K. G. (2000). Growth and ecological impacts of traditional agroforestry tree species in Central Himalaya, India. *Agroforestry Systems*, *48*, 257–272.

33. Murthy, I. K., Gupta, M., Tomar, S., Munsi, M., Tiwari, R., Hegde, G. T., & Ravindranath, N. H. (2013). Carbon sequestration potential of agroforestry systems in India. *Journal of Earth Science & Climate Change*, *4*(1), 1–7.

34. Nair, P. K. R. (1979). Agroforestry Research: A retrospective and prospective appraisal. *Proc. Int. Conf.* International cooperation in agro-forestry. ICRAF, Nairobi, 275–296 pp.

35. Nair, P. K. R., & Nair, V. D. (2003). Carbon storage in North American Agroforestry systems. In: The potential of U.S. forest soils to sequester carbon and mitigate the greenhouse effect, Boca Raton, FL. Edited by Kimble, J., Heath, L.S., Birdsey, R.A., & Lal, R. (eds), CRC Press LLC. PP. 333–346.

36. Nair, P. K. R., & Nair, V. D. (2014). Solid-fluid-gas: the state of knowledge on carbon sequestration potential of agroforestry systems in Africa. *Current Opinion in Environ. Sustain*, *6*, 22–27.

37. Nair, P. K. R., Kumar, B. M., & Nair, V. D. (2009). Agroforestry as a strategy for carbon sequestration. *J. Plant Nutr. Soil Sci.*, *172*, 10–23.

38. Nair, P. K. R., Vimala, D. N., Kumar, B. M., & Showalter, J. M. (2011). Carbon sequestration in agroforestry systems. *Advances in Agronomy*, *108*, 237–307.

39. National Research Centre for Agroforestry (NRCAF) (2007). Vision-2025: NRCAF Perspective Plan. Jhansi, India.

40. NRCAF (2005). *Annual Report*. National Research Centre for Agroforestry, Jhansi.

41. Oelbermann, M., Voroney, R. P., Gordon, A. M., Kass, D. C. L., Schlnvoigt, A. M., & Thevathasan, N. V. (2006). Carbon input, soil carbon pools, turnover and residue stabilization efficiency in tropical and temperate agroforestry systems. *Agroforest. Syst.*, *68*, 27–36.

42. Parihaar, R. S. (2016). Carbon Stock and Carbon Sequestration Potential of different Land-use Systems in Hills and Bhabhar belt of Kumaun Himalaya. PhD Thesis, Kumaun University, Nainital, pp. 350.

43. Parrotta, J. A. (1999). Productivity, nutrient cycling, and succession in single- and mixed-species stands of *Casuarina equisetifolia*, *Eucalyptus robusta* and *Leucaena leucocephala* in Puerto Rico. *For. Ecol. Manage.*, *124*, 45–77.

44. Prasad, R., Saroj, N. K., Ram Newaj, Venkatesh, A., Dhyani, S. K., & Dhanai, C. S. (2010). Atmospheric carbon capturing potential of some agroforestry trees for mitigation of warming effect and climate change. *Indian Journal of Agroforestry*, *12*(2), 37–41.

45. Rai, P., Yadav, R. S., Solanki, K. R., Rao, G. R., & Singh, R. (2001). Growth and pruned biomass production of multipurpose tree species in silvi-pastoral system on degraded lands in semi-arid region of Uttar Pradesh, India. *Forest Tree and Livelihood*, *11*, 347–364.

46. Raj, A., & Singh, L. (2017). Effects of girth class, injury and seasons on Ethephon induced gum exudation in *Acacia nilotica* in Chhattisgarh. *Indian Journal of Agroforestry*, *19*(1), 36–41.

47. Raj, A., & Jhariya, M. K. (2017). *Sustainable Agriculture with Agroforestry: Adoption to Climate Change*. Pp. 287–293. In: Climate Change and Sustainable Agriculture. Edited by Kumar, P.S., Kanwat, M., Meena, P.D., Kumar, V., & Alone, R.A. (Ed.). ISBN: 9789-3855-1672-6. New India Publishing Agency (NIPA), New Delhi, India.

48. Raj, A. (2015a). Evaluation of Gummosis Potential Using Various Concentration of Ethephon. MSc Thesis, I.G.K.V., Raipur (C.G.), p 89.

49. Raj, A. (2015b). Gum exudation in *Acacia nilotica*: effects of temperature and relative humidity. In Proceedings of the National Expo on Assemblage of Innovative ideas/ work of postgraduate agricultural research scholars, Agricultural College and Research Institute, Madurai (Tamil Nadu), pp. 151.

50. Raj, A., Haokip, V., & Chandrawanshi, S. (2015). *Acacia nilotica:* a multipurpose tree and source of Indian gum Arabic. *South Indian Journal of Biological Sciences, 1*(2), 66–69.

51. Raj, A., & Toppo P. (2014). Assessment of floral diversity in Dhamtari district of Chhattisgarh. *Journal of Plant Development Sciences, 6*(4), 631–635.

52. Raj, A., Jhariya, M. K., & Pithoura, F. (2014a). Need of Agroforestry and Impact on ecosystem. *Journal of Plant Development Sciences, 6*(4), 577–581.

53. Raj, A., Jhariya, M. K., & Toppo, P. (2014b). Cow dung for eco-friendly and sustainable productive farming. *International Journal of Scientific Research, 3*(10), 42–43.

54. Rose, S. K., Ahammad, H., Eickhout, B., Fisher, B., Kurosawa, A., Rao, S., Riahi, K., & van Vuuren, D. P. (2012). Land-based mitigation in climate stabilization. *Energy Economics*, 365–380.

55. Roshetko, M., Delaney, M., Hairiah, K., & Purnomosidhi, P. (2002). Carbon stocks in Indonesian homegarden systems: Can smallholder systems be targeted for increased carbon storage? *Am. J. Alt. Agr., 17,* 125–137.

56. Sathaye, J. A., & Ravindranath, N. H. (1998). Climate change mitigation in the energy and forestry sectors of developing countries. *Annual Review of Energy & Environment, 23,* 387–437.

57. Singh, G. (2005). Carbon sequestration under an agrisilvicultural system in the arid region. *Indian Forester, 131*(4), 543–552.

58. Singh, N. R., & Jhariya, M. K. (2016). Agroforestry and agrihorticulture for higher income and resource conservation. Pp. 125–145. In: Innovative Technology for Sustainable Agriculture Development, Edited by Sarju Narain and Sudhir Kumar Rawat (Ed.). ISBN: 978–81–7622–375–1. Biotech Books, New Delhi, India.

59. Singh, N. R, Jhariya, M. K and Raj, A. (2013). Tree crop interaction in agroforestry system. *Readers Shelf, 10*(3), 15–16.

60. Singh, V. S., & Pandey, D. N. (2011). Multifunctional agroforestry systems in India: Science-based policy options. RSPCB Occasional Paper No 4, 2–35.

61. Smith, P., Martino, D., & Cai, Z. (2008). Greenhouse Gas Mitigation in Agriculture. *Philosophical Transactions of the Royal Society B, 363,* 789–813.

62. Sudha, P., Ramprasad, V., Nagendra, M. D. V., Kulkarni, H. D., & Ravindranath, N. H. (2007). Development of an agroforestry carbon sequestration project in Khammam district, India. *Mitigation and Adaptation Strategies for Climate Change, 12,* 1131–1152.

63. Swamy, S. L., & Puri, S. (2005). Biomass production and C-sequestration of *Gmelina arborea* in plantation and agroforestry system in India. *Agroforestry Systems, 64,* 181–195.

64. Swamy, S. L., Puri, S., & Singh, A. K. (2003). Growth, biomass, carbon storage and nutrient distribution in *Gmelina arborea* Roxb. stands on red lateritic soils in central India. *Bioresource Technology*, *90*, 109–126.

65. Takimoto, A., Nair, P. K. R., & Nair, V. D. (2008b). Carbon stock and sequestration potential of traditional and improved agroforestry systems in the West African Sahel. *Agri. Eco. Environ.*, *125*, 159–166.

66. Toppo, P., Raj, A., Harshlata. (2014). Biodiversity of woody perennial flora in Badal Khole sanctuary of Jashpur district in Chhattisgarh. *Journal of Environment and Bio-sciences*, *28*(2), 217–221.

67. Verchot, L. V., Noordwijk, M. V., Kandji, S., Tomich, T., Ong, C., Albrecht, A., Mackensen, J., Bantilan, C., Anupama, K. V., & Palm, C. (2007). Climate change: linking adaptation and mitigation through agroforestry. *Mitg. Adapt. Strat. Glob. Change*, *12*, 901–918.

68. Verma, K. S., Kumar, S., & Bhardwaj, D. R. (2008). Soil organic carbon stocks and carbon sequestration potential of agroforestry systems in H.P. Himalaya Region of India. *Journal of Tree Sciences*, *27*(1), 14–27.

69. World Agroforestry Centre (2010). Transforming lives and landscapes, pp. 1–5.

70. Yadav, R. P., Bisht., J. K., & Pandey, B. M. (2015). Above ground biomass and carbon stock of fruit tree based land use systems in Indian Himalaya. *The Ecoscan*, *9*(3&4), 779–783.

CHAPTER 8

Tree-Soil Interaction Studies on Different Species in Arboretum

BIBEK BIRUA,[1] SANT KUMAR SINGH,[2] and
PRABHAT RANJAN ORAON[3]

[1]*Department of Natural Resource Management, Faculty of Forestry,
Birsa Agricultural University, Ranchi–834006, Jharkhand, India,
Mobile: +00-91-9934958010, E-mail: vivek.forestry@gmail.com*

[2]*Professor & Head, Department of Natural Resource Management,
Faculty of Forestry, Birsa Agricultural University, Ranchi–834006,
Jharkhand, India, Mobile: +00-91-9801334301,
E-mail: sksnrm@gmail.com*

[3]*Junior Scientist-cum-Assistant Professor, Department of Silviculture
and Agroforestry, Faculty of Forestry, Birsa Agricultural University,
Ranchi–834006, Jharkhand, India, Mobile: +00-91-9431326222,
E-mail: prabhat.ranjan.oraon@gmail.com*

ABSTRACT

The present experiment has been conducted in Arboretum of Faculty of
Forestry, Birsa Agricultural University, Ranchi, Jharkhand state, India
planted in the year 2006 with the objective to study the physiochemical
properties of soil in selected tree species, growth parameter of different
tree species, and to know-how the interrelation between soil properties and
tree growth. Seven-year-old plantation in the arboretum was done in the
unequal block and, hence, a random sampling method has been adopted
in carrying the present experiment. The present study includes *Acacia
catechu* (Khair), *Tectona grandis* (Teak), *Emblica officinalis* (Anola),
Eucalyptus tereticornis (Safeda) and *Pongamia pinnata* (Karanj). The soil
physical properties such as soil texture, bulk density, soil porosity, water

holding capacity, and particle density have been analyzed. The chemical properties viz. soil pH, soil organic carbon, and available nitrogen, available phosphorus, and available potassium have been determined. Besides these, the growth parameters like height, diameter, and crown width have been recorded. Statistically, the bulk density has been found significantly higher in *Eucalyptus tereticornis* followed by *Pongamia pinnata, Tectona grandis, Acacia catechu,* and *Emblica officinalis;* while particle density is significantly higher only in *Tectona grandis* and other species are non-significant. The porosity has been found significantly higher in *Acacia catechu* followed by *Pongamia pinnata, Tectona grandis* and rest are non-significant. The water holding capacity is found significantly higher in *Acacia catechu* followed *Pongamia pinnata* and *Tectona grandis,* but others are non-significant. The sand percentage is found significantly higher in *Acacia catechu* and *Tectona grandis;* while others are non-significant; whereas in case of percentage of silt, it has been found significantly higher in *Eucalyptus tereticornis* followed by *Pongamia pinnata* and *Emblica officinalis,* but *Tectona grandis* and *Acacia catechu* are non-significant. The clay percentage has been found significantly higher in *Emblica officinalis;* whereas other species are non-significant. Thus, a physical property of soil has shown improvement in the plated area. In statistical analysis, the soil pH is found significantly higher in *Emblica officinalis* followed by *Acacia catechu, Tectona grandis, Pongamia pinnata,* and *Eucalyptus tereticornis;* while, the soil organic carbon is significantly higher in *Pongamia pinnata* followed by *Eucalyptus tereticornis, Acacia catechu, Tectona grandis,* and *Emblica officinalis.* The available nitrogen is found significantly higher in *Eucalyptus tereticornis* followed by *Acacia catechu, Pongamia pinnata, Emblica officinalis,* and *Tectona grandis;* whereas, the available phosphorous has been found significantly higher in *Eucalyptus tereticornis* followed by *Tectona grandis, Emblica officinalis, Acacia catechu* and *Pongamia pinnata;* while, the available potassium is found significantly higher in *Eucalyptus tereticornis* followed by *Acacia catechu, Emblica officinalis, Tectona grandis,* and *Pongamia pinnata.* The nutrient status in all the species of the planted area has been found in the order like K>N>P. Both diameter and height have been found the maximum in *Eucalyptus tereticornis* followed by *Tectona grandis, Pongamia pinnata, Emblica officinalis* and minimum in *Acacia catechu.* The crown width has been found the maximum in *Pongamia pinnata* followed by *Eucalyptus tereticornis, Tectona grandis,*

Emblica officinalis and minimum in *Acacia catechu*. Results revealed that soil physical properties improved area significantly under plantation. The pH, soil organic carbon (OC), available nitrogen (N), available potassium (K) in the planted area is significantly higher than unplanted area; whereas in case of available phosphorous, it is lower in planted area than unplanted area except *for Eucalyptus tereticornis*. It may be said that soil pH is increasing in the plantation area showing betterment for tree growth.

8.1   INTRODUCTION

The forest, which is a kind of natural resource, is dwindling day by day, particularly in tropical countries majorly due to deforestation and illicit felling. The forest crops are nowadays at the alarming condition and need proper attention and propagation. The total forest and tree cover account for 7,57,010 Km² or more than 22% of the geographical limit in respect to the targeted amount of 33% as per National Forest Policy, 1988. However, the recorded forest area in the country is 7,68,436 Km² or 23.38% of the country's geographical area (Anon, 2011). Therefore, it is the urgent need of the hour to make up the gap planting and growing more and more forest tree species to get healthy and pollution free environment. Regarding ecological importance of forest, it considered that additional 300–600 million ha area is required to be brought under rapidly growing forest plantation in the world to sequester about 2.9 billion tones increment of carbon accumulating in the atmosphere through the burning of fossil fuel and deforestation. Wood is considered as a universal and chief source of energy amongst coal, oil, gas, etc. for thousands of year, until the advent of other sources of energy arises. Even today more than half of the world's consumption of wood is for fuel. Wood remains the major source of domestic fuel in India and most of which is obtained from the forest. Using forest crops as a source of energy is nothing spectacular, because at present most of the energy sources have hazardous by-products and or after effect like radiation from nuclear reactors, water and air pollution from thermal plants, etc. There is a wide gap between production from forest and demand for various usages, i.e., fuel, wood, timber, fodder, green manure, etc. Keeping the view of multifarious usages of trees and its produce, much emphasis is also given to increase forest productivity by plantations.

The importance of forests and wood in alleviating the energy crisis can be judged from the fact that wood is the main product of forest and it contributes 68.5% as a fuel source in rural India. It has been estimated that about 3 MW of power can be generated from wood produced by a fast-growing tree planted in 1000 ha of land (Dayal, 1984 & 1985). Energy crisis and rising wood prices have brought forest productivity under attention, and now the emphasis has been given to raise forest planting majorly timber and fuelwood species.

Decomposers in the form of soil microbiota play an active role in disintegration and material cycling of nutrients. It is suggested by Wardle (2002) that the aboveground and belowground subsystems are closely linked influencing the outcome of ecosystem processes. Trees have a multifaceted of impacts on soil a property which includes variation in nutrient inputs in the form of wet and dry fall of N in the form of fixing of N. In this way, trees affect morphology and chemical characteristic of above and belowground soil. Different plant component significantly influences the nutrient flow in the soil ecosystem through decomposition and nutrient release and soil moisture level which influences the breakdown of litter. The lateral development of roots harbors nutrient from a diverse system under the tree canopy.

The growth of tree depends upon various factors like genetic, climatic and, edaphic conditions as well as biotic interference. Various site factors such as edaphic, climatic, physiographic and biotic factors together interact with each other. They do not work independently but the growth and survival of a plant in the resulting of these all factor interacting with each other and hence, their cumulative effect matter a lot. The physiological background of the growth of trees has a very complex nature because of the interaction of these factors amongst which soil and climate are the most important. In general, trees represent both conduits through which nutrient cycle and sites get the accumulation of the nutrients within a landscape in an ecological perspective. The soil patch found beneath tree canopy is important local and regional nutrient reserve that influence community structure and ecosystem function. Understanding the species-specific difference in tree soil interaction is very important (Rhoades, 1997). Tree species influence on the soil mineralization process can regulate overall nutrient cycling in a forest ecosystem, which may occur through their effect on substrate quality of physiochemical soil properties (Ushio et al., 2010).

Soil is also a kind of renewable natural resource. It is the most important resource of the earth planet, and we cannot imagine life without it. Soil is a component of the lithosphere and biosphere on which life-supporting system-depends. Soil is the complex expression of physical, chemical and biological processes occurring across the spatial and temporal scale. Soil properties at specific location integrate and reflect both past and present condition. Different factors influence differentially depending upon the range of interest (Robertson and Gross, 1994). Globally or at the regional level, climate and edaphic condition are important determinants (Birkeland, 1984). The ability of the land to produce is limited as governed by soil, climate, and landform condition. However the capacity of the soil to produce is limited, and the limits to production are set by intrinsic characteristic, agroecology setting use and management (FAO, 1993a).

The soil is a very important factor playing a vital role in determining the varieties as well as the survival of both plant and animal kingdom. The properties of the soil pertaining different types of zone or area determines different types of plants to be grown in that very area being adapted through the force of natural selection to suit in that very soil properties and survive in that particular area or zone. Accordingly, the animals also living in that very area have to adapt for survival in that area or zone by eating and taking such plants to be a assimilated in their body, which is also adopted by the force of natural selection and getting energy for their survival from the herbivores onwards through the systems of energy-flow and nutrient-cycling. In this way, food-chains and food-webs are maintained through energy-flow mechanisms covering different zones or areas by checking and eliminating unwanted or excess population in the natural system, which is ultimately related to the survival of human being who is part and parcel of the system. Thus, it is very much clear that soil as an edaphic factor, is the important basic factor playing a key role in maintaining a natural balance. Soil ecosystem along with their attributes is an important aspect as a life support system to maintain the natural balance for living creatures leading to their prosperity. The properties of soil developed under a particular region, have a profound influence on the growth and development of plants. Properties such as texture, pH, nutrient status, moisture relations and those related to parent materials are particularly important. Plant growth is influenced by the texture of the soil. Granular structure possesses more volume of air and water than other structure types. Compaction of superficial layers results in poor penetration of soil. Water, air, tree roots are detrimental to plant growth (Dwivedi, 1993).

In a limiting environment, the restrictive factors operate markedly in favor of survival against growth; hence the native vegetation is so specialized and adapted to the site that it is not capable of realizing the maximum productivity of the site. Several species such as *Eucalyptus globulus* and *Cryptomeria japonica* in India and *Pinus petula* in South Africa and *Tectona grandis* in Java have shown at least 4–6 times more growth than in their original home and the native vegetation of site. It means that sites were really capable of supporting vegetation of high productivity, but due to some restrictive factors, the potential was not harvested (Champion and Seth, 1968). In case of decomposition of leaf litter in multipurpose tree species, the highest amount of nitrogen release was found in *Eucalyptus teriticornis* followed by *Dendrocalamus strictus*, and maximum phosphorous in *Cassia fistula* while maximum potassium in *Eucalyptus tereticornis* and *Terminalia balerica* was found (Bhalawe et al., 2013). The nutrient balance of N, P, K, Ca, Mg, Zn, B, Fe or Al, Ca with Mg, Fe with Al provided the key to promote the growth of teak in acid soils, and soil Zn as a primary trace element that are affecting tree growth (Zaizhi, 2016). N, P and K input in soil due to litter decomposition differed significantly between two species *Tectona grandis* and *Shorea robusta* in the order of N>K>P. Both tree species showed a similar pattern of nutrient release during decomposition of their leaf litter (Das, 2016). The non-significant difference was found in the evaluation of tree growth between exotic and indigenous tree species (Chima et al., 2016). In case of nutrient return through litter fall like N, P and K were found highest in the litterfall of *Azadirachta indica* and lowest in *Hardwickia binata*. All the trees of the studies increased organic carbon, organic matter, and available nutrients in the soil as compared to the open crop field. All the tree species appeared to be the most efficient in capturing carbon and removal of carbon dioxide (Desai, 2017). The nitrogen-fixing tree species showed a beneficial effect on available nutrient content in soil and improved organic carbon in the soil. These species increase nitrogen content in soil, and also phosphorus and potassium content showed noticeable increment (Dalvi et al., 2017).

Plants and animals always try to adjust and adapt to the physical environment. They also try to modify the physical environment and reduce the limiting effects of the locality factors. Such happens both at the community and species level. Tree imposes significant influence over soil quality as well as local climatic conditions. The continuous deposition of leaves and other debris create organic mulch on the surface of the forest floor which creates a more active and stable soil providing a better and wider field for soil microorganism. Such activity and leaching of organic acid and other decomposition products give much difference in chemical, physical and biological properties of soil. The most salient addition to the forest soil is organic matter. The surface litter layer contributes to the accumulation of soil organic matter and exerts considerable influence in the underlying mineral soil and the associated population of microorganisms and soil biota. Tree roots penetrate through compacted layers and improve aeration, soil structure, water infiltration and retention; and nutrient supplying capacity of the soil of forest cover. Roots of trees reach deep into the profiles, extract bases and return them to the surface or litterfall.

Forest soil properties need to be investigated properly towards better management of forests resources. Under the process of nutrient cycling among components of the forest stand give a distinctive character to the soil with a forest cover. Because of the close and integral relationship between soils and plant communities and their influence one upon the other help in understanding the soil processes. In this way the forest growth that linked with physical and chemical properties of soil and nutrient status act as a tool in decision making policy. Therefore, numerous factors affecting plants richness, soil play a major role in influencing the composition and structure of terrestrial flora. Heterogeneity of their properties creates a niche with a specific condition affecting the spatial distribution of the relation between plants. Soil properties are very important from ecological as well as a production point of view.

Keeping this view in mind the present work of tree-soil interaction studies has been taken in the Arboretum of Faculty of Forestry, Birsa Agricultural University, Ranchi, Jharkhand, India considering five tree species like Khair (*Acacia catechu*), Teak (*Tectona grandis*), Anola (*Emblica officinalis*), Safeda (*Eucalyptus tereticornis*) and Karanj (*Pongamia pinnata*) with the following objectives:

1. To study the physiochemical properties of soil in selected tree species.
2. To study the growth parameters of different tree species.
3. To study the interrelation between soil properties and tree species

8.2   METHODOLOGY

8.2.1   LOCATION MAP

The present work has been taken in the arboretum of Birsa Agricultural University Ranchi, (Jharkhand) India (Figure 8.1). The plantation has been done row-wise in blocks (7 years of age). Since the blocks are not equal in size, random sampling method has been applied in the present work. To study tree-soil interaction in Arboretum, five tree species like *Tectona grandis* (Teak), *Acacia catechu* (khair), *Eucalyptus tereticornis* (Safeda), *Emblica officinalis* (Amla), *Pongamia pinnata* (Karanj) have been taken. In each tree species at least 10 soil samples (0–30 cm deep) were collected from 10 trees (treating as planted area) by random selection method and accordingly 10 soil samples collected from just opposite of such selected tree from about 5 meters distance having no vegetation (treating as unplanted area) in each species. Further, from the same randomly selected trees in each species; height, diameter and crown width have also been recorded to assess as growth performance of the species.

8.2.2   SOIL PHYSICAL PROPERTIES

8.2.2.1   BULK DENSITY (BD): CORE METHOD

It has been determined by the formula given by Blake (1965).

8.2.2.2   PARTICLE DENSITY (PD)

It has been calculated by the Formula given by Blake (1965).

8.2.2.3   SOIL POROSITY (SP)

It has been estimated by the formula given by Blake and Hartge (1986).

8.2.2.4   WATER HOLDING CAPACITY (WHC)

It has been found out by the formula postulated by Veihmeyer and Hendrickson (1931).

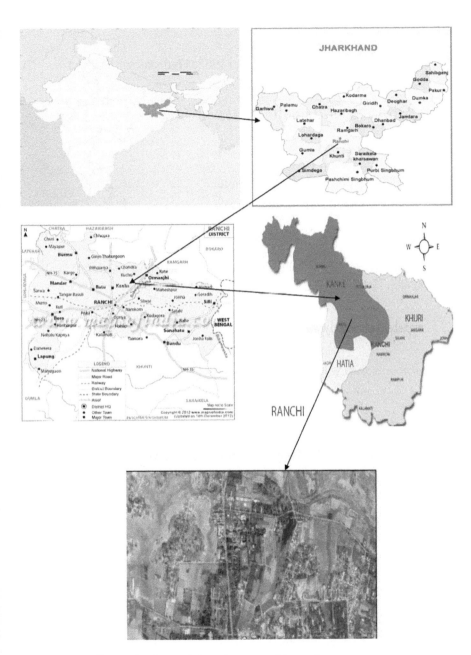

FIGURE 8.1   (See color insert.) Location map of the study area.

8.2.2.5   SOIL TEXTURE (ST)

Soil texture estimated by Bouyoucos (1962) Hydrometer Method.

8.2.3   SOIL CHEMICAL PROPERTIES

8.2.3.1   SOIL PH

The soil pH was observed by a pH meter.

8.2.3.2   SOIL ORGANIC CARBON (OC)

It has been calculated by the volumetric method postulated by Walkley and Black (1934).

8.2.3.3   AVAILABLE NITROGEN

Available nitrogen was calculated by Subbia and Asija method (1956).

8.2.3.4   AVAILABLE PHOSPHORUS

It has been found out by Bray's method No. 1 given by Bray and Kurtz (1945).

8.2.3.5   AVAILABLE POTASSIUM

It has been estimated by Flame photometric method postulated by Toth and Prince (1949).

8.2.4   MEASUREMENT OF TREE PARAMETERS

8.2.4.1   DIAMETER

The diameter was measured at breast height with the help of Calliper.

8.2.4.2 HEIGHT (DBH)

Height was measured of standing tree by the support of instrument Hypsometer.

8.2.4.3 CROWN WIDTH

Crown width has been estimated by ocular estimation method.

8.2.5 STATISTICAL ANALYSIS

Statistical analysis has been estimated by Paired T-tests and is given by Rosie Shier (2004).

8.3 RESULTS

The experiment was carried out in tree-soil interaction on different tree species in the arboretum and found out the extent of influence of soil properties on the growth of five tree species with ten soil parameters and three growth parameters are variables. The results obtained after statistical analysis under the following headings are as follows:

8.3.1 PHYSICAL PROPERTIES OF SOIL

8.3.1.1 BULK DENSITY

The bulk density of soil has been found higher in the unplanted area than planted area in all the species. The highest value of bulk density was found in *Eucalyptus tereticornis* stand followed by *Tectona grandis, Acacia catechu, Emblica officinalis* and minimum in *Pongamia pinnata*. Statistically, it has been found significantly higher in *Eucalyptus tereticornis* followed by *Pongamia pinnata, Tectona grandis Acacia catechu* and *Emblica officinalis* (Table 8.1).

TABLE 8.1 Soil Bulk Densities Under Different Tree Species

Name of species	Unplanted (gm cm^{-3})	Planted (gm cm^{-3})	t-test value
Tectona grandis	1.574	1.548	12.28**
Acacia catechu	1.735	1.541	3.185**
Emblica officinalis	1.546	1.516	6.489**
Pongamia pinnata	1.584	1.506	15.36**
Eucalyptus tereticornis	1.613	1.611	16.13**

t(9)0.05 = 2.262** Highly significant; 9 = Degree of freedom; 0.05 = 5% level of significance; 2.262 = Tabulated value of t-table.

8.3.1.2 PARTICLE DENSITY

The particle density of soil has been found higher in planted area than the unplanted area in all the species. The highest value of particle density is found in *Emblica officinalis* planted stand followed by *Acacia catechu, Eucalyptus tereticornis, Tectona grandis* and minimum in *Pongamia pinnata*. Statistical analysis shows it is highly significant only in *Tectona grandis* whereas other species are non-significant (Table 8.2).

TABLE 8.2 Soil Particle Densities Under Different Tree Species

Name of species	Unplanted (gm cm^{-3})	Planted (gm cm^{-3})	t-test value
Tectona grandis	2.333	2.374	19.52**
Acacia catechu	2.382	2.391	NS
Emblica officinalis	2.464	2.530	NS
Pongamia pinnata	2.290	2.329	NS
Eucalyptus tereticornis	2.371	2.388	NS

t(9)0.05 = 2.262** Highly significant; 9 = Degree of freedom; 0.05 = 5% level of significance; 2.262 = Tabulated value of t-table.

8.3.1.3 POROSITY

The porosity of soil has been found higher in planted area than unplanted area showing the highest value of porosity in *Emblica officinalis* followed by *Eucalyptus tereticornis, Acacia catechu, Pongamia pinnata* and minimum in *Tectona grandis*. In statistical analysis, it is found significantly higher in *Acacia catechu* followed by *Pongamia pinnata, Tectona grandis* while it is non-significant in *Eucalyptus tereticornis* and *Emblica officinalis* (Table 8.3).

TABLE 8.3 Soil Porosity (%) Under Different Tree Species

Name of species	Unplanted	Planted	t-test value
Tectona grandis	30.37	35.35	49.8**
Acacia catechu	29.29	37.30	53.4**
Emblica officinalis	35.72	38.75	NS
Pongamia pinnata	31.23	36.26	50.3**
Eucalyptus tereticornis	33.19	37.36	NS

t(9)0.05 = 2.262** Highly significant; 9 = Degree of freedom; 0.05 = 5% level of significance; 2.262 = Tabulated value of t-table.

8.3.1.4 WATER HOLDING CAPACITY

The water holding capacity of soil was found higher in planted area than unplanted area except *for Emblica officinalis*. The highest value of water holding capacity was found in *Tectona grandis* in the planted area followed by *Acacia catechu, Eucalyptus tereticornis, Pongamia pinnata,* and minimum in *Emblica officinalis*. From a statistical point of view, it is found significantly higher in *Acacia catechu* followed by *Pongamia pinnata,* then *Tectona grandis. Eucalyptus tereticornis,* and *Emblica officinalis* are non-significant (Table 8.4).

TABLE 8.4 Soil Water Holding Capacity Under Different Tree Species

Name of species	Unplanted ml/100cm^3	Planted ml/100cm^{-3}	t-test value
Tectona grandis	35.5	38.6	26.20**
Acacia catechu	32.7	36.2	46.8**
Emblica officinalis	30.6	28.4	NS
Pongamia pinnata	33.8	35.6	34.80**
Eucalyptus tereticornis	33.7	35.6	NS

t(9)0.05 = 2.262** Highly significant; 9 = Degree of freedom; 0.05 = 5% level of significance; 2.262 = Tabulated value of t-table.

8.3.1.5 SOIL TEXTURE

8.3.1.5.1 Sand (%)

The percentage of sand has been found higher in the unplanted area than planted area except *Pongamia pinnata* and *Eucalyptus tereticornis*. The highest value of the percentage of sand was found in *Pongamia pinnata* in the planted area followed by *Eucalyptus tereticornis, Emblica officinalis, Tectona grandis* and minimum in *Acacia catechu.* Statistically, it is found significantly higher in *Acacia catechu* and *Tectona grandis;* but non-significant in *Emblica officinalis, Eucalyptus tereticornis* and *Pongamia pinnata* (Table 8.5).

TABLE 8.5 Percentage of Sand in Soil Under Different Tree Species

Name of species	Unplanted	Planted	t-test value
Tectona grandis	60.3	58.9	22.65**
Acacia catechu	61.3	57.2	51.25**
Emblica officinalis	61.2	61.1	NS
Pongamia pinnata	62.1	62.6	NS
Eucalyptus tereticornis	61.3	61.9	NS

t(9)0.05 = 2.262** Highly significant; 9 = Degree of freedom; 0.05 = 5% level of significance; 2.262 = Tabulated value of t-table.

8.3.1.5.2   Silt (%)

The percentage of silt has been found higher in planted area than unplanted area except *Emblica officinalis* and *Pongamia pinnata.* The highest value of the percentage of silt is found in *Acacia catechu* in the planted area followed by *Emblica officinalis, Eucalyptus tereticornis, Tectona grandis,* and minimum in *Pongamia pinnata.* Statistically, it shows significantly higher in *Eucalyptus tereticornis* following *Pongamia pinnata* and *Emblica officinalis* showing the same value, while *Tectona grandis* and *Acacia catechu* showed non-significant values (Table 8.6).

TABLE 8.6 Percentage of Silt in Soil Under Different Tree Species

Name of species	Unplanted	Planted	t-test value
Tectona grandis	24.3	25.4	NS
Acacia catechu	24.9	28.9	NS
Emblica officinalis	28.7	27.4	22.88**
Pongamia pinnata	26.3	25.3	22.88**
Eucalyptus tereticornis	25.0	25.8	45.55**

t(9)0.05 = 2.262** Highly significant; 9 = Degree of freedom; 0.05 = 5% level of significance; 2.262 = Tabulated value of t-table.

8.3.1.5.3 Clay (%)

The percentage of clay in soil has been found higher in planted area than unplanted area except *Eucalyptus tereticornis*. The highest value of the percentage of clay is found in *Tectona grandis* in the planted area followed by *Acacia catechu, Eucalyptus tereticornis, Pongamia pinnata* and minimum in *Emblica officinalis*. From a statistical point of view, it has been found significantly higher in *Emblica officinalis,* and other species are non-significant (Table 8.7).

TABLE 8.7 Percentage of Clay in Soil Under Different Tree Species

Name of species	Unplanted	Planted	t-test value
Tectona grandis	15.4	15.7	NS
Acacia catechu	13.8	13.9	NS
Emblica officinalis	10.2	11.5	61.90**
Pongamia pinnata	11.6	12.1	NS
Eucalyptus tereticornis	13.7	12.3	NS

t(9)0.05 = 2.262** Highly significant; 9 = Degree of freedom; 0.05 = 5% level of significance; 2.262 = Tabulated value of t-table.

8.3.2 SOIL CHEMICAL PROPERTIES

8.3.2.1 SOIL PH

The soil pH has been found higher in planted area than the unplanted area in all the species. The highest value of soil ph is found in *Pongamia pinnata* in the planted area followed by *Tectona grandis, Emblica officinalis, Eucalyptus tereticornis* and minimum in *Acacia catechu*. Statistically, it has been found significantly higher in *Emblica officinalis* followed by *Acacia catechu, Tectona grandis, Pongamia pinnata* and minimum in *Eucalyptus tereticornis* (Table 8.8).

8.3.2.2 SOIL ORGANIC CARBON

The soil organic carbon has been found higher in planted area than the unplanted area in all the species. Its highest value in the planted area has been found in *Acacia catechu* and *Pongamia pinnata* followed by *Tectona grandis, Emblica officinalis* as well as *Eucalyptus tereticornis*. In statistical analysis, it is found significantly higher in *Pongamia pinnata* followed by *Eucalyptus tereticornis, Acacia catechu, Tectona grandis* and *Emblica officinalis* (Table 8.9).

TABLE 8.8   Soil pH Under Different Tree Species

Name of species	Unplanted	Planted	t-test value
Tectona grandis	5.21	5.64	5.79**
Acacia catechu	4.84	5.24	6.32**
Emblica officinalis	5.27	5.43	10.67**
Pongamia pinnata	4.89	5.74	5.48**
Eucalyptus tereticornis	5.01	5.26	5.08**

$t(9)0.05 = 2.262$** Highly significant; 9 = Degree of freedom; 0.05 = 5% level of significance; 2.262 = Tabulated value of t-table.

TABLE 8.9   Soil Organic Carbon (%) Under Different Tree Species

Name of species	Unplanted	Planted	t-test value
Tectona grandis	0.33	0.39	7.97**
Acacia catechu	0.34	0.40	10.00**
Emblica officinalis	0.34	0.39	4.18**
Pongamia pinnata	0.33	0.40	14.22**
Eucalyptus tereticornis	0.35	0.39	12.75**

$t(9)0.05 = 2.262$** Highly significant; 9 = Degree of freedom; 0.05 = 5% level of significance; 2.262 = Tabulated value of t-table.

8.3.2.3   AVAILABLE NITROGEN

The soil available nitrogen in soil has been found higher in planted area than the unplanted area in all the species. The highest value of soil available nitrogen was found in *Tectona grandis* in the planted area followed by *Acacia catechu, Eucalyptus tereticornis, Pongamia pinnata* and minimum in *Emblica officinalis*. Statistically, it is found significantly higher in *Eucalyptus tereticornis* followed by *Acacia catechu, Pongamia pinnata, Emblica officinalis* and *Tectona grandis* (Table 8.10).

8.3.2.4   AVAILABLE PHOSPHOROUS

The available phosphorous of soil has been found higher in unplanted area than the planted area in all species except *Eucalyptus tereticornis*. The highest value of soil available phosphorous is found the maximum in

Eucalyptus tereticornis in the planted area followed by *Emblica officinalis, Acacia catechu, Tectona grandis* and minimum in *Pongamia pinnata.* From a statistical point of view, it has been found significantly higher in *Eucalyptus tereticornis* followed by *Tectona grandis, Emblica officinalis, Acacia catechu* and *Pongamia pinnata* (Table 8.11).

TABLE 8.10 Available Nitrogen (kg/ha) in Soil Under Different Tree Species

Name of species	Unplanted	Planted	t-test value
Tectona grandis	252.90	273.02	20.20**
Acacia catechu	253.46	269.84	34.10**
Emblica officinalis	248.79	262.48	26.32**
Pongamia pinnata	252.11	268.13	32.84**
Eucalyptus tereticornis	251.71	269.39	41.21**

t(9)0.05 = 2.262** Highly significant; 9 = Degree of freedom; 0.05 = 5% level of significance; 2.262 = Tabulated value of t-table.

TABLE 8.11 Available Phosphorous (kg/ha) in Soil Under Different Tree Species

Name of species	Unplanted	Planted	t-test value
Tectona grandis	7.49	6.96	7.80**
Acacia catechu	7.77	7.21	4.95**
Emblica officinalis	7.60	7.38	6.19**
Pongamia pinnata	7.37	6.91	4.59**
Eucalyptus tereticornis	7.07	10.47	12.48**

t(9)0.05 = 2.262** Highly significant; 9 = Degree of freedom; 0.05 = 5% level of significance; 2.262 = Tabulated value of t-table.

8.3.2.5 AVAILABLE POTASSIUM

The available potassium of soil has been found higher in planted area than the unplanted area in all the species. The highest value of soil available potassium is found in *Pongamia pinnata* in the planted area followed by *Eucalyptus tereticornis, Tectona grandis, Emblica officinalis* and minimum in *Acacia catechu.* Statistically, it has been found significantly higher in *Eucalyptus tereticornis* followed by *Acacia catechu, Emblica officinalis, Tectona grandis* and *Pongamia pinnata* (Table 8.12).

TABLE 8.12 Available Potassium (kg/ha) in Soil Under Different Tree Species

Name of species	Unplanted	Planted	t-test value
Tectona grandis	339.35	476.78	9.68**
Acacia catechu	340.00	445.22	13.36**
Emblica officinalis	345.79	454.53	10.23**
Pongamia pinnata	337.65	504.97	8.44**
Eucalyptus tereticornis	321.56	488.45	13.76**

t(9)0.05 = 2.262** Highly significant; 9 = Degree of freedom; 0.05 = 5% level of significance; 2.262 = Tabulated value of t-table.

8.3.3 GROWTH PARAMETER

8.3.3.1 MEAN DIAMETER

The trend of diameter in the studied species has been found the maximum in *Eucalyptus tereticornis* followed by *Tectona grandis, Pongamia pinnata, Emblica officinalis* and minimum in *Acacia catechu* (Table 8.13).

TABLE 8.13 Mean Diameters of Different Tree Species

Sl. No.	Name of the tree species	Mean diameter (in cm)
1.	*Tectona grandis*	10.51
2.	*Emblica officinalis*	7.43
3.	*Pongamia pinnata*	8.82
4.	*Eucalyptus tereticornis*	16.32
5.	*Acacia catechu*	5.56

8.3.3.2 MEAN HEIGHT

The trend of height among the species has been found the maximum in *Eucalyptus tereticornis* followed by *Tectona grandis, Pongamia pinnata, Emblica officinalis* and minimum in *Acacia catechu* (Table 8.14).

8.3.3.3 CROWN WIDTH (CW)

The trend of crown width among the species has been found the maximum in *Pongamia pinnata* followed by *Eucalyptus tereticornis, Emblica officinalis, Tectona grandis* and minimum in *Acacia catechu* (Table 8.15).

TABLE 8.14 Mean Height of the Different Tree Species

Sl. No.	Name of the tree species	Mean Height (in m)
1.	*Tectona grandis*	8.04
2.	*Emblica officinalis*	4.75
3.	*Pongamia pinnata*	5.25
4.	*Eucalyptus tereticornis*	14.8
5.	*Acacia catechu*	4.01

TABLE 8.15 Mean Crown Width of the Different Tree Species

Sl. No.	Name of the tree species	Mean crown width (in m)
1.	*Tectona grandis*	2.93
2.	*Emblica officinalis*	3.21
3.	*Pongamia pinnata*	4.38
4.	*Eucalyptus tereticornis*	4.26
5.	*Acacia catechu*	1.73

8.4 DISCUSSION

On the basis of data and statistical analysis, the results found in the present investigation are interpreted and discussed under the following headings.

8.4.1 PHYSICAL PROPERTIES OF SOIL

8.4.1.1 BULK DENSITY

As per the results in the present experiment, the bulk density of soil has been found significantly higher in the unplanted area than planted area. The difference between unplanted and planted areas is significantly higher in *Eucalyptus tereticornis* followed by *Pongamia pinnata, Tectona grandis, Acacia catechu,* and *Emblica officinalis.* Zhang et al. (2012) reported the mechanism of alteration in soil quality under *E. grandis* plantation 1–10 years old. As with the recent findings of the present investigation Singh et al. (2011) reported the negative impact of the plantation of multipurpose tree species on the sodic soils of Indo-Gangetic alluvial plains which reflects a reduction in tree density. Soil bulk density did not vary significantly under eucalyptus plantation at Koga watershed, western Amhara region in Ethiopia (Alemie, 2009). Such report is almost alike with the present finding. Hosur and Dasog (1995) studied on the effect of tree species like teak, shisham, and khair on soil properties and found that trees decreased bulk density which is similar to the present investigation. Mishra and Nanda (1985) worked on available water holding capacity of soil textural classes and reported that bulk density influences significantly water holding capacity of the soil texture. In the present experiment, the bulk density in the planted area decreased while water holding capacity is found increase.

8.4.1.2   PARTICLE DENSITY

As per the recorded data of present finding, the particle density of soil has been found significantly higher in planted area than unplanted area in *Tectona grandis*; while Jordan et al. (2003) found that soil particle density significantly decreased seedling height, dry matter production and also affected microbial activity as well as decreased young tree growth along with nitrogen fertilizer uptake.

8.4.1.3   SOIL POROSITY

In the present investigation, it was found that the porosity of soil is significantly higher in planted area than the unplanted area which comparable with the earlier findings of Singh et al. (2011), who studied on sodic soils of Indo-Gangetic alluvial plains of India and found a significant increase in porosity under tree plantations. Woodward (1996) investigated the soil physical and chemical properties and found significant decreased total porosity in both microporosity and water availability.

8.4.1.4   WATER HOLDING CAPACITY

During the present investigation area under plantation reflected higher water holding capacity. Our finding corroborates with earlier findings of Zhang et al. (2012) in the case of Eucalyptus plantation in South Western China. Alemie (2009) found a significant effect on water holding capacity in the undergrowth. Alavi (2002) studied the relationship between tree growth and water availability for a Norway spruce stand during a 22-years-period at a forest in southern Sweden. The results suggested that shortage of soil water has a strong effect on tree growth in an area of southern Sweden with generally high precipitation. In the present studies, the result has shown that water holding capacity has increased in the planted area. Misra and Nanda (1985) worked on the available water holding capacity of the soil textural classes, and found that water holding capacity influenced by bulk density, texture, and organic matter.

8.4.1.5 SOIL TEXTURE

Results of the present investigation reveal sand percentage is higher in amount in the area under without plantation with an exception such as *Eucalyptus tereticornis* and *Pongamia pinnata*; the silt percentage and clay percentage have been found higher in the planted area except *Emblica officinalis* and *Pongamia pinnata* and *Eucalyptus tereticornis*, respectively. A similar trend was reported by Alemie (2009). Present investigation reveals higher particle density, soil porosity and water holding capacity and reduced level of bulk density. Mishra and Nanda (1985) have worked on available water holding a capacity of three soil textured classes from sand to clay and reported that bulk density influence significantly water holding capacity of the soil texture which is almost alike to the present investigation.

8.4.2 CHEMICAL PROPERTIES OF SOIL

8.4.2.1 SOIL pH

As per the present investigation, the soil pH has been found significantly higher in planted area than the unplanted area in all the species. The difference between planted and unplanted areas is found significantly higher in *Emblica officinalis* followed by *Acacia catechu, Tectona grandis, Pongamia pinnata* and minimum in *Eucalyptus tereticornis*. A similar result has been found by Singh et al. (2011). As per their report, elevated pH level was recorded under tree plantation in comparison to the natural fallow system. Alemie (2009) found that there was no pronounced change in soil pH due to growing eucalyptus as hedgerow in farmland such report is almost alike with the present finding. Prevost et al. (1999) conducted an experiment in peatland of eastern Quebec, Canada and reported that the nutrient content of peat soil water was enhanced by drainage. They also found that drainage also produced a pH increase of one unit, a result attributed to increased runoff from the upland part of the treated watershed. Present findings corroborate with the earlier findings of Binkley and Giardina (1998) in case of soil pH. However, Hosur and Dasog (1995) have found decrease pH planted area of teak, shisham, and khair. Sharma et al. (1985) investigated on soil properties of *Alnus nepalensis* from 7 to 56 years old trees in the eastern Himalayas and found that soil pH increased slowly in the surface soil, which is similar to the present investigation.

8.4.2.2   SOIL ORGANIC CARBON

In the present studies, the soil organic carbon has been significantly higher in planted area than the unplanted area. The same findings comparable with the present investigation has been found by Desia et al. (2017) such as all the tree species in the planted area showed higher organic carbon, organic matter and available nutrients in the soil than the open area. The investigation of Dalvi et al. (2017) revealed that nitrogen-fixing trees improve organic carbon in the soil which the similar finding of the present experiment. The similar results have been reported by Singh et al. (2011), who studied multipurpose tree species on sodic soils of Indo-Gangetic alluvial plains of India, stating that significant improvement in organic carbon has been found under tree plantation than natural fallow. Watanabe et al. (2010) studied on the teak plantation and found that site factors influence organic carbon having a significant correlation with site factors. As per Tomar et al. (2003), field investigation revealed gradual buildup of soil organic carbon level for a time period of 9 years had been recorded by 31 species under the semi-arid climatic condition of India. In this connection in our study *Eucalyptus tereticornis* reflected more than 0.4% of increment in soil organic carbon in the upper layer of soil (30 cm). *Acacia auriculiformis, Albizia lebbeck, Bauhinia variegata, Cassia glauca, Syzygium cuminii, Crescentia alata, Samanea saman,* and *Terminalia arjuna* may be comparable with the present findings.

Binkley and Giardina (1998), worked on plant-soil interactions work has concluded that plant species have a significant impact on soil quality for carbon concentration, which is almost similar to the present finding. The high value of organic carbon content in the soils is ascribed to the lower bulk density in the soil is almost alike with the present finding. The higher soil organic carbon value in upper layers of soil profile than the other profiles may have resulted from the dilution of soil matrix with lesser denser material and improvement in soil aggregation. Improvement in aggregation encourages a fluffy and porous condition in soil due to the increase in soil organic carbon (Coote and Ramsay, 1983; Sharma et al., 1995; Lavbhushan, 1998).

8.4.2.3   AVAILABLE NITROGEN

As per the recorded data of present finding, the available nitrogen in soil has been found significantly higher in planted area than the unplanted area. Comparable results have been reported by Dalvi et al. (2017) in case of nitrogen-fixing tree species with increased nitrogen content in the soil in the planted area. The comparable finding has been found with the present experiment by Das et al. (2016) in the pattern of nutrient releases like K>N in *Tectona grandis* and *Shorea robusta*. Singh et al. (2011) carried out work on multipurpose tree species on sodic soils of Indo-Gangetic alluvial plains of India, and found significant improvement in available nitrogen under tree plantation than natural fallow such finding is similar with the present result. An investigation was taken by Watanabe et al. (2010) on teak plantation and found site factors influence on available nitrogen showing significant correlation with site factors. According to Swamy et al. (2004), who studied on a chronosequence of *Gmelina arborea* ranging from 1to 6 years old and found that available nitrogen in the soil improved significantly after plantation which is similar to the present result. A comparable result like present finding has been reported by Binkley and Giardina (1998), who have studied on plant-soil interactions and concluded that plant species have a significant impact on soil quality for nitrogen concentration. Changes in topsoil chemical properties in an agroforestry system in the Philippines using three multipurpose tree legumes (*Gliricidia sepium, Acacia auriculiformis,* and *A. mangium*) intercropped with upland rice, and mung bean (*Vigna radiate*) showed positive variation in agroforestry and monoculture tree treatments. An increment in the level of total N (11.6%) and the soil chemical properties did not vary much across or along the tree rows (Anon, 1997). Woodward (1996) investigated on soil compaction and topsoil removal effect on soil properties and seedling growth in Amazonian Ecuador and found a significant decrease in nitrogen content in the soil. Narain and Singh (1985) worked on the nitrogen status of three kinds of forest soils under Sal (*Shorea robusta*) forest, eucalyptus coppice re-growth, and mixed brushwood at Selakui, Dehradun and higher nitrate concentration was observed in the top 20 cm. Hosur and Dasog (1995) have reported that nutrient status followed the order K>N in *Acacia catechu*, which is comparable with the present finding planted in sandy-loam soil.

8.4.2.4   AVAILABLE PHOSPHOROUS

The result of the present experiment showed that available phosphorous of soil is significantly higher in the unplanted area than the planted area in all species except *Eucalyptus tereticornis*. Comparable results have been reported by Dalvi et al. (2017) in case of nitrogen-fixing tree species with increased phosphorous content in the soil in the planted area. An almost similar result has been observed in case of *Tectona grandis* and *Shorea robusta* by Das et al. (2016) in the pattern of nutrient release like K>P with the slight different than the present investigation which is K>N>P. Zhang et al. (2012) reported no significant impact on soil available phosphorous level as we observed during our present investigation. An experiment was taken by Singh et al. (2011) on multipurpose tree species on sodic soils of Indo-Gangetic alluvial plains of India and found significant improvement in available phosphorous under tree plantation than natural fallow. Such report is almost at par with the present finding except *Eucalyptus tereticornis*. Alemie (2009) found a significant effect available phosphorous in the undergrowth like the present finding. Swamy et al. (2004) reported an elevated level of phosphorous in the soil after plantation. Binkley and Giardina (1998) carried out studies on plant-soil interactions and reported that plant species have a significant impact on soil quality for phosphorous concentration. Woodward (1996) investigated on soil compaction and topsoil removal effect on soil properties and seedling growth in Amazonian Ecuador and found a significant decrease in phosphorous content in the soil. The results found in the present studies are similar to them.

8.4.2.5 AVAILABLE POTASSIUM

In the present investigation, the available potassium of soil has been found significantly higher in planted area than the unplanted area in all the species. Comparable results have been reported by Dalvi et al. (2017) in case of nitrogen-fixing tree species with increased potassium content in the soil in the planted area. The almost comparable finding has been found with the present experiment by Das et al. (2016) in a pattern of nutrient release like K>P>N in *Tectona grandis* and *Shorea robusta*; whereas in the present experiment the pattern is slightly changed like K>N>P. A similar result has been obtained by Singh et al. (2011), who concluded studies on the ameliorative effect of ten years old multipurpose tree species grown on sodic soil of Indo-Gangetic alluvial plain in India and found significant improvement in available potassium under tree plantation than natural fallow. According to Swamy et al. (2004) reported an elevated level of soil potassium under plantation. Hosur and Dasog (1995) reported that nutrient status followed in the order like K>N in *Acacia catechu*. This is similar with the present result planted in sandy-loam soil. Sharma et al. (1985) investigated on soil properties of *Alnus nepalensis* from 7 to 56-years-old trees in the eastern Himalayas and found that exchangeable potassium decreased with soil depth, particularly in the older stands. However, there was no consistent relationship between exchangeable potassium or calcium content and stand age.

8.4.3 TREE GROWTH PARAMETER

8.4.3.1 TREE DIAMETER

The significant trend of diameter in the studied species has been found the maximum in *Eucalyptus tereticornis* followed by *Tectona grandis, Pongamia pinnata, Emblica officinalis* and minimum in *Acacia catechu*. The non-comparable result was found by Chima et al. (2016) in case of exotic and indigenous tree species while during the present investigation exotic species (*E. tereticornis*) deferred significantly in growth in respect of diameter than other indigenous species. Comparable results have been found by Swamy et al. (2003) who reported significant differences in growth parameters of *Gmelina arborea* of 1 to 6 years old among various sites. Tomer et al. (2002) also explained that stem girth of the tree species measured in term of diameter at breast height was equivalent to the gain in their height in the 9 years of the plantation which is comparable with the present study. The *Eucalyptus tereticornis* which is comparable fast-growing exotic species has shown maximum diameter than other species.

8.4.3.2   TREE HEIGHT

The trend of height among the species in the present investigation has been found the maximum in *Eucalyptus tereticornis* followed by *Tectona grandis, Pongamia pinnata, Emblica officinalis* and minimum in *Acacia catechu.* Chima et al. (2016) found a non-comparable result in case of exotic and indigenous tree species while during the present experiment in case of exotic (*E. tereticornis*) differed significantly in growth in respect of height than other indigenous species. Swamy et al. (2003) studied on *Gmelina arborea* stands on red-laterite soil in Chhattisgarh, India and pointed out that growth parameters like total height varied significantly due to the difference in age and site quality. Woodward (1996) has reported in his investigation on soil physical and chemical properties as well as the growth of planted trees in the tropical moist forest that height growth reduced in all species and most consistently the height reduced in compacted subsoil with topsoil removed compared with undisturbed soil. According to Troup (1986), *Eucalyptus tereticornis* which is comparatively fast-growing exotic tree species than other species has shown maximum height. The present data are comparable with the result obtained by him.

8.5   CONCLUSION

It may be concluded that physical properties of soil are found improved in planted area than the unplanted area. The pH, soil organic carbon, available nitrogen, available potassium in planted area is significantly higher than the unplanted area. In case of soil available phosphorous, it is found lower in planted area than unplanted area except the planted area of *Eucalyptus tereticornis* species, where it is higher. It is found that *Eucalyptus tereticornis*, which is fast-growing exotic species, has attained maximum height and diameter followed by *Tectona grandis* in comparison to other species. In case of crown width, it was found the maximum in *Pongamia pinnata* followed by *Eucalyptus tereticornis, Emblica officinalis, Tectona grandis,* and *Acacia catechu.* It is also observed that *Eucalyptus tereticornis and Tectona grandis* have maximum diameter and height which may be due to maximum available nitrogen in the soil. In all the species of the planted area, the nutrient status has followed the order such as K>N>P. On the basis of the findings, it may be recommended to plant forest crops in the barren and wastelands to reclaim soil properties improving its fertility and getting better revenue from forest crops as well as agricultural crops in case of social forestry or agroforestry as output.

KEYWORDS

- **agroforestry**
- **arboretum**
- **available nitrogen**
- **available phosphorus**
- **available potassium**
- **biotic interference**
- **bulk density**
- **crown width**
- **ecological processes**
- **edaphic conditions**

- **natural resource**
- **organic carbon**
- **particle density**
- **social forestry**
- **soil pH**
- **soil porosity**
- **soil texture**
- **tree-soil interaction**
- **water holding capacity**

REFERENCES

1. Alavi, G. (2002). The impact of soil moisture on stem growth of spruce forest during a 22-year period. *Forest Ecology and Management, 166*(1–3), 17–33.
2. Alemie, T. C. (2009). The effect of Eucalyptus on crop productivity and soil properties in the Yoga watershed, Western Amhara Region, Ethiopia.
3. Anon (2011). Methods Manual Soil Testing in India, Department of Agriculture & Cooperation, Ministry of Agriculture, Govt. of India, New Delhi, pp. 77–100.
4. Anon (1997). Effect of legume trees on soil chemical properties under agroforestry system. *Annals of Bangladesh Agriculture, 7*(2), 95–103.
5. Anon (1993a). Framework for Land Evaluation. Soils Bulletin, Food and Agriculture Organization, 32, Rome.
6. Bhalawe, S., Kukadia, M. U., & Nayak, D. (2013). Nutrient Release Pattern of Decomposited Leaf Litter in Different Multipurpose Trees. *The India Forester, 139*(3).

7. Binkley, D., & Giardina, C. (1998). Why do tree species affect soil? The warp and woof of tree-soil interaction. *Biogeochemistry, 42*, 89–106.

8. Birkeland, P. W. (1984). Soil and geomorphology. Oxford University Press, New York, 372 pp.

9. Blake, G. R. (1965). Bulk Density in Methods of Soil Analysis. Agronomy, No. 9, Part 11, C.A. Black, ed. pp. 374–390. Agronomy, No. 9, Part 11, C.A. Black, ed. pp. 374–390.

10. Blake, G. R., & Hartge, K. H. (1986). Bulk density. p. 363–375. In A. Klute (ed.) Methods of soil analysis. Part 1. (2nd ed.). Agron. Monogr. 9. ASA and SSSA, Madison, WI.

11. Bouyoucos, G. J. (1962). Hydrometer method improvement for making particle size analysis of soils. *Agron. J., 54*, 179–186.

12. Bray, R. H., & Kurtz, L. T. (1945). Determination of total organic and available forms of phosphorus in soil. *Soil Science, 59*, 36–46.

13. Champion, H. G., & Seth, S. K. (1968). *Forest types of India*. Govt. of India Publication. New Delhi.

14. Chande, S. R. S. (2003). *A Hand Book of Biostatistics*. Achal Prakashan Mandir, Kanpur, pp. 163–165.

15. Chima, U. D., Akhabue E. F., & Gideon I. K. (2016). Rhizosphere soil properties and growth attributes of four tree species in a four-year arboretum at the University of Port Harcourt, Nigeria. *Nigerian Journal of Agriculture, Food and Environment, 12*(2), 74–80.

16. Coote, R. D., & Ramsey, J. F. (1983). Quantification of the Effect of Over 35 Years of Intensive Cultivation of Four Soils. *Canadian Journal of Soil Science, 63*, 1–74.

17. Das, C., & Mondal, N. K. (2016). Litterfall, decomposition and nutrient release of *Shorea robusta* and *Tectona grandis* in a sub-tropical forest of West Bengal, Eastern India. *Journal of Forestry Research, 27*(5), 1055–1065.

18. Dalvi, V. V., Pawar, P. R., More, S. S., & Shigwan, A. S. (2015). Effect of different nitrogen-fixing tree species on soil chemical properties and primary nutrients in lateritic soil. *Indian Journal of Agroforestry, 17*(2), 31–35.

19. Dayal, M. (1985). Proceedings of Bioenergy Society 2nd Convention. New Delhi, India.

20. Dayal, M. (1984). Proceedings of Bioenergy Society of 1st Convention. New Delhi, India.

21. Desai, N. H., Shakhela, R. R., & Bhatt, P. K. (2014). Nutrient recycling through litter production from different multipurpose trees in agrisilviculture system under rainfed conditions of Gujarat. *Indian Journal of Agroforestry, 16*(2), 36–39.

22. Dwivedi, A. P. (1993). *A Textbook of Silviculture*. IBD, Dehra Dun, India, pp. 505, ISBN: 81-7089-198-1.

23. Hosur, G. C., & Dasog, G. S. (1995). Effect of tree species on soil properties. *Journal of the Indian Society of Soil Science, 43*(2), 256–259.

24. Jordan, D., Ponder, F. Jr., & Hubbard, V. C. (2003). Effects of soil compaction, forest leaf litter and nitrogen fertilizer on two oak species and microbial activity. *Applied Soil Ecology, 23*(1), 33–41.

25. Lavbhusan (1998). Changes in soil physical properties and crop yields with long-term addition of *Lantana camara* biomass in rice-wheat cropping. *PhD Thesis*, Himachal Pradesh Krishi Vishwavidyalaya, Palampur, H.P.

26. Narain, P., & Singh, R. (1985). Behavior of nitrates under different land uses. *Indian Forester, 111*(4), 230–239.

27. Mishra, D. P., & Nanda, S. S. K. (1985). Available water of different textural classes in the Ayacuts of Hirakud Command Area (Orissa). *Journal of the Indian Society of Soil Science, 33*(2), 372–379.

28. Prevost, M., Plamondon, A. P., & Belleau, P. (1999). Effects of drainage of a forested peatland on water quality and quantity. *Journal of Hydrology, 214*(1–4), 130–143.

29. Rhoades, C. C. (1997). Single-tree influence on soil properties in agroforestry lessons from natural forest and savanna ecosystem, agroforestry system. Kluwer academic publisher, Netherland, *35*, 71–94.

30. Robertson, G. P., & Gross, K. L. (1994). Assessing the Heterogeneity of below Ground Resources Quantifying Pattern and Scale in Coldwell MM and Pearcy RW Exploitation of environmental heterogeneity by plant. Academic Press, New York, pp. 237–253.

31. Rosie Shier (2004). Statistics: 1.1 Paired t-tests, United Kingdom. Mathematics Learning Support Centre.

32. Sharma, E., Ambasht, R. S., & Singh, M. P. (1985). Chemical soil Properties under five age series of *Alnus nepalensis* plantations in the Eastern Himalayas. *Plant and Soil, 84*(1), 105–113.

33. Sharma, P. K., Verma, T. S., & Bhagat, R. M. (1995). Soil structure improvement with the addition of *Lantana camara* biomass in rice-wheat cropping. *Soil Use and Management, 11*, 199–203.

34. Singh, Y. P., Singh, G., & Sharma, D. K. (2011). Amelioration effect of multipurpose tree species grown on sodic soils of Indo-Gangetic Alluvial Plain of India. *Arid Land Research and Management, 25*(1), 55–74.

35. Subbia, B. V., & Asija, G. L. (1956). A rapid procedure for estimating of available nitrogen in soil. *Current Science, 25*, 259–260.

36. Swamy, S. L., Puri, S., & Singh, A. K. (2003). Growth, biomass, carbon storage and nutrient distribution in *Gmelina arborea* Roxb. stands on red lateritic soils in central India. *Bioresource Technology, 90*(2), 109–126.

37. Swamy, S. L., Kushwaha, S. K., & Puri, S. (2004). Tree growth, biomass, allometry and nutrient distribution in *Gmelina arborea* stands grown in red lateritic soils of central India. *Biomass and Bioenergy, 26*(4), 305–317.

38. Tomar, O. S., Minhas, P. S., Sharma, V. K., Singh, Y. P., & Gupta, Raj, K. (2003). Performance of 31 tree species and soil conditions in a plantation established with saline irrigation. *Forest Ecology and Management, 177*(1–3), 333–346.

39. Toth, S. J., & Prince, A. L. (1949). Estimation of cation- exchange capacity and exchangeable Ca, K and Na contents of soils by flame photometer techniques. *Soil Sci., 67*, 435–439.

40. Troup, R. S. (1986). *Silviculture of Indian Trees*. FRI Publication, *2*, 556–589.

41. Ushio, M., Kitayama, K., & Balser, T. C. (2010). Tree species effects on soil enzyme activities through effects on soil physicochemical and microbial properties in a tropical montane forest on Mt. Kinabalu, Borneo. *Pedobiologia, 53*, 227–233.

42. Veihmayer, F. J., & Hendrickson, A. H. (1949). Methods of measuring field capacity and permanent wilting percentage of soils. *Soil Sci., 68*, 75–94.

43. Walkley, A., & Black, I. A. (1934). An examination of the Degtjareff method for determining organic carbon in soils: Effect of variations in digestion conditions and of inorganic soil constituents. *Soil Sci.*, *63*, 251–263.

44. Wardle, D. A. (2002). Community and Ecosystems. Linking the aboveground and belowground components. Princeton University Press, Princeton.

45. Watanabe, Y., Sekyere, E. O., Masunaga, T., Buri, M. M., Oladele, O. I., & Wakatsuki, T. (2010). Teak (*Tectona grandis*) growth as influenced by soil physiochemical properties and other site condition an Ashanti region, Ghana. *Journal Food, Agriculture and Environment*, *8*(2), 1040–1045.

46. Woodward, C. L. (1996). Soil compaction and topsoil removal effects on soil properties and seedling growth in Amazonian Ecuador. *Forest Ecology and Management*, *82*(1–3), 197–209.

47. Zhang, D. (2012). Effect of afforestation with *Eucalyptus grandis* on soil physiochemical and microbiological properties Sichuan agricultural university, Wenjiang, China. *Soil Research*, *50*(2), 167–176.

48. Zhou, Z. (2017). Growth and mineral nutrient analysis of teak (*Tectona grandis*) grown on acidic soils in south China. *Journal of Forestry Research*, *28*(3), 503–511.

CHAPTER 9

Phytoremediation of Coal Mine-Based Wastelands: An Approach in the Raniganj Coalfield (RCF)

DEBALINA KAR and DEBNATH PALIT

¹Guest Lecturer in Environmental Science, Michael Madhusudan Memorial College, Durgapur–713216, India, Mobile: 09732128980, E-mail: ka.debalina@gmail.com

²Associate Professor, Department of Botany, Durgapur Government College, Durgapur–713214, India, Mobile: 0983215737, E-mail: debnath_palit@yahoo.com

ABSTRACT

Mining activity especially open cast mining has a great impact on the mining-affected land. The indigenous vegetation of mining sites suffered from a massive damage due to such activities. Phytoremediation technology is an approach to overcome such damages. An attempt was made to repair the degraded land through a proper revegetation technique for reclamation of mine sites of Raniganj coalfield areas in the present study. Characteristic features of spoil of Raniganj coalfield was evaluated to know the condition of the soil. Native vegetations were surveyed during the study period. One hundred species belonging to 28 families were identified. A pot experiment was held in an ex-situ condition in a field of the institution with five tolerant plant species in six different treatment conditions. Different plants showing growth differences in different treatment conditions. Rehabilitation of coal mine spoil can be possible through implementation of proper treatment through pot experiment of ecologically suitable species. The growth of these plants reflects their tolerance to the soil conditions, and supports their use in the restoration process of the degraded wastelands that were subjected to active coal mining activities in the past.

9.1 INTRODUCTION

At the early stages of an ecosystem development, soils constitute a critical controlling component. Without the progress of natural processes of soil development, ecosystems would remain in an under-developed condition. Mine spoil heaps are composed of coarse rocks excavated during the deep coal mining operations and associated coal processing. Landscape, as well as biological communities, faces massive damages due to mining activity (Down and Stock, 1977; Ghose, 1990; Ahmad and Singh, 2004). Improper planning and negligence of regulations of mining activity cause drastic changes in biodiversity due to landscape degradation which also affect some medicinal plants and deteriorate the quality of soil, water and air (Mahalik and Satapathy, 2016). Natural plant communities get disturbed due to mining activities, and following the mining, the habitats become impoverished presenting a very rigorous condition for its growth.

Coal mine spoils are recalcitrant medium for plant and microbial growth because of low organic matter contents, unfavorable pH, either coarse texture or compact structure. Thus the qualities of spoil are very poor in respect to their physical, chemical, nutritional and biological features. Jha and Singh (1993) reported that during revegetation processes some features such as acidity, alkalinity, salinity, sodicity, poor water holding capacity and toxicity causes major problems. Bradshaw (1983) has described the major problems of mine spoils and their short and long-term treatment.

Vegetation is responsible for stabilization, nutrient enrichment and pH maintenance of soils (Tordoff et al., 2000; Conesa et al., 2007). Its development on mine overburdens is essential for ecorestoration, sustenance of biodiversity for moderation of meteorological phenomena and stabilization of environment in the coal mining area. However, these overburdens represent the extremely rigid substrate for plant growth and development. Colonization, establishment, and maintenance of vegetation on these spoils are enormously difficult. In spite of such stressed conditions, vegetation establishment occurs, but the entire process is time-consuming affair (Bradshaw and Chadwick, 1980; Dobson et al., 1997) that cannot keep pace with the amount of deterioration caused by mining activities. The process of revegetation generates the stability of soil condition through restoration and maintenance of landscape with the help of selected plant species which have the capability to stabilize the soil structure.

In the context of increasing mined land degradation, both the ecological and economic imperatives demand that restoration of land be prioritized and for that, it is required to maintain an equilibrium between development and environment through implementation of strategies of ecorestoration. In view of this, the present study aims to formulate a strategy for ecorestoration of opencast coal mining sites through inoculation of appropriate amendments in the mined spoils to transform them into soil and revegetation by appropriate tree species selected through pot culture experimental trials using spoils with and without amendments. Remediation through plant species establishment in a degraded and contaminated landscape improves soil structure is surely an emerging technology (Tian et al., 2007; Amaya-Chavez et al., 2006). Much research has been conducted in this field and is gaining global acceptance because of the possibility of adapting this technology to many different types of ecosystems in both developed and developing countries.

9.2   REVIEW OF LITERATURE

The contemporary environmental issues and challenges have emanation from indispensable developmental activities and direction towards restoration and optimization of environment. One of such issues being spread of wastelands it is essential to conceptualize the subject of wastelands collaterally with the causal developmental activity so that strategies for remediation of the environmental derangements could be worked out. In view of this, various perspectives of wastelands relevant to such a resource essential for development as coal and its open-cast mining have been reviewed keeping the focus initially on the 'state of art' of the subject and subsequently on the theme of the present research work.

Mining cause massive damage to the landscapes and biological communities and impact of severity depends on whether the mine is working or abandoned, mining methods, and the geological conditions (Bell et al., 2001; Singh et al., 2007). The wasteland generally comprises of the bare stripped area, loose soil piles, waste rock, and overburden surfaces, subsided land areas, other degraded lands by mining activities. Among them the waste rocks many often stands as a great threat to the restoration process. The mining wastes visibly destroy soil microbe population and nutritional cycles which are sine qua non to sustain the equilibrium of ecosystem. Hence, as a result, we see the destruction of existing vegetation and soil profile (Kundu and Ghose, 1997). Moreover, the effect of mine waste may gradually culminate in soil erosion, air and water pollution, intoxication, loss of biodiversity and subsequently environmental pollution. (Wong and Bradshaw, 1982; Sheoran et al., 2008) . Hence mine wastes ultimately jeopardize the economic wealth of country (Ghose, 1989, 2004).

Coal mine spoils when freshly tipped has a great range of particle size ranging from large pieces of shale to silt and clay (Molyneux, 1963). These mine spoils represent extremely rigid substrata for plant growth and development. Colonization, establishment, and maintenance of vegetation on these spoils are enormously difficult. Among the factors which hinder the growth of plant species on these spoils, acidity merits special attention. Extreme acidity is caused due to the oxidation of iron pyrites (Chadwick, 1973). Opencast coal mines generate a variety of wastes with differing physicochemical properties and changes with their toxicity level. According to Davcheva (1990) and Gonzalez and Gonzalez-Chavez (2006), it has been recognized that the potentiality of mining operation in their large scale destroy the flora and fauna, contaminate the soil, air, and water in the mining. Mining operations have been a major environmental concern due to overburdening of large areas with spoiled soils or spoils. The condition of the mine spoils was very meticulous with low organic matter content and unfavorable soil condition that affect soil microbes and plant growth (Meyer, 1973; Jha and Singh, 1994). In coal mining area, the spoil has the low potentiality of plants habitat establishment, growth rate, and survivability ratio. Coal mine spoil has very low organic matter content as a consequence. pH and water-holding capacity also seems low in this condition (Vogel, 1982; Chaubey et al., 2012). In the viewpoint of Reddy and Reddy (2001), the scientists always focused on the condition of revegetation optimization through suitable amendment. It is a green plant which naturally converts the spoil to good fertile soil through an autogenic plant succession. It leads the natural succession in a slow and steady way. It was observed that nature restores the alterations through autogenic succession. Mostly plant succession involves animal and microbes under the presence of various factors and environmental changes. Such natural process by plants are considered as phytoremediation.

9.3　MATERIALS AND METHODS

9.3.1　STUDY SITES

The study area encompasses a large stretches of old Opencast Coal Pit (OCPs) in the area of Eastern Coalfield Limited (ECL) which are situated in the Raniganj Coalfield division of Burdwan district of West Bengal of India. Raniganj Coalfield with latitude 23.7500° and longitude 86.8333°, which falls under Eastern Coalfield Ltd., is the birthplace of coal mining in the Country (Figure 9.1). As many as four collieries (mine generated wasteland) were selected to depict the nature of the spoil and their improvement through proper technique. Physic-chemical properties of selected overburden spoils were sampled and analyzed with different physicochemical parameters. The climate condition that prevails in the study area provides three seasons in a year, i.e., pre-monsoon (March to June), monsoon (July to October) and post-monsoon (November to February). Soil samples were collected seasonally. The overall study was conducted in the year of 2014 to 2016.

9.3.2　STUDY OF INDIGENOUS VEGETATION

The study of vegetation is most important phases to understand and analyze the status. The composition of vegetation on mine spoil dumps of different ages were studied by laying randomly placed quadrats during the study period 2014 to 2015. Quadrates of 1 m × 1 m were laid out for herbs, 5 m × 5 m for shrubs and 10 m × 10 m for tree species by using standard quadrat method (Srivastava, 2001). Ten quadrates were laid on different aspects of each site for the analysis of vegetation. To study the detailed of different aspects of any community, a number of characters (parameters) are taken into consideration. The various characters are analytical characters viz. frequency, density, abundance.

9.3.3　POT CULTURE TECHNIQUE

Long term monitoring field experiments were set up at an experimental garden in 2015 to 2016. In pot culture experiments suitable tree species for revegetation of mine spoils were screened. Characteristics such as

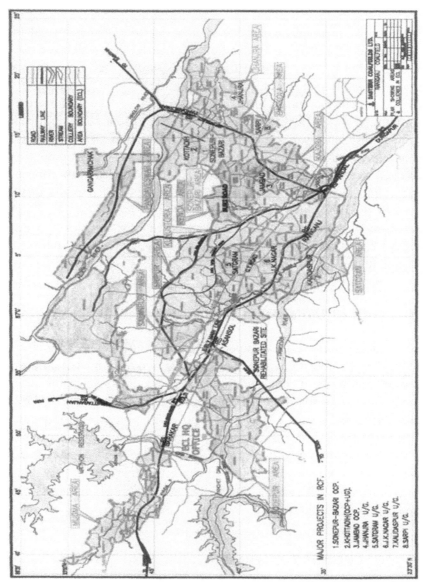

FIGURE 9.1 **(See color insert.)** Location of study sites in Raniganj Coalfield areas.

adaptability of plant species to the particular physical and chemical condition of mine spoil and longevity of established plants were considered for selection of suitable plant species. Choice of multipurpose trees can be of economic and social significance. The availability, growth, adaptability, and longevity are important parameters in selecting suitable tree species for revegetation purpose.

9.3.4 PREPARATION OF PLOT FOR POT CULTURE EXPERIMENT

In pot culture techniques each pot was filled with mine spoil, garden soil with different treatment manures, fertilizers, fly ash and husk separately, and three-month-old saplings were transplanted in the pot. Healthy saplings of five tree species were planted in monsoon in 6 cm depth in experimental garden. Each pot was measured with the height 14 cm and diameter 1.3 ft. Experiments were set up in completely Randomized Block Design (RBD) with five replications.

9.3.5 SELECTION OF SPECIES FOR POT CULTURE EXPERIMENT

For selection of suitable plant species for revegetation of mine spoils, species characteristics such as adaptability of plant species to the particular physical and chemical conditions of mine spoils, and to macro and micro-climatic conditions, drought tolerance, rooting depth, hardiness, palatability of seeds, stabilization ability, aesthetic appeal, ease of propagation, and longevity of established plants were considered (Singh et al., 1995). Choice of multipurpose trees can be of economic and social significance. Therefore, multipurpose trees should be given preference (Jha and Singh, 1994). The plant architecture and geometry, morphology, phenology, root growth, architecture, and activity are important parameters in selecting suitable tree species for revegetation purpose (Table 9.1).

9.3.6 COLLECTION OF SAPLING

Saplings of five selected tree species were collected from Forest Nursery with due permission. Healthy sapling of each species was collected from nursery and transported to experimental garden. Then each species are planted in each pot for experiment. Five replicate of each selected tree sapling was set up under the treatment of spoil with fly ash. Overall 150 saplings were planted for screening out to determine the growth pattern and to assess their potentialities for coal mined wastelands.

TABLE 9.1   List of Selected Species Used for Revegetation/Pot Culture Experiment

Sl. No.	Plant Name	Local Name	Family
1	*Cassia fistula* L.	Bandarlathi	Fabaceae
2	*Emblica officinalis* Gaertn.	Amlaki	Euphorbiaceae
3	*Dalbergia sissoo* DC.	Sisoo	Fabaceae
4	*Azadirachta indica* A. Juss.	Neem	Meliaceae
5	*Pongamia glabra* Vent.	Karanj	Fabaceae
6	*Albizia lebbeck (*L.) Benth.	Siris	Fabaceae
7	*Holoptelea integrifolia* Planch.	Challa	Moraceae
8	*Acacia auriculiformis* Benth.	Akashmoni	Fabaceae
9	*Swietenia macrophylla* King	Mehogoni	Meliaceae

9.3.7   MATERIALS USED IN POT CULTURE EXPERIMENT

Based upon the availability and cost and suitability in the field condition four treatment components such as compost, husk, fly ash, and chemical fertilizer was selected. Following treatments were given, such as a) mine spoil, b) garden soil (for control), c) mine spoil + compost, d) mine spoil + husk, e) mine spoil + fly ash and f) mine spoil + chemical fertilizer (Table 9.2).

TABLE 9.2   Composition of Pot Culture Experiment

Treatment Name	Treatment	Material	Mixing ratio
T_1	Spoil	-	-
T_2	Soil	-	-
T_3	Spoil + Compost	Compost	3:1
T_4	Spoil + Husk	Husk	3:1
T_5	Spoil + Fly ash	Fly ash	3:1
T_6	Soil + Chemical Fertilizer	Chemical Fertilizer	10 gm NPK mixed in 1000 ml of water and 10 ml of solution was added in each pot

9.3.8 MAINTENANCE OF EXPERIMENTAL GARDEN

An experimental garden was maintained by regular watering, removal of weeds from a different pot and periodic cleaning.

9.3.9 MEASUREMENT OF GROWTH

Plant growth comprises of three distinct phenomena, i.e., increase of cell, longitudinal growth of the cell (maturation stage), and horizontal growth of the cell. To access the growth through morphological expression, some vegetative growth parameters were taken into consideration for measurement of plant growth. The observations for the morphological growth attributes were taken every month. Data for shoot length, number of node, number of internode, number of auxiliary bud, number of branches were collected recorded on monthly basis.

9.4 RESULTS AND DISCUSSION

9.4.1 CHARACTERISTICS OF MINE SPOIL

The soil features of the wastelands of Raniganj Coalfields differed considerably in terms of the seasons (Table 9.3) reflecting the dynamic nature of the soil that varies with the seasons. All parameters of the coal mine spoil were differing considerably in respect of controlled area. In case of four studied sites increasing pH of mine spoil was analyzed with the range of 6.7 to 8. Bulk density, particle density, and water holding capacity decreased considerably due to the lack of vegetation cover. Nutrient content of all sites was decreasing in nature. As a consequence of the differences in the soil characteristics, the corresponding changes in the habitat quality can be assumed that may provide a clue for the observed differences in the flora. The results appeared to be similar to the earlier studies on the wasteland conditions where the land mass was subjected to coal mine activities in the past (Ghose, 2004).

The observations on the plant diversity and the soil features in the selected sites of the wastelands of Raniganj Coalfields, West Bengal, revealed the presence of at least 100 different angiosperms belonging to 28 families (Table 9.4). Considerable differences in the relative abundance of the plant species were observed that substantiates the differences in the colonization ability and the niche requirements of the species.

TABLE 9.3   The Characteristic Features of the Soil in the Selected Sites of the Wastelands of Raniganj Colliery, West Bengal, India

PARAMETERS	Pre-monsoon	Monsoon	Post-monsoon
pH	0.53–0.96	0.93–1.21	0.97–1.17
	0.66 ± 0.12	1.02 ± 0.09	1.03 ± 0.07
Bulk Density (BD)	0.94 - 1.21	0.93–1.3	1.08–1.37
	1.10 ± 0.02	1.10 ± 0.13	1.17 ± 0.10
Particle Density (PD)	0.58 - 1.32	0.71–1.22	1.04–1.16
	1.34 ± 0.13	0.98 ± 0.20	1.10 ± 0.04
Soil Porosity (SP)	0.23–0.95	0.43–1.64	0.44–1.07
	0.89 ± 0.31	0.86 ± 0.41	0.82 ± 0.25
Water Holding Capacity (WHC)	0.36 - 1.63	0.48–1.40	0.68–1.32
	1.00 ± 0.17	0.91 ± 0.36	0.97 ± 0.22
Conductivity (CON)	0.04–1.14	0.07–3	0.28–1.5
	0.40 ± 0.38	0.97 ± 1.06	0.62 ± 0.40
Organic Carbon (OC)	0.07 - 1.19	0.11–5	0.07–0.5
	0.27 ± 0.52	1.72 ± 1.98	0.27 ± 0.17
Available Nitrogen (AN)	0.20 - 0.34	0.12–2	0.19–1.33
	0.19 ± 0.08	0.81 ± 0.70	0.71 ± 0.47
Available Potassium (AP)	0.11–1.02	0.63–1.36	0.31–0.99
	0.50 ± 0.16	0.94 ± 0.26	0.64 ± 0.27
Available Sodium (AS)	0.10–3.56	0.09–1.38	0.05–0.88
	1.62 ± 2.33	0.74 ± 0.60	0.41 ± 0.34
Available Phosphate Phosphorous (APP)	0.11 - 1.20	0.83–1.97	0.25–3.37
	0.83 ± 0.27	1.40 ± 0.45	1.64 ± 1.22

TABLE 9.4   The Plant Species Observed in the Course of Sampling the Selected Sites of the Wastelands Abandoned Colliery of Raniganj, West Bengal, India

Family	Plant Name
Acanthaceae	*Andrographis echioides, A. paniculata, Ruellia tuberosa*
Agavaceae	*Agave sisalana*
Aizoaceae	*Trianthema portulacastrum*
Alangiaceae	*Alangium lamarckii*
Amaranthaceae	*Alternanthera sessilis, A. tenella, Amaranthus spinosus, A. viridis, Gomphrena celosioides*
Apocynaceae	*Alstonia scholaris, Thevetia neriifolia*

TABLE 9.4 *(Continued)*

Family	Plant Name
Arecaceae	*Phoenix sylvestris*
Asclepiadaceae	*Calotropis gigantean, C. procera, Hemidesmus indicus, Pergularia daemia*
Asteraceae	*Cnicus wallichii, Eclipta alba, Mikania scandens, Spilanthes paniculata, Tridax procumbens, Vernonia cinerea, Xanthium strumarium, Eupatorium odoratum*
Capparaceae	*Cleome gynandra, C. viscosa*
Combretaceae	*Terminalia arjuna*
Commelinaceae	*Commelina benghalensis*
Convolvulaceae	*Ipomoea maxima, I. pinnata, I. pes-tigridis*
Cucurbitaceae	*Coccinia cordifolia, Trichosanthes cucumerina*
Cyperaceae	*Cyperus rotundus, Kyllinga monocephala*
Euphorbiaceae	*Acalypha indica, Croton bonplandianus, Emblica officinalis, Euphorbia antiquorum, E. hirta, Jatropha curcas, J. gossypiifolia, Phyllanthus amarus*
Fabaceae	*Acacia Arabica, A. auriculaeformis, A. nilotica, Atylosia scarabaeoides, Butea monosperma, Cassia alata, C. fistula, C. obtusifolia, C. siamea, C. sophera, C. tora, Crotalaria pallida, Dalbergia sissoo, Desmodium gangeticum, Pongamia glabra, Tephrosia purpurea, T. villosa, Teramnus labialis*
Lamiaceae	*Clerodendrum viscosum, Gmelina arborea, Hyptis suaveolens, Leonurus sibiricus, Leucas aspera, Ocimum canescens, Vitex negundo*
Malvaceae	*Abutilon indicum, Sida acuta, S. cordata, S. cordifolia, Urena lobata*
Meliaceae	*Azadirachta indica*
Mimosaceae	*Albizia lebbeck*
Moraceae	*Ficus benghalensis, F. cunea, F. religiosa*
Papaveraceae	*Argemone mexicana*
Poaceae	*Aristida adscensionis, Chloris barbata, Cynodon dactylon, Eragrostis coarctata, Eulaliopsis binate, Heteropogon contortus, Oplismenus compositus, Panicum maximum, Poa annua, Saccharum munja, S. spontaneum*
Polygonaceae	*Polygonum barbatum*
Scrophulariaceae	*Scoparia dulcis*
Simaroubaceae	*Ailanthus excelsa*
Solanaceae	*Datura metel, Physalis minima, Solanum nigrum, Solanum sisymbriifolium, S. surattense*

Similar observations were found in previous work which deals with the variations in the plant species assemblages on different coal mining sites of India (Maiti and Ghose, 2005; Nath, 2009; Rai et al., 2010; Hazarika et al., 2006; Borpujari, 2008; Dowarah et al., 2009; Singh, 2011 & 2012; Chaubey et al., 2012; Singh, 2013).

9.4.2 *TREATMENT EFFECT DUE TO POT EXPERIMENT*

All the plants selected for the study exhibited growth reflected through the changes in the morphological features, monitored over the entire observation period of one year. The morphological features varied considerably against the treatments as shown in Table 9.5. For all the plant species, *Cassia fistula, Emblica officinalis, Dalbergia sissoo, Azadirachta indica, Pongamia glabra, Albizia lebbeck, Holoptelea integrifolia, Acacia auriculiformis,* and *Swietenia macrophylla,* the morphological features were observed to change with time (Table 9.6) indicating the growth and thus the tolerance against the treatments considered in the study. As shown in Tables 9.5 and 9.6, the changes in the morphological features like: (i) length of shoot; (ii) number of node; (iii) average length of internode; (iv) number of auxiliary bud; and (v) number of branches, were observed similar to the studies made elsewhere using similar conditions.

TABLE 9.5 Data on the Different Variables (Plant morphometric features) As Observed in Different Treatments (Treatments–spoil, soil, spoil+compost, spoil+husk, spoil+fly ash, spoil+chemical fertilizer) Irrespective of Months Conditions ($n = 5$ different plant individuals; 1 plant per pot)

	TREATMENT 1 (SPOIL)				
	Morphometric features				
	SL	NN	ALIN	NAB	NB
CAF	0–49	0–19	0–16	0–3	0–8
	27.23 ± 3.31	4.30 ± 0.59	5.12 ± 0.48	0.56 ± 0.20	0.19 ± 0.12
EMO	0–69.30	0–43	0–4.10	0–18	0–6
	43.87 ± 2.43	20.03 ± 1.51	2.27 ± 0.13	2 ± 0.35	0.25 ± 0.07
DAS	0–188	0–41	0–8	0–7	0–9
	86.59 ± 4.88	2.63 ± 1.23	3.32 ± 0.11	0.90 ± 0.13	2 ± 0.19

TABLE 9.5 *(Continued)*

		TREATMENT 1 (SPOIL)			
		Morphometric features			
	SL	**NN**	**ALIN**	**NAB**	**NB**
AZI	0–59.60	0–19	0–43	0–2	0–7
	22.62 ± 2.43	3.38 ± 0.42	5.48 ± 0.83	0.58 ± 0.10	0.60 ± 0.14
POG	0–132	0–26	0–9.02	0–8	0–2
	66.33 ± 3.42	15.42 ± 0.76	4.31 ± 0.21	2.10 ± 0.27	0.38 ± 0.08
ALL	0–83	0–23	0–9.33	0–6	0–3
	46.68 ± 2.66	10.72 ± 0.69	4.43 ± 0.26	1.40 ± 0.15	1.28 ± 0.11
HOI	0–75.20	0–25	0–8.60	0–11	0–5
	42.54 + 2.94	11.32 ± 0.73	3.59 ± 0.26	1.62 ± 0.24	1.12 ± 0.14
ACA	0–177	0–61	0–6.09	0–19	0–20
	60.14 ± 6.21	20.70 ± 0.95	2.98 + 0.13	4.85 ± 0.37	4.97 ± 0.58
SWM	0–80	0–25	0–7.30	0–9	0–12
	30.15 ± 4.12	9.23 ± 0.76	4.11 ± 0.42	0.90 ± 0.16	1.14 ± 0.28

		TREATMENT 2 (SOIL)			
		Morphometric features			
	SL	**NN**	**ALIN**	**NAB**	**NB**
CAF	0–90	0–21	0–18	0–7	0–3
	21.20 ± 3.45	5.17 ± 0.56	5.67 ± 0.78	0.37 ± 0.13	0.07 ± 0.04
EMO	0–90	0–41	0–6.09	0–19	0–5
	35.63 ± 4.45	16.90 ± 1.66	2.98 ± 0.14	2.09 ± 0.44	0.84 ± 0.15
DAS	0–140.13	0–45	0–13.67	0–19	0–25.13
	79.12 ± 6.17	18.13 ± 1.14	7 ± 0.45	2.34 ± 0.37	9.23 ± 0.80
AZI	0–142	0–20	0–20	0–7	0–5
	30.60 ± 6.23	8.78 ± 0.87	6 ± 0.60	0.79 ± 0.16	0.69 ± 0.30
POG	0–245	0–34	0–8.63	0–8	0–3
	107 ± 7.99	19.38 ± 1.15	4.80 ± 0.34	3.67 ± 0.30	0.87 ± 0.17
ALL	0–120.30	0–19	0–51.30	0–8	0–4
	46.53 ± 4.67	8.90 ± 0.89	6.50 ± 1.14	1.67 ± 0.23	1.16 ± 0.20
HOI	14–106	4–26	1.92–5.95	0–2	0–3
	60.23 ± 3.30	17.70 ± 0.66	3.50 ± 0.15	1.25 ± 0.12	1.65 ± 0.12

TABLE 9.5   *(Continued)*

TREATMENT 2 (SOIL)				
Morphometric features				
ACA	0–147	0–47	0–13.86	0–16	0–8
	62.88 ± 4.60	22.12 ± 1.60	3.34 ± 0.12	5.12 ± 0.50	3.14 ± 0.34
SWM	0–73	0–21	0–8.15	0–2	0–2
	12.32 ± 3.19	3.24 ± 1.23	0.87 ± 0.30	0.23 ± 0.09	0.23 ± 0.12

TREATMENT 3 (SPOIL+COMPOST)				
Morphometric features				
	SL	**NN**	**ALIN**	**NAB**	**NB**
CAF	0–60.13	0–18	0–10.12	0–6	0–9
	21.32 ± 4.12	3.40 ± 0.60	3.13 ± 0.47	0.23 ± 0.11	0.63 ± 0.15
EMO	26.23–70.72	7–34	1.67–10	0–16	0–25
	55.90 ± 1.46	23.13 ± 0.84	2.83 ± 0.19	3.56 ± 0.42	5.90 ± 0.97
DAS	15–170.45	7–38	1.38–61.38	0–5	0–22
	103.71 ± 4.76	23.11–1.06	6.60 ± 1.00	2.02 ± 0.19	11.22 ± 0.83
AZI	0–78.72	0–27	0–16.56	0–9	0–15
	45.19 ± 2.33	9.32 ± 0.63	6.21 ± 0.50	1 ± 0.15	3.12 ± 0.45
POG	22–118.30	6–26	2.54–9.48	0–4	0–5
	79.72 ± 2.84	15.48 ± 0.53	5.37 ± 0.23	2.12 ± 0.13	1.32 ± 0.17
ALL	13–112.50	3–21	2.48–28.86	0–5	0–4
	67.34 ± 3.76	10.25 ± 0.52	7.17 ± 0.55	1.70 ± 0.16	1.53 ± 0.13
HOI	22–91.10	2–29	1.58–12.50	0–5	0–10
	59.89 ± 2.81	17.33 ± 0.78	3.73 ± 0.21	1.53 ± 0.16	3.50 ± 0.30
ACA	16–149.20	6–44	1.54–13.50	0–19	0–15
	87.74 ± 4.95	24.50 ± 1.48	4.19 ± 0.35	5.82 ± 0.67	5.83 ± 0.56
SWM	0–64.50	0–22	0–3.73	0–5	0–4
	27.49 ± 3.06	9.45 ± 1.07	1.75 ± 0.19	0.95 ± 0.15	0.92 ± 0.16

TREATMENT 4 (SPOIL+HUSK)				
Morphometric features				
	SL	**NN**	**ALIN**	**NAB**	**NB**
CAF	0–40	0–10	0 –18.40	0–2	0–0
	20.29 ± 1.60	4.60 ± 0.41	4.51 ± 0.48	0.42 ± 0.07	0 ± 0
EMO	15–69.70	0–42	0–59.30	0–20	0–3
	55.45 ± 1.43	21.10 ± 1.47	5.30 ± 1.21	3.05 ± 0.52	0.38 ± 0.08

TABLE 9.5 *(Continued)*

| | | TREATMENT 4 (SPOIL+HUSK) | | | |
| | | Morphometric features | | | |
	SL	NN	ALIN	NAB	NB
DAS	15–196	8–32	1.83–13.73	0–6	0–11
	110.03 ± 5.02	19.48 ± 0.75	6.22 ± 0.43	2.08 ± 0.20	5.85 ± 0.42
AZI	0–78.70	0–14	0–15.85	0–5	0–5
	34.25 ± 2.92	5.47 ± 0.51	5.98 ± 0.48	0.83 ± 0.13	1.38 ± 0.19
POG	0–112.30	0–21	0–45	0–6	0–2
	69.21 ± 5.05	11.87 ± 0.95	4.99 ± 0.75	1.72 ± 0.19	0.28 ± 0.07
ALL	0–88.80	0–18	0–17.76	0–6	0–3
	58.68 ± 2.98	10.53 ± 0.56	6.12 ± 0.52	1.23 ± 0.17	1.48 ± 0.14
HOI	0–53.30	0–17	0–8.47	0–4	0–5
	23.68 ± 2.42	8 ± 0.80	2.25 ± 0.26	0.73 ± 0.13	0.87 ± 0.19
ACA	15– 92.80	7–39	0.51–5.01	1–15	0 12
	62.31 ± 2.85	22.77 ± 0.99	2.79 ± 0.12	5.82 ± 0.41	4.53 ± 0.31
SWM	0–69.90	0–16	0 –10.30	0–2	0–1
	29.22 ± 3.07	7.50 ± 0.73	2.89 ± 0.35	0.53 ± 0.08	0.10 ± 0.04

| | | TREATMENT 5 (SPOIL+FLY ASH) | | | |
| | | Morphometric features | | | |
	SL	NN	ALIN	NAB	NB
CAF	0–44	0–13	0–18	0–2	0–3
	22.76 ± 1.81	3.67 ± 0.37	5.48 ± 0.57	0.35 ± 0.07	0.52 ± 0.13
EMO	0–63.30	0–30	0–4.87	0–7	0–5
	32.44 ± 2.82	12.23 ± 1.06	2.06 ± 0.18	1.30 ± 0.22	0.33 ± 0.13
DAS	0–182	0–35	0–19.61	0–5	0–12
	104.06 ± 5.85	18.27 ± 1.15	6.51 ± 0.59	1.72 ± 0.17	3.95 ± 0.45
AZI	0–28.40	0–8	0–7.86	0–2	0–4
	10.09 ± 1.36	2.47 ± 0.34	2.23 ± 0.32	0.37 ± 0.07	0.67 ± 0.18
POG	0–122	0–27	0–8.15	0–5	0–2
	63.66 ± 3.53	14.77 ± 0.82	4.31 ± 0.21	1.63 ± 0.13	0.55 ± 0.10
ALL	0–135	0–23	0–12.14	0–4	0–6
	56.25 ± 3.74	10.42 ± 0.65	5.72 ± 0.33	1.12 ± 0.10	0.32 ± 0.11

TABLE 9.5 *(Continued)*

	SL	NN	ALIN	NAB	NB
	TREATMENT 5 (SPOIL+FLY ASH)				
	Morphometric features				
HOI	0–67.10	0–21	0–5.40	0–5	0–4
	35.38 ± 2.79	11.08 ± 0.84	2.60 ± 0.20	1.10 ± 0.13	0.75 ± 0.19
ACA	0–170	0–40	0–5.07	0–10	0–8
	56.60 ± 4.74	17.72 ± 1.21	2.89 ± 0.17	3 ± 0.31	3.10 ± 0.38
SWM	0–62	0–22	0–18	0–3	0–4
	23.10 ± 2.74	7.95 ± 0.87	2.09 ± 0.33	0.63 ± 0.10	0.95 ± 0.15

	SL	NN	ALIN	NAB	NB
	TREATMENT 6 (SPOIL+CHEMICAL FERTILIZER)				
	Morphometric features				
CAF	0–65	0–10	0–41	0–2	0–2
	24.26 ± 2.61	2.95 ± 0.34	6.71 ± 0.95	0.40 ± 0.08	0.32 ± 0.09
EMO	0–55	0–28	0–44	0–8	0–8
	41.27 ± 1.72	11.90 ± 0.10	4.95 ± 0.85	2.22 ± 0.23	2.42 ± 0.39
DAS	0–195	0–40	0–14.02	0–6	0–18
	92.03 ± 7.74	19.95 ± 1.70	4.23 ± 0.42	2.18 ± 0.26	7.47 ± 0.70
AZI	0–144	0–35	0–14.06	0–6	0–5
	47.74 ± 4.93	11.32 ± 1.33	4 ± 0.39	1.15 ± 0.19	2.23 ± 0.21
POG	0–152	0–26	0–12.37	0–6	0–3
	71.43 ± 3.26	12.47 ± 0.63	5.91 ± 0.29	1.57 ± 0.17	0.87 ± 0.13
ALL	0–110	0–22	0–22.20	0–7	0–4
	53.75 ± 2.60	9.22 ± 0.58	7.26 ± 0.63	1.97 ± 0.18	0.73 ± 0.14
HOI	0–95.20	0 –31	0–5.87	0–6	0–3
	49.73 ± 4.16	14.72 ± 1.11	2.73 ± 0.23	1.17 ± 0.17	0.73 ± 0.14
ACA	0–198	0–55	0–23.40	0–16	0–17
	90.48 ± 7.11	24.70 ± 2.03	4.42 ± 0.61	6.05 ± 0.63	5.70 ± 0.60
SWM	0–55	0–18	0–5.56	0–4	0–3
	28.83 ± 2.83	8.80 ± 0.85	2.30 ± 0.23	0.88 ±0.14	0.38 ± 0.09

Abbreviations: CAF–*Cassia fistula*, EMO–*Emblica officinalis,* DAS–*Dalbergia sissoo*, AZI–*Azadirachta indica*, POG–*Pongamia glabra*, ALL–*Albizia lebbeck*, HOI–*Holoptelia integrifolia*, ACA–*Acacia auriculiformis*, SWM–*Swetenia macrophyla*, SL–Shoot Length, NN–Number of Node, ALIN–Average Length of Internode, NAB–Number of Auxiliary Bud, NB–Number of Branches.

TABLE 9.6 Data on the Different Variables (Plant morphometric features) As Observed in Different Months (January through December) Irrespective of Treatment Conditions (n = 5 different plant individuals; 1 plant per pot)

	JANUARY				
	Morphometric features				
	SL	**NN**	**ALIN**	**NAB**	**NB**
CAF	0–54.10	0–10	0–20	0–0	0–3
	23.64 ± 3.38	3.70 ± 0.64	5.43 ± 0.97	0 ± 0	0.30 ± 0.14
EMO	0–69.10	0–37	0–59.30	0–8	0–20
	51.19 ± 2.67	15.57 ± 1.86	5.49 ± 1.90	1.27 ± 0.33	1.87 ± 0.73
DAS	0–169.30	0–33	0–19.61	0–5	0–18
	113.86 ± 5.39	18 ± 1.58	7.27 ± 0.72	1 ± 0.20	7.80 ± 0.98
AZI	0–116.20	0–30	0–15.10	0–3	0–8
	38.14 ± 5.37	6.70 ± 1.18	5.30 ± 0..78	0.57 ± 0.13	1.90 ± 0.41
POG	0–155.10	0–30	0–9.86	0–7	0–2
	88.63 ± 5.67	15.03 ± 1.08	6.05 ± 0.38	1.43 ± 0.29	0.73 ± 0.14
ALL	0–107	0–18	0–25.35	0–4	0–6
	61.92 ± 4.08	8.33 ± 0.80	8.83 ± 0.97	1.13 ± 0.18	1.40 ± 0.24
HOI	0–93.60	0–25	0–7.54	0–3	0–8
	50.34 ± 4.97	11.70 ± 1.23	3.75 ± 0.37	0.73 ± 0.14	1.60 ± 0.34
ACΛ	54.80–158.30	5–50	2.33–23.40	1–15	0–16
	90.86 ± 5.21	24.20 ± 2.03	4.93 ± 0.79	5.53 ± 0.76	6.13 ± 0.69
SWM	0–66.70	0–20	0–7.27	0–2	0–3
	29.39 ± 4.91	7.47 ± 1.31	2.37 ± 0.42	0.67 ± 0.15	0.60 ± 0.17

	FEBRUARY				
	Morphometric features				
	SL	**NN**	**ALIN**	**NAB**	**NB**
CAF	0–55.30	0–11	0–41.20	0–2	0–3
	24.85 ± 3.55	4.10 ± 0.70	5.76 ± 1.45	0.17 ± 0..08	0.40 ± 0.14
EMO	0–72.30	0–36	0–44	0–8	0–21
	50.27 ± 3.23	15.10 ± 1.93	6.08 ± 1.66	1.47 ± 0.32	2 ± 0.78
DAS	0– 175	0–35	0–61.38	0–4	0–20
	118.44 ± 5.71	18.77 ± 1.56	9.09 ± 1.93	1.10 ± 0.21	8.23 ± 1.02
AZI	0–127	0–28	0–18.02	0–3	0–8
	40.06 ± 5.71	6.80 ± 1.15	5.37 ± 0.79	0.67 ± 0.17	1.90 ± 0.39
POG	0–157.30	0–31	0–12.37	0–6	0–3
	87.73 ± 6.41	15.07 ± 1.25	5.85 ± 0.46	1.40 ± 0.29	1 ± 0.19

TABLE 9.6 *(Continued)*

	SL	NN	ALIN	NAB	NB
	FEBRUARY				
	Morphometric features				
ALL	0–113.50	0–19	0–51.30	0–4	0–3
	62.16 ± 4.66	8.47 ± 0.87	9.09 ± 1.63	1 ± 0.17	1.33 ± 0.20
HOI	0–94	0–24	0–6.79	0–3	0–9
	51.79 ± 5.08	12.20 ± 1.27	3.68 ± 0.35	0.87 ± 0.16	1.73 ± 0.37
ACA	57–167	7–48	2.55–17.90	1–15	0–17
	95 ± 5.54	24.87 ± 1.98	4.67 ± 0.59	5.37 ± 0.74	6.30 ± 0.71
SWM	0–68.20	0–20	0–10.30	0–3	0–3
	28.46± 5.08	6.63 ± 1.29	2.63 ± 0.54	0.57 ± 0.16	0.53 ± 0.16

	SL	NN	ALIN	NAB	NB
	MARCH				
	Morphometric features				
CAF	0–56	0–12	0–14.17	0–2	0–3
	24.10 ± 3.80	4.40 ± 0.75	4.17 ± 0.72	0.27 ± 0.11	0.40 ± 0.14
EMO	0–78	0–36	0–32.50	0–8	0–21
	50.06 ± 3.76	15.27 ± 2.01	5.09 ± 1.21	1.80 ± 0.31	2.10 ± 0.78
DAS	0–186.60	0–37	0–13.33	0–5	0–20
	116.07 ± 8.45	19.33 ± 1.84	6.21 ± 0.64	1.57 ± 0.29	8.27 ± 1.09
AZI	0–136.60	0–27	0–15.06	0–3	0–9
	38.58 ± 6.44	6.53 ± 1.25	4.48 ± 0.73	0.83 ± 0.18	1.90 ± 0.43
POG	0–165.50	0–30	0–11.09	0–7	0–3
	91.60 ± 6.66	15.87 ± 1.26	5.70 ± 0.41	1.83 ± 0.27	1.17 ± 0.17
ALL	0–117.20	0–19	0–26.50	0–5	0–3
	64.51 ± 4.81	9.27 ± 0.86	7.70 ± 0.92	1.37 ± 0.23	1.43 ± 0.21
HOI	0–94.40	0–25	0–6.99	0–5	0–9
	53.56 ± 5.24	12.63 ± 1.27	3.64 ± 0.34	0.90 ± 0.22	1.83 ± 0.36
ACA	58.80–175.30	7–51	2.60–18.06	1–16	0–17
	99.45 ± 6.06	26 ± 2.07	4.65 ± 0.59	5.97 ± 0.67	6.67 ± 0.75
SWM	0–69.90	0–21	0–9.99	0–4	0–3
	24.35 ± 5.25	6.43 ± 1.45	1.81 ± 0.44	0.67 ± 0.19	0.60 ± 0.17

TABLE 9.6 *(Continued)*

	APRIL				
	Morphometric features				
	SL	**NN**	**ALIN**	**NAB**	**NB**
CAF	0–63	0–13	0–10.75	0–3	0–3
	21.20 ± 4.20	4 ± 0.83	2.88 ± 0.60	0.40 ± 0.13	0.40 ± 0.15
EMO	0–81.20	0–37	0–22.23	0–9	0–21
	41.49 ± 5.11	13.97 ± 2.27	3.26 ± 0.78	1.80 ± 0.36	1.87 ± 0.74
DAS	0–192.20	0–39	0–13.73	0–6	0–21
	114.23 ± 10.50	18.90 ± 2.07	5.69 ± 0.66	2.07 ± 0.33	8.37 ± 1.15
AZI	0–140.70	0–29	0–11.43	0–3	0–10
	29.14 ± 6.87	5.27 ± 1.35	2.90 ± 0.63	0.87 ± 0.20	1.57 ± 0.45
POG	0–192.30	0–32	0–7.80	0–7	0–4
	87.88 ± 9.07	15.90 ± 1.61	4.73 ± 0.44	2 ± 0.27	1.30 ± 0.21
ALL	0–124	0–20	0–17.76	0–5	0–4
	63.96 ± 5.95	9.67 ± 0.99	6.24 ± 0.67	1.67 ± 0.27	1.50 ± 0.24
HOI	0–94.80	0–27	0–5.71	0–5	0–5
	55.23 ± 5.41	13.87 ± 1.36	3.38 ± 0.31	1.20 ± 0.24	1.77 ± 0.29
ACA	0–186.60	0–54	0–8.45	0–17	0–18
	87.96 ± 9.39	24.73 ± 2.83	3.25 ± 0.37	5.27 ± 0.83	6.20 ± 0.88
SWM	0–70.60	0–21	0–4.38	0–4	0–4
	17.16 ± 4.92	5.27 ± 1.53	0.10 ± 0.29	0.67 ± 0.22	0.60 ± 0.21

	MAY				
	Morphometric features				
	SL	**NN**	**ALIN**	**NAB**	**NB**
CAF	0–72	0–15	0–8.80	0–3	0–3
	18.49 ± 4.47	4.13 ± 1	1.89 ± 0.47	0.57 ± 0.17	0.33 ± 0.14
EMO	0–86	0–38	0–16.98	0–11	0–23
	36.42 ± 5.90	13.13 ± 2.53	2.33 ± 0.61	2.47 ± 0.51	2.13 ± 0.80
DAS	0–196	0–41	0–13.07	0–6	0–22
	112.29 ± 12.36	19.50 ± 2.30	5.33 ± 0.65	2.37 ± 0.36	8.93 ± 1.20
AZI	0–144	0–35	0–9.03	0–6	0–11
	23.82 ± 7.17	4.93 ± 1.56	1.96 ± 0.52	1.10 ± 0.32	1.30 ± 0.42
POG	0–233	0–34	0–8.63	0–8	0–5
	93.30 ± 10.58	17.17 ± 1.82	4.41 ± 0.46	2.63 ± 0.36	1.70 ± 0.26

TABLE 9.6 *(Continued)*

	SL	NN	ALIN	NAB	NB
			MAY		
			Morphometric features		
ALL	0–135	0–22	0–11.88	0–7	0–4
	60.37 ± 7.75	9.97 ± 1.29	4.44 ± 0.60	1.73 ± 0.31	1.47 ± 0.27
HOI	0–101	0–31	0–5.13	0–6	0–10
	51.08 ± 6.39	14.33 ± 1.75	2.64 ± 0.32	1.37 ± 0.30	2.10 ± 0.42
ACA	0–198	0–55	0–7.06	0–19	0–18
	84.69 ± 11.84	23.83 ± 3.36	2.62 ± 0.37	5.53 ± 0.10	6.37 ± 0.99
SWM	0–72	0–22	0–3.93	0–5	0–4
	13.66 ± 4.66	4.57 ± 1.55	0.70 ± 0.24	0.63 ± 0.25	0.53 ± 0.21

	SL	NN	ALIN	NAB	NB
			JUNE		
			Morphometric features		
CAF	0–35	0–7	0–18	0–2	0–1
	16.24 ± 1.77	3.1 ± 0.39	5.11 ± 0.88	0.50 ± 0.12	0.03 ± 0.03
EMO	15–45.40	6–25	1.62–3.60	0–14	0–3
	31.33 ± 1.30	14.03 ± 0.85	2.36 ± 0.10	4 ± 0.53	0.20 ± 0.11
DAS	5.60–55	2–20	1.38–6.60	0–5	0–2
	27.50 ± 1.99	10.90 ± 0.79	2.79 ± 0.21	1.60 ± 0.23	0.37 ± 0.12
AZI	0–61	0–13	0–37	0–2	0–3
	27.59 ± 2.67	5.07 ± 0.57	7.04 ± 1.27	0.33 ± 0.11	0.33 ± 0.15
POG	21–56	0–13	0–45	0–6	0–0
	34.39 ± 1.51	7.67 ± 0.55	5.62 ± 1.39	1.60 ± 0.26	0 ± 0
ALL	12.40–38.20	2–14	1.14–9.55	0–5	0–3
	21.10 ± 1.17	6.73 ± 0.50	3.61 ± 0.32	1.17 ± 0.23	0.37 ± 0.13
HOI	0–27	0–13	0–12.50	0–4	0–3
	19.63 ± 1.14	7.70 ± 0.63	3.22 ± 0.47	1.20 ± 0.21	0.37 ± 0.15
ACA	11–30	5–21	1.09–3.88	0–11	0–5
	19.97 ± 0.75	10.17 ± 0.68	2.13 ± 0.12	1.33 ± 0.38	0.77 ± 0.24
SWM	0–31	0–10	0–18	0–2	0–2
	17.65 ± 1.37	5.67 ± 0.57	2.97 ± 0.58	0.20 ± 0.09	0.23 ± 0.10

TABLE 9.6 *(Continued)*

	JULY				
	Morphometric features				
	SL	**NN**	**ALIN**	**NAB**	**NB**
CAF	0–61.20	0–9	0–17.50	0–4	0–1
	20.66 ± 2.73	4.37 ± 0.53	3.94 ± 0.66	0.77 ± 0.20	0.10 ± 0.06
EMO	26–64	8–38	1.31–3.62	2–16	0–11
	43.61 ± 1.76	20.70 ± 1.29	2.27 ± 0.12	5.53 ± 0.79	0.97 ± 0.41
DAS	0–101	0–35	0–4.59	0–6	0–6
	55.64 ± 3.83	20.07 ± 1.22	2.73 ± 0.17	2.83 ± 0.24	2.43 ± 0.30
AZI	0–66	0–16	0–10.25	0–2	0–5
	30.92 ± 3.39	7.07 ± 0.81	4.21 ± 0.47	0.80 ± 0.13	1.07 ± 0.25
POG	0–72.10	0–21	0–7.43	0–8	0–1
	45.75 ± 2.32	11.73 ± 0.77	3.65 ± 0.27	2.53 ± 0.25	0.07 ± 0.05
ALL	22–61	6–15	2.30–6	1–6	0–3
	35.27 ± 1.54	10.53 ± 0.48	3.49 ± 0.17	2.13 ± 0.27	0.63 ± 0.14
HOI	0–46.10	0–20	0–4.52	0–7	0–3
	29.91 ± 2.09	12.50 ± 0.88	2.32 ± 0.17	1.60 ± 0.27	0.80 ± 0.19
ACA	19–50	9–27	1.52–3.70	0–10	0–5
	31.60 ± 1.36	15.10 ± 0.77	2.16 ± 0.09	3.77 ± 0.47	1.97 ± 0.25
SWM	0–36.10	0–15	0–3.40	0–3	0–3
	19.42 ± 2.30	8.13 ± 0.10	1.78 ± 0.21	0.80 ± 0.15	0.37 ± 0.14

	AUGUST				
	Morphometric features				
	SL	**NN**	**ALIN**	**NAB**	**NB**
CAF	0–62.60	0–10	0–18.05	0–1	0–2
	20.99 ± 3.23	4.50 ± 0.66	3.78 ± 0.70	0.33 ± 0.09	0.17 ± 0.10
EMO	26.20–67	8–38	1.36–3.45	1–20	0–20
	46.85 ± 1.93	24 ± 1.44	2.08 ± 0.10	3.90 ± 0.73	1.67 ± 0.72
DAS	0–128	0–40	0–5	0–6	0–11
	77.52 ± 4.51	25–1.27	3.01 ± 0.16	2.93 ± 0.26	4 ± 0.49
AZI	0–71	0–22	0–15.50	0–3	0–7
	31.80 ± 3.82	7.60 ± 1.10	4.44 ± 0.69	0.73 ± 0.14	1.53 ± 0.35
POG	0–83.10	0–22	2.29–5.09	0–7	0–1
	55.47 ± 2.68	15.80 ± 0.66	3.54 ± 0.13	2.53 ± 0.25	0.07 ± 0.05

TABLE 9.6 *(Continued)*

| | AUGUST | | | | |
| | Morphometric features | | | | |
	SL	NN	ALIN	NAB	NB
ALL	27–66	7–21	2.06–5.50	1–4	0–3
	45.34 ± 2.07	12.83 ± 0.61	3.63 ± 0.16	2.07 ± 0.17	0.80 ± 0.16
HOI	0–62	0–26	0–4.40	0–8	0–3
	35.88 ± 3.04	15.07 ± 1.23	2.21 ± 0.18	1.50 ± 0.30	0.93 ± 0.20
ACA	26–66.10	13–32	0.51–3.24	1–12	0–5
	40.69 ± 1.91	19.93 ± 0.86	2.05 ± 0.10	4.57 ± 0.52	2.60 ± 0.27
SWM	0–45	0–18	0–5.04	0–4	0–4
	22.74 ± 2.77	9.63 ± 1.16	1.80 ± 0.23	0.77 ± 0.21	0.67 ± 0.20

| | SEPTEMBER | | | | |
| | Morphometric features | | | | |
	SL	NN	ALIN	NAB	NB
CAF	0–63.40	0–10	0 –9.60	0–2	0–2
	22.65 ± 3.32	4.37 ± 0.65	4.06 ± 0.62	0.60 ± 0.10	0.17 ± 0.10
EMO	27–67.20	9–42	1.46–3.38	0–6	0–20
	48.61 ± 2.01	23.63 ± 1.50	2.20 ± 0.11	2.80 ± 0.28	1.93 ± 0.74
DAS	0–144	0–37	0–8.03	0–7	0–17
	93.59 ± 4.61	26.27 ± 1.35	3.58 ± 0.23	3 ± 0.29	6.03 ± 0.77
AZI	0–95	0–32	0–17.06	0–4	0–4
	34.56 ± 4.60	9.17 ± 1.49	4.02 ± 0.66	1.07 ± 0.22	1.33 ± 0.27
POG	0–110	0–26	0–6.27	0–7	0–2
	70 43 ± 3.72	17.43 ± 1.03	3.67 ± 0.24	2.63 ± 0.30	0.50 ± 0.13
ALL	33.20–100	6–22	2.62–10.14	0–4	0–3
	55.39 ± 3.02	13.23 ± 0.75	4.46 ± 0.31	1.60 ± 0.18	0.87 ± 0.16
HOI	0–75	0–29	0–5.40	0–9	0–8
	44.34 ± 3.74	16.23 ± 1.36	2.54 ± 0.20	1.97 ± 0.33	1.40 ± 0.32
ACA	35.60–86	17–33	1.69–3.67	1–15	0–7
	57.97 ± 2.23	24 ± 0.80	2.46 ± 0.10	5.97 ± 0.64	3.33 ± 0.32
SWM	0–49	0–19	0–8.15	0–2	0–4
	28.69 ± 3.44	10.73 ± 1.33	2.13 ±0.31	0.87 ± 0.14	0.77 ± 0.20

TABLE 9.6 *(Continued)*

	OCTOBER				
	Morphometric features				
	SL	**NN**	**ALIN**	**NAB**	**NB**
CAF	0–64	0–10	0–12.80	0–2	0–2
	23.82 ± 3.37	3.90 ± 0.60	4.91 ± 0.74	0.33 ± 0.10	0.20 ± 0.10
EMO	0–67.60	0–41	0–5.20	0–4	0–20
	48.26 ± 2.52	20.43 ± 1.71	2.55 ± 0.18	1.60 ± 0.22	1.77 ± 0.73
DAS	0–146	0–37	0–12.74	0–4	0–17
	104.66 ± 4.78	22.63 ± 1.49	4.68 ± 0.41	1.77 ± 0.20	6.73 ± 0.88
AZI	0–104	0–32	0–17.70	0–4	0–8
	35.19 ± 4.74	8.33 ± 1.37	4.48 ± 0.75	0.93 ± 0.15	1.63 ± 0.37
POG	0–134	0–31	0–8.31	0–3	0–2
	82.51 ± 4.76	17.10 ± 0.99	4.79 ± 0.28	1.87 ± 0.16	0.40 ± 0.11
ALL	35.40–103	4–23	3.15–13.15	0–4	0–3
	59.15 ± 3.18	11.23 ± 0.79	5.77 ± 0.42	1.33 ± 0.16	0.97 ± 0.16
HOI	0–92	0–27	0–5.35	0–3	0–8
	47.98 ± 4.46	15.17 ± 1.33	2.78 ± 0.24	1.17 ± 0.15	1.40 ± 0.32
ACA	46.60–114.40	15–38	2.03–5.26	1–16	0–9
	73.37 ± 3.01	23.97 ± 1.14	3.15 ± 0.13	6.07 ± 7.01	3.97 ± 0.36
SWM	0–55.20	0–19	0–4	0–2	0–4
	29.76 ± 3.86	10.20 ± 1.30	2.05 ± 0.26	0.83 ± 0.14	0.77 ± 0.21

	NOVEMBER				
	Morphometric features				
	SL	**NN**	**ALIN**	**NAB**	**NB**
CAF	0–65	0–10	0–18.50	0–1	0–3
	24.48 ± 3.42	3.90 ± 0.65	5.77 ± 0.10	0.20 ± 0.07	0.27 ± 0.14
EMO	0–68	0–40	0–16	0–2	0–20
	49 ± 2.55	18.83 ± 1.75	3.40 ± 0.51	0.27 ± 0.10	1.77 ± 0.73
DAS	0–146.70	0–38	0–16.20	0–5	0–17
	106.27 ± 4.84	20.10 ± 1.76	6.38 ± 0.68	0.77 ± 0.20	7.13 ± 0.90
AZI	0–105	0–33	0–15.75	0–5	0–8
	37.22 ± 4.90	7.07 ± 1.31	5.43 ± 0.73	0.63 ± 0.18	1.63 ± 0.37
POG	0–148.60	0–33	0–8.65	0–3	0–2
	86.44 ± 5.49	15.77 ± 1.19	5.59 ± 0.32	1.60 ± 0.20	0.57 ± 0.12

TABLE 9.6 *(Continued)*

			NOVEMBER		
			Morphometric features		
	SL	NN	ALIN	NAB	NB
ALL	36.20–103.40	3–23	3.61–28.86	0–3	0–3
	61.15 ± 3.40	9.23 ± 0.90	8.39 ± 0.96	1.07 ± 0.14	1.03 ± 0.17
HOI	0–92.70	0–27	0–7	0–3	0–8
	49.31 ± 4.60	14.33 ± 1.32	3.07 ± 0.28	0.73 ± 0.14	1.43 ± 0.33
ACA	50.20–127	10–49	2.17–9.17	0–16	0–11
	81.50 ± 4.03	23.20 ± 1.88	3.97 ± 0.31	6.13 ± 0.83	4.43 ± 0.42
SWM	0–57.30	0–20	0–5.73	0–3	0–4
	31.94 ± 4.15	9.57 ± 1.26	2.40 ± 0.32	0.77 ±0.14	0.77 ± 0.21

			DECEMBER		
			Morphometric features		
	SL	NN	ALIN	NAB	NB
CAF	0–53.70	0–10	0–19.80	0–1	0–3
	23.04 ± 3.32	3.50 ± 0.64	6.08 ± 1.14	0.10 ± 0.06	0.30 ± 0.14
EMO	0–68.20	0–39	0–10.20	0–7	0–20
	50.18 ± 2.61	15.90 ± 1.90	3.39 ± 0.43	1.17 ± 0.29	1.77 ± 0.73
DAS	0–147	0–34	0–16.43	0–4	0–18
	109.79 ± 5.03	18.23 ± 1.56	6.96 ± 0.66	0.67 ± 0.15	7.50 ± 0.97
AZI	0–105.90	0–32	0–17.97	0–2	0–8
	37.92 ± 4.96	6.83 ± 1.22	5.74 ± 0.87	0.67 ± 0.10	1.70 ± 0.38
POG	0–154	0–30	0–9.02	0–6	0–2
	87.44 ± 5.57	14.90 ± 1.04	5.95 ± 0.34	1.90 ± 0.26	0.60 ± 0.13
ALL	36.90–104	2–19	3.65–21.50	0–8	0–3
	62.02 ± 3.41	8.47 ± 0.82	9.11 ± 0.84	1.50 ± 0.27	1.07 ± 0.17
HOI	0–93.20	0–28	0–7.28	0–4	0–8
	50.12 ± 4.67	12.40 ± 1.22	3.67 ± 0.34	0.90 ± 0.14	1.53 ± 0.33
ACA	52–148	5–51	1.93–22.60	0–14	0–16
	86.73 ± 4.87	24.03 ± 2.11	4.98 ± 0.82	5.70 ± 0.75	5.77 ± 0.71
SWM	0–63.60	0–19	0–7.14	0–2	0–4
	32.10 ± 4.30	8.83 ± 1.18	2.74 ± 0.38	0.73 ± 0.13	0.77 ± 0.21

Abbreviations: CAF – *Cassia fistula,* EMO – *Emblica officinalis,* DAS – *Dalbergia sissoo,* AZI – *Azadirachta indica,* POG – *Pongamia glabra,* ALL – *Albizia lebbeck,* HOI – *Holoptelia integrifolia,* ACA – *Acacia auriculiformis,* SWM – *Swetenia macrophyla,* SL – Shoot Length, NN – Number of Node, ALIN – Average Length of Internode, NAB – Number of Auxiliary Bud, NB – Number of Branches.

TABLE 9.7   Individual Plant Species-Based Variations Judged Through Univariate ANOVA A–CAF: *Cassia fistula*

		A. Cassia fistula			
Variable	Source	SS	df	MS	F
SL	Soil condition	1101.97	5	220.39	0.624
	Error	125001.42	354	353.11	
	Total	126103.39	359		
NN	Soil condition	484.48	5	96.90	**7.628**
	Error	4496.52	354	12.70	
	Total	4981.00	359		
ALIN	Soil condition	673.10	5	134.62	**6.152**
	Error	7746.22	354	21.88	
	Total	8419.32	359		
NAB	Soil condition	3.05	5	0.61	1.466
	Error	147.15	354	0.42	
	Total	150.20	359		
NB	Soil condition	19.56	5	3.91	**10.415**
	Error	132.93	354	0.38	
	Total	152.49	359		

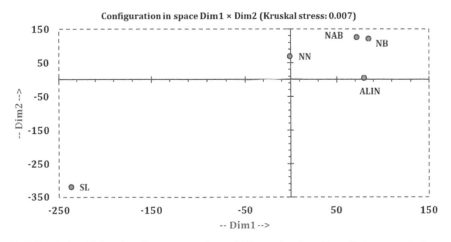

FIGURE 9.2   Biplot showing non-metric multidimensional scaling of plant morphology of different features of the plant.

TABLE 9.7   *(Continued)*

B–EMO: *Emblica officinalis*

B. *Emblica officinalis*					
Variable	Source	SS	df	MS	F
SL	Soil condition	23925.58	5	4785.117	17.340
	Error	97687.13	354	275.9524	
	Total	121612.7	359		
NN	Soil condition	5996.147	5	1199.229	**12.940**
	Error	32807.05	354	92.67528	
	Total	38803.2	359		
ALIN	Soil condition	582.8152	5	116.563	**5.044**
	Error	8180.498	354	23.10875	
	Total	8763.313	359		
NAB	Soil condition	180.3556	5	36.07111	**4.832**
	Error	2642.3	354	7.464124	
	Total	2822.656	359		
NB	Soil condition	1436.347	5	287.2694	**27.624**
	Error	3681.317	354	10.3992	
	Total	5117.664	359		

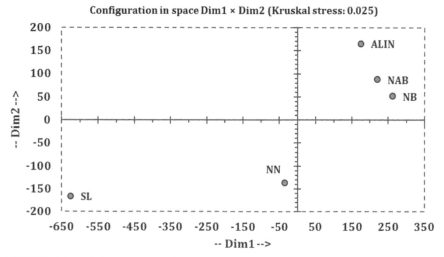

FIGURE 9.3   Biplot for the ordination of the different features of the plant (plant morphology, based on the non-metric multidimensional scaling).

TABLE 9.7 *(Continued)*

C–DAS: *Dalbergia sissoo*

		C. *Dalbergia sissoo*			
Variable	**Source**	**SS**	**df**	**MS**	**F**
SL	Soil condition	43890.28	5	8778.06	4.546
	Error	683498.54	354	1930.79	
	Total	727388.82	359		
NN	Soil condition	2411.69	5	482.34	**5.768**
	Error	29600.08	354	83.62	
	Total	32011.78	359		
ALIN	Soil condition	543.43	5	108.69	**5.783**
	Error	6653.01	354	18.79	
	Total	7196.44	359		
NAB	Soil condition	69.32	5	13.86	**5.877**
	Error	835.07	354	2.36	
	Total	904.39	359		
NB	Soil condition	2980.13	5	596.03	**27.880**
	Error	7567.77	354	21.38	
	Total	10547.90	359		

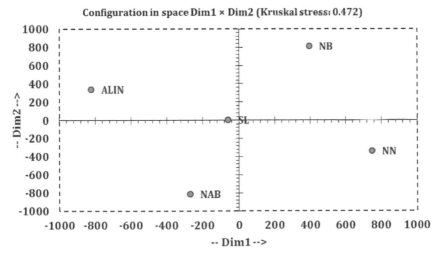

FIGURE 9.4 Biplot for the ordination of the different features of the plant (plant morphology, based on the non-metric multidimensional scaling).

TABLE 9.7   *(Continued)*

D–AZI: *Azadirachta indica*

		D. Azadirachta indica			
Variable	**Source**	**SS**	**df**	**MS**	**F**
SL	Soil condition	65742.99	5	13148.60	20.478
	Error	227299.54	354	642.09	
	Total	293042.53	359		
NN	Soil condition	3755.38	5	751.08	**21.352**
	Error	12452.28	354	35.18	
	Total	16207.66	359		
ALIN	Soil condition	658.35	5	131.67	**7.7519**
	Error	6012.83	354	16.99	
	Total	6671.17	359		
NAB	Soil condition	24.57	5	4.91	**5.404**
	Error	321.83	354	0.91	
	Total	346.40	359		
NB	Soil condition	292.93	5	58.59	**18.051**
	Error	1148.97	354	3.25	
	Total	1441.90	359		

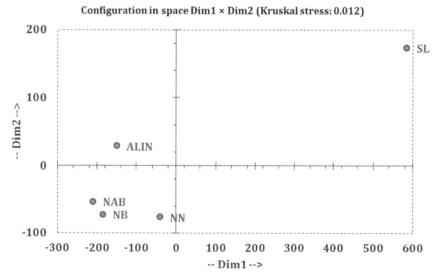

FIGURE 9.5   Biplot for the ordination of the different features of the plant (plant morphology, based on the non-metric multidimensional scaling).

TABLE 9.7 *(Continued)*

E–POG: *Pongamia glabra*

		E. *Pongamia glabra*			
Variable	**Source**	**SS**	**df**	**MS**	**F**
SL	Soil condition	71606.21	5	14321.24	11.839
	Error	428231.61	354	1209.69	
	Total	499837.83	359		
NN	Soil condition	2335.85	5	467.17	**11.689**
	Error	14148.35	354	39.97	
	Total	16484.20	359		
ALIN	Soil condition	114.48	5	22.90	**2.657**
	Error	3050.70	354	8.62	
	Total	3165.19	359		
NAB	Soil condition	68.91	5	13 78	**6.068**
	Error	804.08	354	2.27	
	Total	873.00	359		
NB	Soil condition	42.19	5	8.44	**10.639**
	Error	280.78	354	0.79	
	Total	322.98	359		

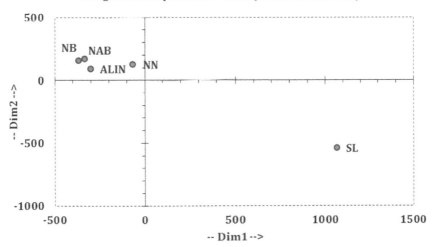

FIGURE 9.6 Biplot for the ordination of the different features of the plant (plant morphology, based on the non-metric multidimensional scaling).

TABLE 9.7　　*(Continued)*

F–ALL: *Albizia lebbeck*

		F. *Albizia lebbeck*			
Variable	**Source**	**SS**	**df**	**MS**	**F**
SL	Soil condition	22123.41	5	4424.68	7.140
	Error	219361.94	354	619.67	
	Total	241485.35	359		
NN	Soil condition	365.88	5	73.18	**3.221**
	Error	8042.78	354	22.72	
	Total	8408.66	359		
ALIN	Soil condition	340.98	5	68.20	**3.106**
	Error	7771.58	354	21.95	
	Total	8112.56	359		
NAB	Soil condition	29.08	5	5.82	**3.894**
	Error	528.78	354	1.49	
	Total	557.86	359		
NB	Soil condition	66.72	5	13.34	**12.999**
	Error	363.40	354	1.03	
	Total	430.12	359		

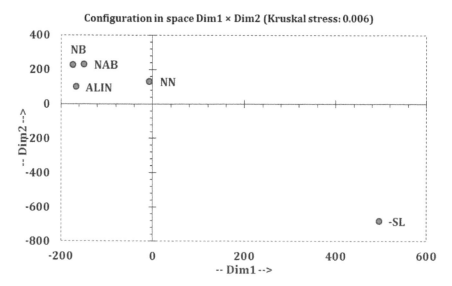

FIGURE 9.7　Biplot for the ordination of the different features of the plant (plant morphology, based on the non-metric multidimensional scaling).

TABLE 9.7   *(Continued)*

G–HOI: *Holoptelia integrifolia*

		G. *Holoptelea integrifolia*			
Variable	Source	SS	df	MS	F
SL	Soil condition	58542.77	5	11708.55	21.903
	Error	189233.60	354	534.56	
	Total	247776.36	359		
NN	Soil condition	3967.36	5	793.47	**19.607**
	Error	14325.27	354	40.47	
	Total	18292.62	359		
ALIN	Soil condition	116.51	5	23.30	**8.282**
	Error	995.87	354	2.81	
	Total	1112.38	359		
NAB	Soil condition	35.46	5	7.09	**4.587**
	Error	547.17	354	1.55	
	Total	582.62	359		
NB	Soil condition	338.89	5	67.78	**31.734**
	Error	756.08	354	2.14	
	Total	1094.98	359		

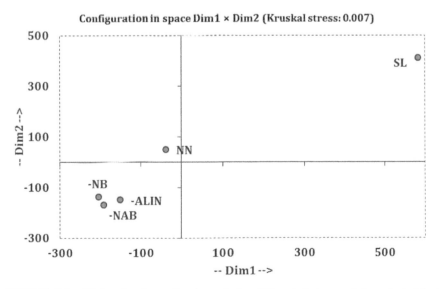

FIGURE 9.8   Biplot for the ordination of the different features of the plant (plant morphology, based on the non-metric multidimensional scaling).

TABLE 9.7 *(Continued)*

H–ACA: *Acacia auriculiformis*

		H. *Acacia auriculiformis*			
Variable	**Source**	**SS**	**df**	**MS**	**F**
SL	Soil condition	62832.32	5	12566.46	8.791
	Error	506004.25	354	1429.39	
	Total	568836.57	359		
NN	Soil condition	2057.95	5	411.59	**3.389**
	Error	42987.05	354	121.43	
	Total	45045.00	359		
ALIN	Soil condition	149.33	5	29.87	**4.386**
	Error	2410.06	354	6.81	
	Total	2559.39	359		
NAB	Soil condition	384.20	5	76.84	**4.959**
	Error	5484.20	354	15.49	
	Total	5868.40	359		
NB	Soil condition	437.99	5	87.60	**6.587**
	Error	4707.38	354	13.30	
	Total	5145.38	359		

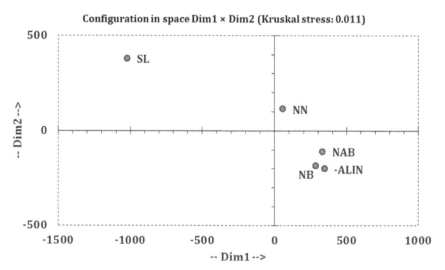

FIGURE 9.9 Biplot for the ordination of the different features of the plant (plant morphology, based on the non-metric multidimensional scaling).

TABLE 9.7   *(Continued)*

I–SWM: *Swietenia macrophylla*

		I. Swietenia macrophylla			
Variable	**Source**	**SS**	**df**	**MS**	**F**
SL	Soil condition	13982.83	5	2796.57	5.674
	Error	174452.88	354	492.80	
	Total	188435.71	359		
NN	Soil condition	1223.82	5	244.76	**5.090**
	Error	17019.63	354	48.08	
	Total	18243.46	359		
ALIN	Soil condition	137.75	5	27.55	**6.722**
	Error	1450.80	354	4.10	
	Total	1588.54	359		
NAB	Soil condition	24.58	5	4.92	**6.134**
	Error	283.68	354	0.80	
	Total	308.26	359		
NB	Soil condition	56.47	5	11.29	**12.816**
	Error	311.93	354	0.88	
	Total	368.40	359		

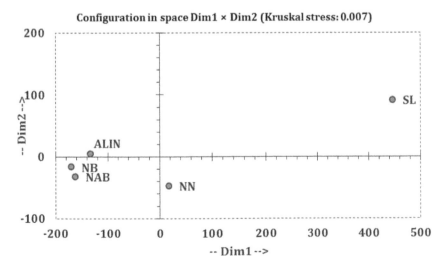

FIGURE 9.10   Biplot for the ordination of the different features of the plant (plant morphology, based on the non-metric multidimensional scaling).

In order to decipher the effects of the soil treatments on the plants at the individual level, the one-way ANOVA was carried out using the data on the morphological features and the soil conditions as the source of variation (explanatory variable). As shown in Table 9.7A through I, in all instances, all the morphological features of the plants were affected by the soil conditions. The variations in the morphological features were obvious, due to the specific growth and anatomical features of the plant species which could be presented through the non-metric multidimensional scaling as shown in the biplots (Tables 9.7A through I; Figures 9.2–9.10). Thus the results reflect that the soil conditions influenced the plant growth considerably, though the plants remained tolerant to the different conditions provided for growth. Under normal conditions, the stability of the soil remains unperturbed than under situation where the land usage pattern is subjected to mining activities. The growth of the plants was unhindered though varied with contrast to the five treatments and the normal soil condition (S), reflecting the tolerance and ability to sustain the perturbed conditions of the environment.

As observed in the present instance, the growth pattern of the plants is quite different from one another, which can partly be attributed to the species-specific growth pattern, irrespective of the treatment conditions. Adaptations of the particular species can be found due to species-specific differences in spite of prominent soil treatment. Thus the growth of *A. indica* was not same as that of *C. fistula* or *D. sissoo*, which however reflects the varying level of importance of these species in organizing the community. Nonetheless, the plants observed in the present study were capable of exhibiting growth in different soil conditions, mimicking the continuum of the soil quality observed in wastelands subjected to open cast coal mining in the past. Earlier studies on these plants have shown prospect of the use in revegetation of the altered soil conditions in abandoned coal fields of Jharia in India (Sheoran et al., 2010). Globally, land mass subjected to coal mining activities in the past has been reclaimed through revegetation, though the selection of the native species has been given priority. The process reinstates the seral stages in the continuum of community development, and therefore the use of the grasses and herbs are less likely being useful. Instance from Poland (Woch et al., 2013; Piekarska-Stachowiak et al., 2014), England (Rostanski, 2005), China (Cheng and Lu, 2005; Donggan et al., 2011; Huang et al., 2015; Zhang et al., 2015) suggests that the reclamation of the wastelands formed due to past mining activity can be augmented through the plantation of selected and desired plant species. Selection of the plant species is crucial in order to enhance the structuring of the community. The tolerance to the existing soil conditions remains an important criterion for selection of the plant species suitable for the restoration of the degraded landscape. In the present instance, the selected plant species, *Cassia fistula, Emblica officinalis, Dalbergia sissoo, Azadirachta indica, Pongamia glabra, Albizia lebbeck, Holoptelea integrifolia, Acacia auriculiformis,* and *Swietenia macrophylla* qualified as tolerant species, owing to their growth under the different treatment conditions. The growth of these plants reflects their tolerance to the soil conditions, and supports their use in the restoration process of the degraded wastelands that were subjected to active coal mining activities in the past. The present study revealed the use of studied plant species in the phytoremediation activity to restore the coal mine wastelands through active plantation.

9.5 CONCLUSION

It can be concluded that coalmine spoil can be rehabilitated by implement of proper treatment through pot experiment of ecologically suitable species. The growth of these plants reflects their tolerance to the soil conditions, and supports their use in the restoration process of the degraded wastelands that were subjected to active coal mining activities in the past. As reflected in the results of the present study, the use of these plant species in the phytoremediation of the wastelands through active plantation may restore the wastelands created through coal mining activities in the past.

Rehabilitation of lands disturbed by mining has become obligatory for both; the state-owned corporations as well as private companies in view of environmental concerns and socio-economic compulsions. As newer and more areas will be brought undermining the need for rehabilitation of these lands will become more obvious. Realizing this, protocols of rehabilitation will have to be implemented as per bioclimatic conditions.

KEYWORDS

- **characteristics features**
- **coal mine**
- **coal mine spoil**
- **indigenous vegetation**
- **mine activity**
- **opencast**
- **phytoremediation**

- **pot experiment**
- **Raniganj**
- **rehabilitation**
- **revegetation**
- **spoil treatment**
- **wasteland**

REFERENCES

1. Ahmad, I., & Singh, S. K. (2004). Seasonal variation in certain chemical properties of the soil of a freshwater pond of Dholi (Muzaffarpur), Bihar, India. *Journal of Applied Biology, 14*(1), 53–55.
2. Amaya-Chavez, A., Martinez-Tabche, L., Lopez-Lopez, E., & Galar-Martinez, M. (2006). Methyl parathion toxicity to and removal of efficiency by *Typha latifolia* in water and artificial sediments. *Chemosphere, 63,* 1124–1129.
3. Bell, F. G., Bullock, S. E. T., Halbich, T. F. J., & Lindsey, P. (2001). Environmental impacts associated with an abandoned mine in the Witbank Coalfield, South Africa. *International Journal of Coal Geology, 45*(2–3), 195–216.
4. Borpujari, D. (2008). Studies on the occurrence and distribution of some tolerant plant species in different spoil dumps of Tikak opencast mine. *The Ecoscan, 2,* 255–260.
5. Bradshaw, A. D., & Chadwick, M. J. (1980). The Restoration of Land: The Ecology and Reclamation of Derelict and Degraded Land. Blackwell Scientific Publication, Oxford, London.
6. Bradshaw, A. D. (1983). The reconstruction of Ecosystems. *Journal of Applied Ecology, 20,* 1–17.
7. Chadwick, K. J. (1973). Methods of assessment of acid colliery spoils as a medium for plant growth. In: Hutnik, R. J., & Davis, G. (eds.), Ecology and reclamation of devastated land. Vol. I. Gordon and Breach. New York. pp. 81–91.
8. Chaubey, O. P., Bohre, P., & Singhal, P. K. (2012). Impact of Bio-reclamation of Coal Mine Spoil on Nutritional and Microbial Characteristics: A Case Study. *International Journal of Bio-Science and Bio-Technology, 4*(3), 69–80.
9. Cheng, J. L., & Lu, Z. H. (2005). Natural vegetation recovery on waste dump in open cast coal mine area. *Journal of Forest Research, 16*(1), 55–57.
10. Conesa, H. M., Faz, A., & Arnaldos, R. (2007). Initial studies for the phytostabilization of a mine tailing from the Cartagena-La Union Mining District (SE Spain). *Chemosphere, 66*(1), 38–44.

11. Davcheva, I. (1990). Coal mine enterprise: an article component in natural ecosystems. In: A. K. M. Rainbow (ed.) Proc. of the Third Int. Symp. On Reclamation, Treatment, and utilization of coal mining Wastes, Glasgow: 119–124.

12. Dobson, A. P., Bradshaw, A. D., & Baker, A. J. M. (1997). Hopes for the future: restoration ecology and conservation biology. *Science, 277,* 515–522.

13. Donggan, G., Zhongke, B., Tieliang, S., Hongbo, S., & Wen, Q. (2011). Impacts of coal mining on aboveground vegetation and soil quality: a case study of Qinxin coal mine in Shanxi Province, China. *Clean- Soil, Air, Water, 39*(3), 219–225.

14. Dowarah, J., Deka Boruah, H. P., Gogoi, J., Pathak, N., Saikia, N., & Handique, A. K. (2009). Ecorestoration of a high sulfur coal mine overburden dumping site in northeast India: a case study. *Journal of Earth System Science, 118,* 597–608.

15. Down, C. G., & Stocks, J. (1977). Environmental impact of mining. Applied Science Publishers Ltd, London.

16. Ghose, A. K. (1990). Mining in 2000 AD: challenges for India. *Journal of the Institute of Engineers* (India), *39*(ii), 1–11.

17. Ghose, M. K. (1989). Land reclamation and protection of environment from the effect of coal mining operation. *Minetech., 10*(5), 35–39.

18. Ghose, M. K. (2004). Effect of opencast mining on soil fertility. *Journal of Scientific and Industrial Research, 63*(12), 1006–1009.

19. Gonzalez, R. C., & Gonzalez-Chavez, M. C. A. (2006). Metal accumulation in wild plants surrounding mining wastes: Soil and sediment remediation (SSR). *Environ. Poll., 144,* 84–92.

20. Hazarika, P., Talukdar, N. C., & Singh, Y. P. (2006). Natural colonization of plant species on coal mine spoils at Tikak Colliery, Assam. *Tropical Ecology, 47*(1), 37–46.

21. Huang, L., Zhang, P., Hu, Y., & Zhao, Y. (2015). Vegetation succession and soil infiltration characteristics under different aged refuse dumps at the Heidaigou opencast coal mine. *Global Ecology and Conservation, 4,* 255–263.

22. Jha, A. K., & Singh, J. S. (1993). Rehabilitation of mine spoils. In: Restoration of Degraded Land: Concepts and Strategies. Ed. Singh, J. S., Rastogi Publications, Meerut. 210–254pp.

23. Jha, A. K., & Singh, J. S. (1994). Rehabilitation of mine spoils with particular reference to multipurpose trees. In: Agroforestry Systems for Sustainable Land Use. Eds. Singh, P., Pathak, P. S., Roy, M. M. Oxford University Press and IBH Publishing Co., New Delhi. 237–249pp.

24. Kundu, N. K., & Ghose, M. K. (1997). Impact of coal mining on land use and its reclamation- A case study. *Trans MGMI, 93,* 97–104.

25. Mahalik, G., & Satapathy, K. B. (2016). Environmental impacts of mining on biodiversity of Angul - Talcher open mining site, Odisha, India. *Scholars Academic Journal of Biosciences (SAJB), Scholars Academic and Scientific Publisher* (An International Publisher for Academic and Scientific Resources), *4,* 224–227.

26. Maiti, S. K., & Ghose, M. K. (2005). Ecological restoration of acidic coal mine overburden dumps- an Indian case study. *Land Contamination and Reclamation, 13*(4), 361–369.

27. Meyer, F. H. (1973). Distribution of ectomycorrhizae in native and manmade forests. In: G. C. Marks and T. T.Klozlowski (eds). Ectomycorrhizae: The ecology physiology. Academic Press, NY: *119,* 79–105.

28. Molyneux, J. K. (1963). Some ecological aspects of colliery waste heaps around Wiga, South Lancashire. *Journal of Ecology*, *51*, 315–321.
29. Nath, S. (2009). Ecosystem approach for mined land rehabilitation and present rehabilitation scenario in Jharkhand coal mines. In: Sustainable Rehabilitation of Degraded Ecosystems. Eds. Chaubey OP, Bahadu V, Shukla PK. Aavishkar publishers, distributors Jaipur, Rajasthan–302003 India. 46–66.
30. Piekarska-Stachowiak, A., Szary, M., Ziemer, B., Besenyei, L., & Wozniak, G. (2014). An application of the plant functional group concept to restoration practice on coal mine spoil heaps. *Ecological Research*, *29*, 843–853.
31. Rai, A. K., Paul, B., & Singh, G. (2010). A Study on the Bulk Density and its Effect on the Growth of selected Grasses in Coal Mine Overburden Dumps, Jharkhand, India. *International Journal of Environmental Sciences*, *1*(4), 677–684.
32. Reddy, S. R., & Reddy, S. M. (2001). Revegetation of coal mine disturbed lands with micorrhizae-problems and prospects. In: Wastelands Management and Environment. Scientific Publishers (India), Jodhpur: 17–25pp.
33. Rostański, A. (2005). Specific features of the flora of colliery spoil heaps in selected European regions. *Polish Botanical Studies*, *19*, 97–103.
34. Sheoran, A. S., Sheoran, V., & Poonia, P. (2008). Rehabilitation of mine degraded land by metallophytes. *Mining Engineers Journal*, *10*(3), 11–16.
35. Sheoran, V., Sheoran, A. S., & Poonia, P. (2010). Soil Reclamation of Abandoned Mine Land by Revegetation: A Review. *International Journal of Soil, Sediment, and Water*, *3*(2), 1–20.
36. Singh, A. (2011). Vascular flora on coal mine spoils of Singrauli coalfields, India. *Journal of Ecology and the Natural Environment*, *3*(9), 309–318.
37. Singh, A. (2012). Pioneer Flora on Naturally Revegetated Coal Mine Spoil in a Dry Tropical Environment. *Bulletin of Environment, Pharmacology and Life Sciences*, *1*(3), 72–73.
38. Singh, A. (2013). Evaluation of the growth performance of planted tree species on coal mine spoil in Singrauli coalfields, India. *Indian Journal of Plant Sciences*, *2*(4), 118–123.
39. Singh, J. S., Singh, K. P., & Jha, A. K. (1995). An Integrated Ecological Study on Revegetation of Mine Spoils: Concepts and Research Highlights. An Interim Report Report of S & T project sponsored by the Ministry of Coal, Govt. of India, through CMPDI, Ranchi (EE-8/92).
40. Singh, M. P., Singh, J. K., Mohonka, K., & Sah, R. B. (2007). Forest Environment and Biodiversity, Daya Publishing House, Delhi. pp 568.
41. Srivastva, H. N. (2001). *Practical Botany*. Pradeep Publications, Jalandhar.
42. Tian, D. L., Xiang, W. H., Yan, W. D., Kang, W. X., Deng, X. W., & Fan, Z. (2007). Biological Cycles of Mineral Elements in a Young Mixed Stand in Abandoned Mining Soils. *Journal of Integrative Plant Biology*, *49*(9), 1284–1293.
43. Tordoff, G. M., Baker, A. J. M., & Willis, A. J. (2000). Current approaches to the revegetation and reclamation of metalliferous mine wastes. *Chemosphere*, *41*(1–2), 219–228.
44. Vogel, W. G. (1982). A guideline for revegetating coal mine soils in the Eastern United States. In: USDA for. Ser. Tech. Rep. NE-68. Northeast for. Exp. Stn. Broomhall, P.A.: 1–190.

45. Woch, M. W., Radwańska, M., & Stefanowicz, A. M. (2013). Flora of spoil Caps after hard coal mining in Trzebinia (southern Poland): effect of substrate properties. *Acta Botanica Croatica, 72*(2), 237–256.
46. Wong, M. H., & Bradshaw, A. D. (1982). A comparison of the toxicity of heavy metals using root elongation of Ryegrass. *Lolium perenne. New Phytol, 91,* 255–261.
47. Zhang, L., Wang, J., Bai, Z., & Lv, C. (2015). Effects of vegetation on runoff and soil erosion on reclaimed land in an opencast coal mine dump in a loss area. *Catena, 128,* 44–53.

CHAPTER 10

Non-Timber Forest Products: Constraints, Prospects and Management, Implications for Combating Climate Change, and Livelihood Development

ABHISHEK RAJ[1], M. K. JHARIYA[2], A. BANERJEE[3], and D. K. YADAV[2]

[1]*PhD Scholar, Department of Forestry, College of Agriculture, I.G.K.V., Raipur–492012 (C.G.), India, Mobile: +00-91-8269718066, Email: ranger0392@gmail.com*

[2]*Assistant Professor, University Teaching Department, Department of Farm Forestry, Sarguja Vishwavidyalaya, Ambikapur–497001 (C.G.), India, Mobile: +00-91-9407004814 (M. K. Jhariya), +00-91-9926615061 (D. K. Yadav), E-mail: manu9589@gmail.com (M. K. Jhariya), dheeraj_forestry@yahoo.com (D. K. Yadav)*

[3]*Assistant Professor, University Teaching Department, Department of Environmental Science, Sarguja Vishwavidyalaya, Ambikapur–497001 (C.G.), India, Mobile: +00-91-9926470656, E-mail: arnabenvsc@yahoo.co.in*

ABSTRACT

Forest plays a significant role in terms of providing economic such as timber and non-timber products and environmental protection through conservation of natural resources, maintaining ecological functioning and supports the entire biosphere. Among the tangible and direct benefits, NTFPs deserve a foremost position in our culture which is based on both plant (flora) and animal (fauna) origin of either natural forest and manmade plantations, plays essential sources of nutrition for health and provides possible incomes to farmers and many forest-based communities around the world. Today, NTFPs get threatened due to overexploitation, unsustainable and unscientific methods of harvesting. Although, the species diversity and productivity of NTFPs gets affected by raising the temperature, uncertain rainfall, and humidity that leads to climate change. As per the present scenario, the impact of climate change happens to impose its impact on NTFPs in a variable way. Climate change can alter the species distribution, their diversity and cause mortality of some important tree species producing minor forest products. Therefore, better management practices for proper utilization of NTFPs and intervention of scientific technology for harvesting and processing can not only make the availability of desired products but also can enhance the productivity, diversity and improve the incomes of residing peoples around and in the forest.

10.1   INTRODUCTION

Around the world, peoples are depends upon nature for many things such as food, fuel, fodder, fruits, craft, gums, resins, honey, bee-wax, medicinal and aromatic plants, dying and tanning materials, fiber, fungi, resin, shelter, small wood for poles, etc. for their sustenance. These all products are gathered and collected by forest as timber and non-timber forest products. Although, NTFPs is not something new, but this is a biological resource based on both plant and animal origin which is harvested and collected from natural forest by tribal and other poor's for their own consumption and considered as a source of income (FAO, 2003; Shackleton and Shackleton, 2004; McLain and Jones, 2005). NTFPs provide multidimensional products in the form of food, maintaining daily livelihood, goods for beautification and many other products of non-wood origin (Chamberlain et al., 1998; Thandani, 2001; Morse, 2003).

Many authors and organizations have modified and defined the term NTFPs in different ways (Belcher, 2003; Gary and Kristin, 2005; Rajesh, 2006). Therefore it is a very typical task to defined NTFPs, and their definition can be changeable as per the need and requirement of non-timber forest produces. Around 350 million people are living in or around the dense forest for their sustenance, and 25% of people depend on the forest for their lives out of 6.2 billion people globally (Killman, 2003). NTFPs such as medicinal herbs and fruits are the source of nutrition and medicine for 80 percent of the people living in developing countries (FAO, 2008) and therefore, fuelwood, medicine, and edible plant and their products are the most used NTFPs (Bouri and Mukharjee, 2013). Therefore, the benefits of NTFPs are not confined only to developing countries. Although, several NTFPs such as berries, honey, and other products are collected by approximately 60% of urban dwellers in Switzerland and they visit forests frequently for the same (Kilchling et al., 2009). Therefore, numbers of literatures are available on the importance of NTFPs and their significant contribution towards household health, nutrition and income generation including livelihoods security of forest-dependent communities. Among the all contribution, household health, nutrition and wealth, i.e. income generation are quite possible benefits through NTFPs. Therefore, various authors have been worked on the role of NTFPs towards household incomes (Figure 10.1). Globally, 1.4–1.6 billion people and forest dwellers are used of at least some NTFPs (FAO, 2001), and different categories of forest users and their numbers are depicted in Table 10.1 (Scherr et al., 2003).

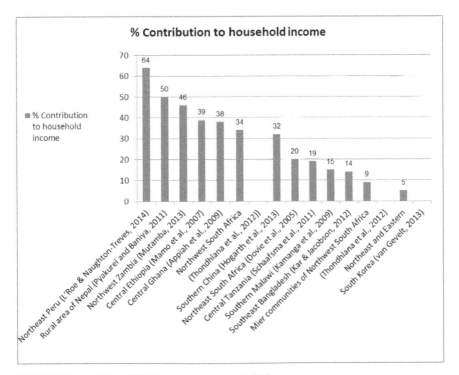

FIGURE 10.1   Role of NTFPs towards household incomes.

TABLE 10.1 Category and Their Estimated Population of Forest Users in World (Modified from Scherr et al., 2003)

Category of user	Estimated population of users
Forest dwellers (depend on natural forests for their Livelihoods)	60 million
Rural people that reside in or around the forest (depend on natural forests for their Livelihoods)	350 million
Small-holder farmers (depends on farm trees or manage remains forests for their Livelihoods)	0.5–1.0 billion
Craftsperson and employees in forest-based small and large enterprises	45 million
Estimated total	1.0–1.4 billion

Land use change, i.e. diversion of forest land to other land use practices and unsustainable harvesting practices results in depletion of NTFPs and pose a great threat to food, nutrition and income security to the tribal peoples and forest. Besides, climate change is another a big issues that is affecting overall production, species distribution, morphology, and phenology of some important NTFPs. In the above point of view, this chapter includes diversity, status, the potential of NTFPs, their opportunities and contribution towards household income, agroforestry role in NTFPs, threats to NTFPs due to climate change and other mismanagement harvesting techniques/practices and their effective conservation and management strategies.

10.2   DIVERSITY OF NON-TIMBER FOREST PRODUCTS

As we know, India is one of the diverse countries that comprise important flora and fauna including various non-timber forest produce. These are the source of nutritive foods, medicine, and incomes for tribals and rural poor's. Shiva and Mathur (1996) have classified NTFPs in two groups (Figure 10.2) of both plant and animal origin which shows a diverse product and their utilization. As we know, India is the rich source of plant diversity, and as per one estimate round 3000 NTFPs species are found out of 45,000 plant species which is spread over sixteen agroclimatic zones and only 126 NTFPs species are potentially recognized marketability (Maithani, 1994; FAO, 2002 & 2005). These include herbs, medicinal and aromatic plants, edible plants, gum, resins, oleoresin, tannins, dye, starch, mucilage's, katha, catechu, essential oils, fats, spices, drugs, insecticides, bamboos and canes, fibers and flosses, grasses, tendu leaves, animal products. Similarly, *Acacia nilotica* tree exudates gums as NTFPs which is an excellent NTFPs which is an excellent source of food and medicine, widely used for an industrial purposes in India and can improve the livelihood of farmers with maintaining biodiversity and ecosystem (Das et al., 2014; Raj, 2015a, 2015b; Raj and Singh, 2017). Although, the poor rural farmers depend on nature for food, fuel, shelter, and medicine (Jhariya and Raj, 2014). Therefore, conserving biodiversity and eliminating poverty are linked global challenges (Raj et al., 2015). Moreover, the utilization trends of medicinal plants are not confining to only human care, also useful for treating disease in domesticated animals.

NTFP CLASSIFICATION

GROUP-I

NTFPs of Plant Origin

- Edible Plant Products
- Spices and Condiments
- Medicinal Plants
- Aromatic Plants
- Fatty Oil Yield Plants
- Gum & Resin Exuding Plants
- Tan Yield Plants
- Dye & Colour Yield Plants
- Fibre & Floss Yield Plants
- Bamboo-Canes
- Fodder & Forage
- Fuelwood, Charcoal Making
- Bidi Wrapper Leaves
- Other Leaves for Plates
- Beads for Ornaments
- Saponin & Marking Nut Plant-Others

GROUP-II

NTFPs of Animal Origin

- Honey
- Lac
- Tussar and Other Silk
- Insects and Animal-Hides, Skins and Feathers
- Horns, Bones and Shellac-Ivory and Musk

FIGURE 10.2 Classifications of NTFPs (Adapted and modified from Shiva and Mathur (1996).

10.3   STATUS OF NTFPs IN INDIA

India is a very diverse country not only in flora and fauna but also in term of society, culture, and religion. A population census report (2011) reveals 8.6% of the population belongs to ST (Scheduled Tribe) categories which are mostly represented by central India and northeast India. These populations are dependent on forest and make their close relationship with nature for their social, culture and daily needs as timber and NTFPs. India has shared 42% in total removal of NTFPs such as tendu leaves and lac followed by other countries like Mexico and Brazil (FRA, 2005). Similarly, the share of different states in lac production in India is depicted in Figure 10.3 (IINRG, 2011). As per Namdeo and Pant (1994) tendu leaves are the sources of employment to nearly 4 million people annually in bidi manufacturing. Tribal belts of the country contribute 70% of NTFPs collections (Mitchell et al., 2003) and this sector alone contributes 55% in employment generation to rural poor's (Joshi, 2003). Therefore, NTFPs contributions of 20.1 to 34.1% and 26.5 to 55.5% in total household's income are reported in the case of Gujarat and West Bengal (Kant, 1997).

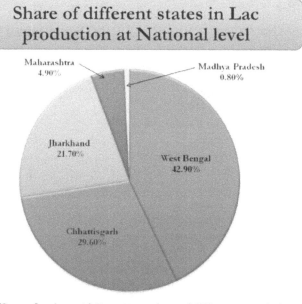

FIGURE 10.3   (See color insert.) Percentage share of different states in lac production in India (IINRG, 2011).

10.4 OPPORTUNITIES AND POTENTIALS

The potential of NTFPs sector are inevitable and plays a great role in creating opportunities for forest dwellers in term of employment. In India, a survey conducted by Planning Commission (2011) reveals up to 55% of household income for forest-dwelling communities comes mostly from food commodities (25–50%). Therefore, this sector provides 10 million workdays per annum basis where women play a significant role in NTFPs marketing and selling. The collection of NTFPs and their analysis is depicted in Figure 10.4 (Keystone Annual Report, 1998).

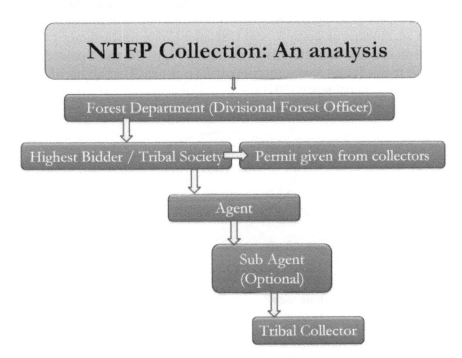

FIGURE 10.4 Collection of NTFPs and their analysis (Keystone Annual Report, 1998).

10.5 CONTRIBUTION OF NTFPS TO NATIONAL INCOME

Indian forest presents a rich diversity of NTFPs, and this makes a subtle contribution in term of forest revenues and of forest export income as 50% and 70% (Shiva and Mathur, 1996). Moreover, collection, processing, and sale of these diverse NTFPs are the potential source of both employment and income generation of poor's in India. Other figures such as the contribution of NTFPs in forest department revenues, net export from forest sector revenue and net export from forest produce are 40%, 75% and 75% in the year of 1986. Similarly, NTFPs based small scale enterprise contributes 50 % of income for 20–30% of the rural labor force in India and contributes 70% of the total wage employment in forestry sector (Bag et al., 2010). In India, NTFPS plays a vital role for 50 million people (Hedge et al., 1996) contribute 50% in forest revenue and 70% of export in forest-based products (ICCF, 2005). Figure 10.5 represents the export-import potential of some medicinal plants in India (Ramawat and Goyal, 2008). Although, India represents the second rank in export value up to 36,750 tons of worth 57,400 US$ after China which shows 1st rank in medicinal and aromatic plants (Lange, 2002). The trade potential of NTFPs in South East Asia comprising India is depicted in Figures 10.6 and 10.7 (NTFP annual report, 2011).

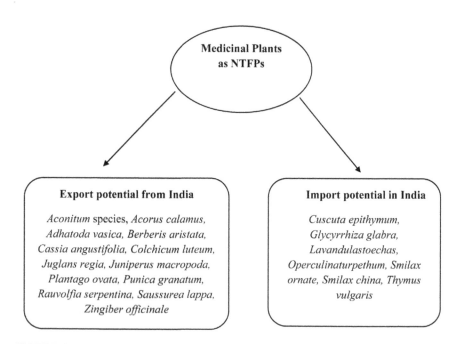

FIGURE 10.5   Export and Import potential of some medicinal plants in India (Ramawat and Goyal, 2008).

SOUTH EAST ASIA - NTFPs TRADE

COUNTRY	NUMBER OF ENTERPRISES	NUMBER OF PERSONS INVOLVED	NUMBER OF VILLAGES	SALES
India	27	2,232	370	€1,81,596
Indonesia	29	1,452	58	€ 99,838
Philippines	64	1,946	68	€ 88,417.8
Cambodia	38	1,649	81	€ 38,344.2
Total	163	7,400	600	€4,08,196

FIGURE 10.6   NTFPs trade in South East Asia (NTFPs Annual report, 2011).

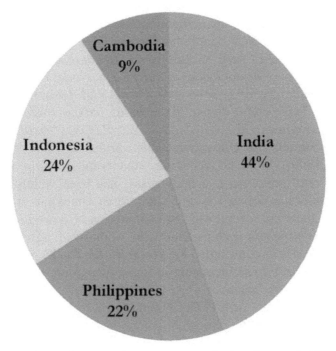

FIGURE 10.7   (See color insert.) Percentage contribution of country in NTFPs trade in South East Asia (NTFPs Annual report, 2011).

10.6   ROLE OF AGROFORESTRY IN NTFPs PRODUCTION

Agroforestry is a practice which incorporates tree plantation in the agroecosystem as well as in the rangeland. This, therefore, makes the entire process more eco-friendly, diversified nature, socio-economically positive for the community stakeholders and most important more practically acceptable at the grass root level (Leakey, 1996). Multifaceted of benefits from agroforestry relies on the protection of soil from an erosive factor of water and wind, maintaining ecological fertility and simultaneously production of wood and non-wood products for economic gain (Cooper et al., 1996). Therefore, agroforestry can produce NTFPs in a sustainable way which depends upon type and nature of management practices. NTFPs form the basis for maintaining the daily livelihood of forest dwellers and therefore are very much essential for the existence of them. Forest dwellers are benefited multi-dimensionally in terms of food, fodder and other NTFPs products. Under such circumstances tree species having multiple uses can provide multifaceted of product which would give economic and environmental benefits (ICRAF, 1993 & 1997). Under tropical conditions farmers give top priority to some specific fruit species such as *Adansonia digitata, Azanza garckeana, Bactris gasipaes, Chrysophyllum albidum, Dacryodes edulis, Garcinia kola, Inga dulis, Irvingia gabonensis, Parkia biglobosa, Ricinodendron heudelottii, Sclerocarya birrea, Tamarindus indica, Uapaca kirkiana, Vangueria infausta, Vitellaria paradoxa, Ziziphus mauritiana*, etc. which are suitable for cultivation under various agroforestry schemes (Leakey, 1999). From an economic standpoint *Irvingia* seed (cure for cholesterol reduction and body mass) has been widely grown by farmers under agroforestry system (Ike, 2008; Omokhua et al., 2012). Therefore, there are several ways of exploitation of *Irvingia gabonensis* and *Irvingia wombolu* such as from wild, homestead, agroforestry and *Irvingia* plantations contributed 89% (major), 85.7%, 83.5% and 39.6% (Chah et al., 2014). The increment in forest plantation, kitchen gardens under agroforestry practices varies significantly in terms of growth and reproduction rate of NTFPs in comparison to mismanaged natural forest ecosystem due to ecological competition (Ticktin et al., 2003), photoperiod (Velasquez-Runk, 1998) and varied environmental factors (Martinez-Balleste et al., 2002).

10.7   CLIMATE CHANGE AND NtFPs PRODUCTION

In the present context, climate change has its disastrous impact over forest ecosystem throughout the world and therefore throws a bigger challenge for human civilization (Parmesan and Yohe, 2003; Kirilenko and Sedjo, 2007). Fewer reports are available on issues such as climate change modification on NTFPs productivity. Changing the pattern of rainfall and temperature results in strong impacts on diversity pattern of important NTFPs which can affect the outcome edible and non-edible materials and their quality. Moreover, climate change results shift in vegetation types, phenology, reproductive biology of forest trees and shrubs and medicinally important herbs. This impact is not confined to only forest ecosystem but also can alter the fires duration, drought occurrence, insect and pathogens outbreaks (Dale et al., 2001, Ravindranath et al., 2006, Mukhopadhyay, 2009). For *Madhuca indica* plant the phenological phenomenon changes its time frame from mid-March to mid-February, therefore, altering an ecological condition of forests (Sushant, 2013). Similar reports were recorded by Bhattacharya and Prasad (2009) in terms of 90% reduction in NTFPs production under changing climatic conditions in Wayanad district, Kerala (WWF, 2011). Therefore, climate change reduced the availability of some NTFPs such as food, fodder, fuel and medicinal herbs that affect income and healthcare (Basu, 2009). However, NTFPs are considered the best strategy by reducing the emission of greenhouse gases (GHGs) through carbon absorption capacity of world forest (Huston and Marland, 2002; Namayanga, 2002; Gorte, 2009).

10.8   ECOLOGICAL IMPACT OF HARVESTING OF NtFPs

Over and unsustainable harvesting of NTFPs results in certain impacts on ecosystems at multiple scales. From an ecological point of view, the term sustainability for NTFPs requires not only long term persistence of their populations but also the practices of harvesting should not affect negatively to other community and ecosystem structure and functions. Although, harvesting practices of NTFPs cause direct impacts on growth, survival, and reproduction of individuals. This impact may be negative as decrease reproductive output in many palms and ferns due to removal of leaves which is a good source of photosynthetic material and nutrients (Milton, 1987;

Endress et al., 2006) and may be positive, i.e. foliage removal promotes growth traits on short term basis because of stored resources reallocation (Endress et al., 2006). Similarly, the harvesting practices of bark, gum, and resin can affect morphology, physiology, phenology, and survival rate of species due to wound appearance. This can be justified by Rijkers et al. (2006), and as per him heavy harvesting of *Boswellia papyrifera* trees for frankincense resin in Ethiopia caused three times less healthy and filled seeds than unharvested trees. Moreover, harvesting of NTFP species may alter the interaction between some organism and NTFP species. This can be justified by Moegenburg and Levey (2002), and as per him the diversity of avian frugivore was reduced by 22% due to practices of overharvesting of acai palm fruit (*Euterpe oleracea*) in the Brazilian Amazon. Moreover, species richness, structure, and plant composition were altered due to high-intensity harvesting practices of NTFPs in dry deciduous forests of India (Murali et al., 1996; Ganeshaiah et al., 1998). Another impact was seen as an enhancement in the cover of invasive species through overharvesting of NTFPs, and this can be mitigating by some good and reliable harvesting technology (Cunningham, 1993; Ticktin et al., 2006).

10.9   CONSERVATION AND MANAGEMENT IMPLICATIONS

The importance, scope and potential utilization of NTFPs well known and cannot be denied. Hall and Bawa (1993) reported that sustainable harvesting practice should be taken care of so that ecosystem resilience process should not be hampered. Moreover, sustainable harvesting can enhance the well being of peoples, maintaining ecosystem function and conserving the biodiversity (Uma Shankar et al., 1996). Therefore, natural distribution, species richness, abundance, population structure, and dynamics are the important factors which can help in assessing the sustainability of NTFPs harvesting (Hall and Bawa, 1996). As per Karki (2001) conservation and proper management of NTFPs should be based on certain aspects of ecology, economics, social, legal, technical, and political. The ecological aspects comprise the type and nature of species that managed for overall production while appropriates methods/ technology of harvesting and processing comes under a technical aspect. Similarly, the social aspect based on cultures, belief, attitude and behavior of peoples.

10.10   CONCLUSION

There is a strong need to recommend for implementation of an appropriate action plan with development of viable strategy behind the management and conservation of NTFPs and their sustainable utilization without damaging their habitat. Moreover, successfully implanted policies should be customized at ecological, political and socio-economic level. Likewise, the inclusion of NTFPs in national strategy and planning of good forest governance would help in the betterment of non-timber forest resources. Additionally, it is also recommended for continuum research of NTFPs in the direction of sustainable forest management. In this direction community participation and their regular monitoring and adopting suitable technology of harvesting and processing can enhance the life of NTFPs.

The scope and potential of NTFPs are well known, and these forest products are sustaining the life of millions of rural peoples and forest dwellers in term of providing both tangible (food, fodder, fruits and other products) and intangible benefits through maintaining biodiversity. But nowadays, the quality, quantity, and availability of NTFPs are affecting due to climate change. Similarly, the unscientific method of harvesting, lack of sufficient tools, mismanagement, etc. accelerate the availability and productive potential of forest species producing non-timber products. Therefore, there is an urgent need to adopt good management practices for harvesting and conservation techniques for full utilization of NTFPs in a sustainable way. A government institution, forest department, NGOs, R&D, and policymakers must take a good decision in this direction for maintaining the status of NTFPs which is a viable option of livelihood support for forest dwellers.

KEYWORDS

- bark
- bee-wax
- dry deciduous forests
- dye
- employment generation
- fiber
- fodder
- food
- forest dwellers
- forest policy
- fruits
- fuelwood

- **gums**
- **honey**
- **intangible benefits**
- **management practices**
- **MPTs**

- **natural forest**
- **natural resources**
- **research and development**
- **resin**
- **tangible benefits**

REFERENCES

1. Appiah, M. B., Blay, D., Lawrence, E., Damnyag, L., Dwomoh, F. K., Pappinen, A., & Luukkanen, O. (2009). Dependence on forest resources and tropical deforestation in Ghana. *Environment, Development, and Sustainability, 11,* 471–487.
2. Bag, H., Ojha, N., & Rath, B. (2010). NTFP Policy Regime after FRA: A study in select states of India. Regional Centre for Development Cooperation, Bhubaneswar, pp. 214.
3. Basu, J. P. (2009). Adaptation, non-timber forest products, and rural livelihood: an empirical study in West Bengal, India. *Earth and Environmental Science, 6,* 3–8.
4. Belcher, B. M. (2003). What isn't an NTFP? *International Forestry Review, 5,* 161–168.
5. Bhattacharya, P., & Prasad, R. (2009). Initial Observation on Impact of Changing Climate on NTFP Resources and Livelihood Opportunities in Sheopur District of Madhya Pradesh (Central India). A Paper presented at XIII World Forestry Congress, Buenos Aires. Argentina, October 18–23.
6. Bouri, T., & Mukherjee, A. (2013). Documentation of traditional knowledge and Indigenous use of non- timber forest products in Durgapur forest range of Burdwan district, West Bengal. Paper presented in National Seminar on Ecology, Environment & Development 25–27 January 2013, organized by Department of Environmental Sciences, Sambalpur University, Sambalpur.
7. Census of India (2011). http://www.censusindia.gov.in/2011-Common/CensusData 2011.html
8. Chah, J. M., Ani, N. A., Irohibe, J. I., & Agwu, A. E. (2014). Exploitation of bush mango (*Irvingia wombolu* and *Irvingia gabonensis)* among rural household in Enugu state, Nigeria. *Journal of Agricultural Extension, 18*(2), 44–56.
9. Chamberlain, J. L., Bush, R. J., & Hammett, A. L. (1998). Non-timber forest products: the other forest products. *Forest Product Journal, 48*(10), 2–12.
10. Cooper, P. J. M., Leakey, R. R. B., Rao, M. R., & Reynolds, L. (1996). Agroforestry and the mitigation of land degradation in the humid and sub-humid tropics of Africa. *Experimental Agriculture, 32,* 235–290.
11. Cunningham, A. B. (1993). African medicinal plants. Setting priorities at the interface between conservation and primary healthcare. People and Plants Working Paper 1. UNESCO, Paris
12. Dale, V. H. (2001). Climate change and forest disturbances. *BioScience, 51,* 723–734.

13. Das, I., Katiyar, P. and, Raj, A. (2014). Effects of temperature and relative humidity on ethephon induced gum exudation in *Acacia nilotica. Asian Journal of Multidisciplinary Studies, 2*(10), 114–116.
14. Dovie, B. K., Witkowski, E. T. F., & Shackleton, C. M. (2005). Monetary valuation of livelihoods indicator for understanding the composition and complexity of rural households. *Agriculture and Human Values, 22,* 87–103.
15. Endress, B. A., Gorchov, D. L., & Berry, E. J. (2006). Sustainability of a non-timber forest product: effects of alternative leaf harvest practices over 6 years on yield and demography of the palm Chamaedorea radicalis. *For Ecol Manage, 234,* 181–191.
16. FAO (2001). How Forests Can Reduce Poverty. FAO, Rome.
17. FAO (2003). Forestry outlook study for Africa: Regional report: Opportunities and challenges towards 2020. FAO Forestry Paper 141. FAO, Rome.
18. FAO (2005). Accessed from http://www.fao.org/documents/en/detail/200714 on 04.03.2013.
19. FAO (2002). Non-wood Forest Products in 15 countries of Tropical Asia: An overview, P. Vantomme, A. Markkula & R. N. Leslie, (eds). Bangkok (available at www.fao.org).
20. FAO (2008). Non-Wood Forest Products, Rome, Italy,
21. FRA (2005). Downloaded from http://www.fao.org/forestry/fra/fra2010/en/
22. Ganeshaiah, K. N., Shaanker, R. U., Murali, K. S., Shankar, U., Bawa, K. S. (1998). Extraction of non-timber forest products in the forest of Bilingiri Rangan Hills, India. 5. Influence of dispersal mode on species response to anthropogenic pressures. *Econ Bot, 52,* 316–319.
23. Gary, L., & Kristin, K. (2005). Creating an Indicator for Non-Timber Forest Products.
24. Gorte, R. W. (2009). Carbon Sequestration in Forests. Congressional Research Service, USA.
25. Hall, P., & Bawa, K. S. (1993). Methods to assess the impact of extraction of non-timber forest products on plant populations. *Economic Botany, 47,* 234–247.
26. Hegde, R., Suryaprakash, S., Achoth, L., & Bawa, K. S. (1996). Extraction of non-timber forest products in the forests of Biligiri Rangan Hills, India: 1. Contribution to rural income. *Economic Botany, 50,* 243–51.
27. Hogarth, N. J., Belcher, B., Campbell, B., & Stacey, N. (2013). The role of forest-related income in household economies and rural livelihoods in the border region of southern China. *World Development, 43,* 111–123.
28. Huston, M. A., & Marland, G. (2002). Carbon management and Biodiversity. *Journal of Environmental Management, 67,* 77–86.
29. ICCF (2005). Stakeholder Organization in the NWFP Sector: Need for a sustainable business model. In Proceeding of National Expert Consultation on NWFP Business Model 25–26 July 2005.
30. ICRAF (1993). *Annual Report 1992.* International Centre for Research in Agroforestry, Nairobi.
31. ICRAF (1997). *ICRAF Medium-Term Plan 1998–2000.* International Centre for Research in Agroforestry, Nairobi, 73p.
32. IINRG (2011). Lac, Plant Resins and Gums Statistics at a Glance, Indian Council of Agricultural Research. Pp. 34.

33. Ike, P. C. (2008). Exploratory survey of the production of non-timber forest products for sustainability of livelihoods: the case of *Irvingia* species in Nsukka agricultural zone of Enugu state, Nigeria. *Extension Farming Systems Journal, 4,* 1.

34. Jhariya, M. K., & Raj, A. (2014). Human welfare from biodiversity. *Agrobios Newsletter, 12*(9), 89–91.

35. Joshi, S. (2003). Supermarket, secretive. Exploitative, is the market in the minor forest produce unmanageable? *Down to Earth, 28,* 27–34.

36. Kamanga, P., Vedeld, P., & Sjaastad, E. (2009). Forest incomes and rural livelihoods in Chiradzulu District, Malawi. *Ecological Economics, 68,* 613–624.

37. Kant, S. (1997). Integration of biodiversity conservation in tropical forest and economic development of local communities. *Journal of Sustainable Forestry, 4*(1/2), 33–61.

38. Kar, S. P., & Jacobson, M. G. (2012). NTFP income contribution to household economy and related socio-economic factors: lessons from Bangladesh. *Forest Policy and Economics, 14,* 136–142.

39. Karki, M. (2001). Institutional & Socioeconomic Factors and Enabling Policies for Non-Timber Forest Products Based Development in Northeast India. Paper presented in the Pre-identification Workshop for NTFP-led Development in NE India, Report No.1145.

40. Keystone annual report (1998). *Keystone Team,* Kotagiri, Nilgiris. Pp. 27.

41. Kilchling, P., Hansmann, R., & Seeland, K. (2009). Demand for non-timber forest products: surveys of urban consumers and sellers in Switzerland. *For Policy Econ, 11,* 294–300.

42. Killman, W. (2003). "Non-wood News." NO.10, March 2003, p.1.

43. Kirilenko, A. P., & Sedjo, R. A. (2007). Climate change impacts on forestry. *Proceedings of the National Academy of Sciences, 104,* 19697–19702.

44. L'Roe, J., & Naughton-Treves, L. (2014). Effects of a policy-induced income shock on forest-dependent households in the Peruvian Amazon. *Ecological Economics, 97,* 1–9.

45. Lange, D. (2002). The role of east and southeast Europe in the medicinal and aromatic plants' trade. *Medicinal Plant Conservation, 8,* 14–18.

46. Leakey, R. R. B. (1996). Definition of agroforestry revisited. *Agroforestry Today,* 8 (1), 5–7.

47. Leakey, R. R. B. (1999). Potential for novel food products from agroforestry trees: a review. *Food Chemistry, 66,* 1–14.

48. Maithani, G. P. (1994). Management perspectives of Minor Forest Produce. MFP News, October–December, 1994. Dehradun.

49. Mamo, G., Sjaastad, E., & Vedeld, P. (2007). Economic dependence on forest resources: a case from Dendi District, Ethiopia. *Forest Policy and Economics, 9,* 916–927.

50. Martinez-Balleste, A., Caballero, J., Gama, V., Flores, S., & Martorell, C. (2002). Sustainability of the traditional management of Xa'an palms by the lowland Maya of Yucatán, México. In J.R. Stepp, F.S. Wyndham and R. Zarger (eds.) *Ethnobiology and Biocultural Diversity.* University of Georgia Press, Athens, pp. 381–388.

51. McLain, R. J., & Jones, E. T. (2005). Nontimber Forest Products Management on National Forests in the United States. General Technical Report PNW-GTR-655. Portland.

52. Milton, S. (1987). Effects of harvesting on four species of forest ferns in South Africa. *Biol. Conserv, 41,* 133–146.

53. Mitchell, C. P., Corbridge, S. E., Jewit, S. L., Mahapatra, A. K., & Kumar, S. (2003). Nontimber forest products: Availability, production, consumption, management, and marketing in Eastern India. Project Report (DFID RNRRS Programme for Forestry Project Reference No. R6916), Pp. 1–278.

54. Moegenburg, S. M., & Levey, D. J. (2002). Prospects for conserving biodiversity in Amazonian extractive reserves. *Ecol Lett, 5,* 320–324.

55. Morse, R. (2003). Preface. P. ii. Proceedings of Hidden forest values: The first Alaskaside non-timber forest products conference and tour, Alaska Boreal Forest Council, comps. USDA For. Serv., Pacific Northwest Res. Stn., Gen. Tech. Rep. GTR-PNW-579. 150 p.

56. Mukhopadhyay, D. (2009). Impact of climate change on forest ecosystem and forest fire in India. *Earth and Environmental Science, 6*(3), 20–27.

57. Murali, K. S., Shankar, U., Shaanker, R., Ganeshaiah, K. N., & Bawa, K. S. (1996). Extraction of non-timber forest products in the forests of Biligiri Rangan Hills, India. 2. Impact of NTFP extraction on regeneration, population structure, and species composition. *Econ Bot, 50,* 252–269.

58. Mutamba, M. (2013). Rural livelihoods, forest products, and poverty alleviation: the role of markets. PhD thesis, Rhodes University, Grahamstown.

59. Namayanga, L. N. (2002). Estimating Terrestrial Carbon Sequestered In Aboveground Woody Biomass from Remotely Sensed Data. Serowe, Botswana.

60. Namdeo, R. K., & Pant, N. C. (1994). Role of minor forest products in tribal economy. *Journal of Tropical Forestry, 10*(1), 36–44.

61. NTFPs Annual Report (2011). Exchange programme for South and South East Asia. pp. 32. https://ntfp.org/wp-content/uploads/2016/07/NTFP-EP-2011-Annual-Report.pdf

62. Omokhua, G. E., Ukoima, H. N., & Aiyeloja, A. A. (2012). Fruits and seeds production of *Irvingia gabonensis* (O' Rorke) and its economic importance in Edo Central, Nigeria. *Journal of Agriculture and Social Research* (JASR), *12*(1), 149–155.

63. Parmesan, C., & Yohe, G. (2003). A globally coherent fingerprint of climate change impacts across natural systems. *Nature, 421,* 37–42.

64. Planning Commission (2011). Report of Sub-Group II on NTFP and their Sustainable Management in the 12th 5-Year Plan. Planning Commission's Working Group on Forests & Natural Resource Management. 2 pp.

65. Pyakurel, D., & Baniya, A. (2011). NTFPs: Impetus for Conservation and Livelihood Support in Nepal–A Reference Book on Ecology, Conservation, Product Development and Economic Analysis of Selected NTFPs of Langtang Area in the Sacred Himalayan Landscape, WWF, Nepal.

66. Raj, A. (2015a). *Evaluation of Gummosis Potential Using Various Concentration of Ethephon.* M.Sc. Thesis, I.G.K.V., Raipur (C.G.), pp. 89.

67. Raj, A. (2015b). Gum exudation in Acacia nilotica: effects of temperature and relative humidity. In Proceedings of the National Expo on Assemblage of Innovative ideas/ work of postgraduate agricultural research scholars, Agricultural College and Research Institute, Madurai (Tamil Nadu), pp. 151.

68. Raj, A., & Singh, L. (2017). Effects of girth class, injury and seasons on Ethephon induced gum exudation in *Acacia nilotica* in Chhattisgarh. *Indian Journal of Agroforestry, 19*(1), 36–41.

69. Raj, A., Haokip, V., & Chandrawanshi, S. (2015). *Acacia nilotica:* a multipurpose tree and source of Indian gum Arabic. *South Indian Journal of Biological Sciences, 1*(2), 66–69.

70. Rajesh, R. (2006). Analytical Review of the Definitions of Non-Timber Forest Product. Institute of Forestry Pokhar.

71. Ramawat, K. G., & Goyal, S. (2008). The Indian Herbal Drugs Scenario in Global Perspectives. In Book: *Bioactive Molecules and Medicinal Plants;* Ramawat, K. G., Merillon, J. M. (eds.) Springer, pp. 323–345.

72. Ravindranath, N. H, Joshi, N. V., Sukumar, R., & Saxena, A. (2006). Impact of climate change on forest in India. *Curr Sci, 90*(3), 354–361.

73. Rijkers, T., Ogbazghi, W., Wessel, M., & Bongers, F. (2006). The effect of tapping for frankincense on sexual reproduction in *Boswellia papyrifera. J Appl Ecol, 43,* 1188–1195.

74. Schaafsma, M., Morse-Jones, S., Posen, P., Swetnam, R. D., Balmford, A., Bateman, I. J., Burgess, N., Chamshama, S. A., Fisher, B., Freeman, T., Geoffrey, V., Green, R., Hepelwa, A. S., Hernández Sirvent, A., Hess, S., Kajembe, G. C., Kayharara, G., Kilonzo, M., Kulindwa, K., Lund, J. F., Madoffe, S. S., Mbwambo, L., Meilby, H., Ngaga, Y. M., Theilade, I., Treue, T., van Beukering, P., Vyamana, V. G., & Turner, R. K. (2011). The importance of local forest benefits: valuation of non-timber forest products in the Eastern Arc Mountains in Tanzania. *Global Environmental Change, 24,* 295–305.

75. Scherr, S. J., White, A., & Kaimowitz, D. (2003) A new agenda for forest conservation and poverty reduction. Making markets work for low-income producers. Forest Trends and Centre for International Forestry Research, Forest Trends, Washington DC.

76. Shackleton, C., & Shackleton, S. (2004). The importance of non-timber forest products in rural livelihood security and as safety nets: a review of evidence from South Africa. *South African Journal of Science, 100,* 658–664.

77. Shiva, M. P., & Mathur, R. B. (1996). Management of Minor Forest Produce for Sustainability, Oxford and IBH Publishing Co. Pvt. Ltd., New Delhi.

78. Sushant (2013). Impact of Climate Change in Eastern Madhya Pradesh, India. *Tropical Conservation Science-Special, 6*(3), 338–364.

79. Thandani, R. (2001). International non-timber forest product issues. In M.R. Emery and R.J. McLain (eds.) *Non-timber Forest Products: Medicinal Herbs, Fungi, Edible Fruits and Nuts, and Other Natural Products from the Forest.* Food Products Press, New York.

80. Thondhlana, G., Vedeld, P., & Shackleton, S. E. (2012). Natural resource use, income, and dependence among San and Mier communities bordering Kgalagadi Transfrontier Park, southern Kalahari, South Africa. *International Journal of Sustainable Development and World Ecology, 19,* 460–470.

81. Ticktin, T., Johns, T., & Chapol Xoca, V. (2003). Patterns of growth in *Aechmea magdalenae* and its potential as a forest crop and conservation strategy. *Agriculture, Ecosystems and Environment, 94,* 123–139.

82. Ticktin, T., Whitehead, A. N., & Fraiola, H. (2006). Traditional gathering of native hula plants in alien invaded Hawaiian forests: adaptive practices, impacts on alien invasive species, and conservation implications. *Environ Conserv, 33,* 185–194.

83. Uma Shankar, Murali, K. S., Urea Shaanker, R., Ganeshaiah, K. N., & Bawa, K. S. (1996). Extraction of non-timber forest products in the forests of Biligiri Rangan

Hills, India. 2. Impact of NTFP extraction on regeneration, population structure, and species composition. *Economic Botany, 50*(3), 252–269.

84. Van Gevelt, T. (2013). The economic contribution of non-timber forest products to South Korean mountain villager livelihoods. *Forests, Trees and Livelihoods, 22,* 156–169.

85. Velasquez-Runk, J. (1998). Productivity and sustainability of a vegetable ivory palm (*Phytelephas aequatorialis*, Araeceae) under three management regimes in northwestern Ecuador. *Economic Botany, 52,* 168–182.

86. WWF (2011). Climate Change in India: A Case Study of Orissa. http://zeenews.india.com/myearth2011/orissa.aspx

CHAPTER 11

Mulberry-Based Agroforestry System: An Effective Way of Maintaining Livelihood Security and Climate Change Mitigation

MUSHTAQ RASOOL MIR,[1] IRFAN LATIF KHAN,[1] M. F. BAQUAL,[1] MUNEESA BANDAY,[2] RAMEEZ RAJA,[3] and SAIMA KHURSHEED[4]

[1]*Associate Professor, Temperate Sericulture Research Institute, Mirgund, SKUAST-K, India, Mobile: +919596024573; E-mail: drrasool@rediffmail.com*

[2]*Junior Research Fellow, Temperate Sericulture Research Institute, Mirgund, SKUAST-K, India*

[3]*Project Assistant, Temperate Sericulture Research Institute, Mirgund, SKUAST-K, India*

[4]*PhD Scholar, Temperate Sericulture Research Institute, Mirgund, SKUAST-K, India*

ABSTRACT

Mulberry (*Morus spp.*) species happens to be the only food items for the rearing of silkworm (*Bombyx mori*) which are cultivated in the form of tree, bush and dwarf condition. Mulberry tree because of its higher canopy and wider planting space provides options to use this plant for agroforestry systems wherein it can be grown with vegetables and other short duration crops. Research report reveals that during Rabi and Kharif season the combined cultivation of mulberry with vegetable crops does not reflect any adverse result in terms of productivity. The farmer has been in a position to get around the year employment and cash flow by way of silkworm cocoons and vegetables. This, in a long way, can go towards enhanced cocoon production, effective resource utilization and economic upliftment of rural folk. The remunerative nature of this agrisilvicultural system will motivate more and more people to go for mulberry plantation which will help in climate change mitigation; since mulberry is a high biomass production tree species utilizing large quantities of CO_2 from the atmosphere and its leaf has quick decomposition after its fall, helping in rapid recycling of carbon back to the soil.

11.1 INTRODUCTION

The major aim towards mulberry cultivation (*Morus spp*) to be used as rearing material for silkworm species (*Bombyx mori L*). It grows in different climatic conditions ranging from tropical to temperate and is unique in the expression of genetic characters under different environmental conditions. Mulberry plantation has wide ecological amplitude to be grown under varied agroclimatic condition ranging from temperate to the tropical climate. This may probably be due to the rich diversity found in this genus wherein different varieties are bestowed with features and characteristics that enable them to grow in a wide range of environments. Mulberry can be trained as bush, dwarf or tree and all these types can be maintained with different cultural practices.

The systematic position of mulberry is as under:

Division:	*Spermatophyta*
Class:	*Angiospermae*
Subclass:	*Dicotyledoneae*
Order:	*Urticales*
Family:	*Moraceae*
Genus:	*Morus*

In the genus *Morus,* there is a wide genetic base with various species coming from a wide climatic range and having the potential to thrive well in different agro-climatic conditions, and the range could even further be extended through further trials and selective breeding. Mulberry is known for being a rugged survivor and is rumored to have been used to push back deserts in China (Jamison, 2014). This is the kind of plant which can be used in agroforestry programmes aimed at fighting the looming issues of malnutrition, deforestation and climate change which are intimately inter-related. It is a miraculously versatile plant which can fulfill the 5 Fs – fruit, foliage, fodder, fuel, and fencing material.

It is a sun-loving, perennial, woody, deep-rooted plant, either monoecious or dioecious. Mulberry plant after four to five years of inception continuously shows full yielding potential up to 17 years without a reduction in leaf yield (Kumereson et al., 1994). However, under temperate conditions, mulberry grows luxuriantly for 50–60 years. The productivity and profitability of sericulture depend mainly on the maximization of leaf yield per unit area at an economically viable cost.

The leaf of mulberry is rich in the nutrients like proteins, sugar, and minerals; Protein molecules are the building block for silk material biosynthesis. In addition to this, the leaf contains some other factors that include attracting factors that attracts the silkworm to the leaf; biting factors which motivates the silkworm to bite and chew the leaf and the swallowing factors which make it to swallow the chewed food. The nutritive composition of mulberry leaf seems to be the secret behind the fact that silkworm (*Bombyx mori* L.) is highly selective and does not eat anything other than the mulberry leaf. The Proverb that "all mulberries will not yield cocoons" is true in the sense that mulberry leaves of bad quality will damage the silkworm crop. Since silkworm is monophagous and depends solely upon mulberry for nutrition, the quality of mulberry has a direct influence on the health and growth of the silkworm.

Mulberry grows in all types of soils and thrives well in slightly acidic soils with pH ranging from 6.2–7.0. The most suitable range of temperature for ideal growth of mulberry is 18–30°C. In the whole of North Indian states representing both temperate and sub-tropical conditions, mulberry grows as scattered trees on roadsides, river bunds, boundaries of kitchen gardens, farms, parks, and school grounds.

Mulberry leaf has its importance significantly in sericulture due to both qualitative and quantitative role in relation to cocoon harvesting. Amongst many other factors, the feed of silkworm, i.e. the mulberry leaf has been proved to contribute 38.2 percent to the success of silkworm rearing, followed by climate (37%), rearing (9.3%), silkworm (7.3%) and others (8.2%) (Miyashita, 1986; Figure 11.1).

Sericulture has become an important figure in the agro-forestry sector under Indian subcontinent as it has been adopted by most of the states. The area under mulberry cultivation is 2,82,244 hectare (Ravindran et al., 1988).

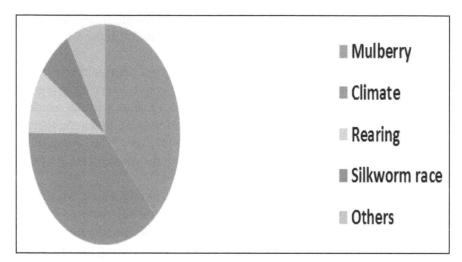

FIGURE 11.1 **(See color insert.)** Regulating factors for efficient silkworm rearing.

As per Vijayan (2011) 68 species have been reported in the genus *Morus* spp. With most of these species occurring in Asia. In China, four species, *Morus alba*, *M. atropurpurea*, *M. multicaulis*, and *M. mizuho* were cultivated among the available 15 species for the rearing of silkworm (Huo, 2000).

India horbors a rich collection of mulberry species (*Morus*, of which *Morus alba*, *M. indica*, *M. laevigata* and *M. serrata*, etc.) represented both in captivity and under wild condition Other species such as *M. multicaulis*, *M. nigra*, *M. sinensis*, and *M. Philippinensis* have been introduced in addition to indigenous plantation (Sastry, 1984).

Central Sericulture Germplasm Research Centre (GSGRC), Hosur, Tamil Nadu has been actively engaged in collecting and conserving the mulberry germplasm and has a collection of 842 mulberry accessions (Sinha, 2001). Some of the popular region specific mulberry varieties are listed in Table 11.1.

Sericulture acts as subside for more than 25,000 rural people of the state. These rural communities mostly belong to economically poor sections of the society Silkworm rearers earns Rs. 2026.00 Lac by producing 1022 MT of cocoons per annum leading to the creation of employment opportunities of 3.5 Lac man-days (on-farm income of Rs. 3.0 Lac and 0.50 Lac 11 off-farm) (Anonymous, 2015). During the 12th plan, there is great thrust on increasing bivoltine cocoon production in the country and the state of Jammu and Kashmir because of its ideal conditions holds promise. Though an easy, less time consuming but a good revenue generating activity, the

TABLE 11.1 Varieties of Mulberry in India (modified from Ravindran et al., 1997)

Variety/Race	Region/area	Center	Source/Method
Kanva-2	Irrigated region of South India	CSRTI, Mysore	Natural selection
S-36	Irrigated region of South India	CSRTI, Mysore	Berhampore local through EMS treatment
S-54	Irrigated region of South India	CSRTI, Mysore	Berhampore local through EMS treatment
Victoria-1	Irrigated region of South India	CSRTI, Mysore	Hybridization between S30 and Berc 776
DD	Irrigated region of South India	KSSRDI, Thalaghattapura	Selection of clone
S-13	Rain dependent region of South India	CSRTI, Mysore	Polycross progeny selection
S-34	Rain dependent region of South India	CSRTI, Mysore	Polycross progeny selection
MR-2	Rain dependent region of South India	CSRTI, Mysore	Open pollinated hybrids selection
S-1	Irrigated region of Eastern and NE India	CSRTI, Berhampore	Domesticated from Myanmar (Mandalaya)
S-7999	Irrigated region of Eastern and NE India	CSRTI, Berhampore	Open pollinated hybrids selection
S-1635	Irrigated region of Eastern and NE India	CSRTI, Berhampore	Selection of triploid progeny
S-146	Irrigated valley region of J&K. and North India	CSRTI, Berhampore	Open pollinated hybrids selection
Tr-10	Valley of East India	CSRTI, Berhampore	Ber. S1Triploid progeny
BC-259	Valley of East India	CSRTI, Berhampore	Hybrid back-crossing (Matigare local and Kosen then with Kosen twice)
Goshoerami	Cold climatic belt	CSRTI, Pampore	Domesticated from Japan.
Chak Majra	Moderately cold climatic belt	RSRS, Jammu	Natural selection
China White	Cold region	CSRTI, Pampore	Selection of clone

number of farmers does not increase to the extent it ought to because of many factors amongst them one being the non-availability of the quality leaf with the farmers which is due to:

1. Less land with farmers, which forces them to go for cultivation of more important crops like food crops, vegetables, etc.
2. Short growing season because of dormancy of most of the plant species during the prolonged winter season.
3. Only one cocoon crop at farmer's level which hence makes the farmers reluctant to devote their land exclusively for mulberry cultivation.
4. Low and inferior varieties in the field.
5. Least attention for mulberry plantation in terms of inputs etc.

11.2   AGROFORESTRY

Agroforestry is the integrated scheme of cultivating tree along with agricultural crops within a specific spatial and temporal regime. Nair (1979) has described agroforestry as an integrated unit of trees, crops, and animals in an environment-friendly way which widely accepted in the farming community. To put it in simpler terms in an agroforestry system:

1. There must be a significant interaction (positive/negative) between the woody and the nonwoody components of the system either ecological and/or economical.
2. This is a production system harmonizing the production of various components and maximization of the total production per unit area of land.

Agroforestry is advantageous as it leads to:

1. Reduction in the insect/mite pest and pathogen population as it provides economically less important alternate host like grasses, etc., to the pests.
2. Reduces hillside erosion and protects topsoil, due to less surface runoff and deep root system.
3. Diversification of farm produce.
4. Economically more viable than monoculture.

5. Insurance against total crop failure under aberrant conditions or pest/disease epidemics.

11.3 AGROFORESTRY AND CLIMATE CHANGE

In December 2015, world leaders from 195 countries met in Paris and agreed on a landmark plan to address global climate change by holding countries accountable for their commitments to carbon emissions reductions and other actions. The agreement calls for all nations to pledge to limits on carbon (C) emissions necessary to keep average global warming less than 2 degrees Celsius above pre-industrial levels. This historic agreement recognizes the importance of "carbon emission reduction from the removal of forest cover and degeneration of forest area, and functional approach for sustainable forestry and increase of carbon stock in the forest." The success of the plan depends on the ability of individual countries to simultaneously reduce emissions and capture carbon from the atmosphere and keep it out of the atmosphere (Anonymous, 2015).

Carbon sequestration (CS) is the long-term storage of carbon in oceans, soils, vegetation, and geologic formations. Storage of carbon in vegetation occurs through uptake of atmospheric carbon dioxide (CO_2) during photosynthesis and its conversion to plant biomass. Soil storage of carbon occurs when soil microbes convert dead plant matter into soil organic matter through decomposition. To meet the ambitious goals of the Paris Agreement, the world needs enhanced CS techniques that provide social, environmental, and economic benefits while reducing atmospheric CO_2 concentrations. Agroforestry systems can contribute to slowing down those increases and, thus, contribute to climate change mitigation. As per Lorenz and Lal (2014) agroforestry practices have been found effective in storing carbon. The areas under tree intercropping, multistrata systems, protective systems, silvopasture, and tree woodlots are estimated at 700, 100, 300, 450, and 50 Mha, respectively all over the world. It has been reported that around 2.2 Pg C can be sequestered in biomass in a half century. The SOC storage in agroforestry systems may amount up to 300 Mg C ha^{-1} to 1 m depth.

Agroforestry (AGF) practices have been approved as a strategy for soil CS by both afforestation and reforestation programs and under the Clean Development Mechanisms of the Kyoto Protocol. This is primarily due to the use of perennial vegetation in AGF practices. Perennial vegetation provides opportunities for above- and below-ground biomass production and potential for greater soil carbon creation. Perennial trees, shrubs, and grasses used in AGF are more efficient in CS than annual monoculture crops and pasture vegetation because they allocate a higher percentage of carbon to belowground biomass and often extend growing seasons. Higher accumulation of carbon through agroforestry practices have been reported in comparison to the presence of forests and pastures as Agroforestry acts as an integrated unit between forest and grassland biome.

Perennial vegetation in agroforestry has many advantages over agricultural monocultures not just in terms of storing carbon in the soil, but also in above and belowground biomass. Perennial trees, shrubs, and grasses have greater carbon storage capacity than annual monoculture crops because they allocate higher percentages of carbon to belowground biomass and often have longer growing seasons. Perennial warm-season grasses, used as forage in silvipasture systems and as biomass for harvest in alley cropping systems, can provide hydrological, water and soil quality, carbon sequestration, feed, and renewable energy benefits.

Extreme weather events such as severe droughts and frequent floods are becoming regular events in India. Increasing flood and drought tolerance of agricultural crops will be essential for adaptation to these new norms. Promising biomass crop species, for deployment in marginal lands along the floodplains, include trees such as cottonwood (*Populus spp.*) mulberry (*Morus spp.*) and willow (*Salix spp.*), grasses such as high biomass sorghum (*Sorghum bicolor* L.) or vegetable crops.

During the present era, numerous works related to carbon sequestration potential and carbon stocks of agroforestry systems have been reported globally. In the West African Sahel, improved agroforestry practices such as live fence and fodder banks sequestered more carbon than traditional parklands. In northwest India, under a popular-based agroforestry system, the soil organic carbon concentration and pools were higher in soils under agroforestry and increased with tree age. Tree density and plant-stand characteristics such as species richness and age of south Indian home gardens also affected soil carbon sequestration. These case studies add to the growing body of literature indicating that agroforestry systems have the potential to sequester more above and below-ground carbon compared to traditional farming systems (Anonymous, 2015).

For climate change mitigation agroforestry systems is a boon in terms of increasing the level of C stock in the aboveground and soil component. The average potential value of agroforestry seems to be 25 t C ha^{-1} over 96 m ha land area. Such a rate of C stock in agroforestry has the capability to store C up to 2400 mt from the Indian perspective. C sequestering potential varies regionally and is often dependent upon some regulatory factors such as the growth rate and nature of tree species (Tiwari et al., 2017).

The contribution of agroforestry in soil carbon sequestration varied between 0.003 to 3.98 Mg C/ha/yr. The total C sequestered in each component differs greatly depending on the region, types of species, system, site quality, and previous land-use. Present review work reveals the fact that besides sequestering carbon agroforestry also aids in fulfilling the target of establishing 33% forest cover in India as per the guidelines of National Forest Policy (Dhyani et al., 2016).

11.4 MULBERRY-BASED AGROFORESTRY FOR BETTER RETURNS

Mulberry-based agroforestry system if practiced seems to be a suitable module with the twin objectives of livelihood security and climate change mitigation. Growing mulberry as a companion crop with other agricultural and fodder crops will open doors for effective land utilization, round the year employment and revenue generation through sericulture and vegetable cultivation. Integrating mulberry and vegetable cultivation is a practical, easy, eco-friendly and efficient way of land utilization. This will make mulberry sericulture more remunerative and ultimately attract more and more farmers to take up this venture.

Mulberry (*Morus* spp) is a hardy plant which can be grown under varied agroclimatic conditions ranging from tropical to temperate climatic conditions (Figure 11.2). Mulberry plant reflects wider adaptability towards various climatic conditions and can be grown under rainfed and irrigated conditions. Given timely pruning, the plant remains mostly free from pests and diseases. It can be grown under various social forestry programmes on forestland, wasteland and other public lands. The plant can be trained in three different forms-bush, dwarf and tree but the latter proves to be better than the other two forms as for as its integration with other agricultural crops is concerned owing to wider spacing and more canopy height. Further mulberry has the edge over other tree species in the temperate region in that it has a very good regeneration capacity. Under the circumstances of tropical climatic condition, mulberry leaves can be harvested for, but the situation changes under the temperate climatic condition of Kashmir mulberry plants foliage during April to November. In spite of this short leaf bearing period, and unlike other tree species of the region, two flushes of a leaf can be harvested in a year from the same plant in this short span of just 5 months. A brief outline of the annual leaf bearing/leaf availability in mulberry under Kashmir conditions is given below:

- Mulberry in Kashmir remains dormant during the winter.
- The winter buds start sprouting from the 4th week of March onwards.
- The leaves increase in size gradually and are fed to silkworm during May–June, i.e., spring crop, which is the main crop in this region.
- This is followed by a pruning of the shoots right from the crown base during the 2nd week of June.
- The buds sprout again and shoots attain a very good height up to the end of September besides bearing luxuriant leaf.
- The leaf is either plucked again to feed the livestock animals, and silkworm (2nd Crop) or is left as such to undergo senescence and natural fall during November.
- The plants remain dormant and bereft of any leaf up to March next.

FIGURE 11.2 **(See color insert.)** Mulberry tree growing in Ladakh (Cold desert).

The aforesaid discussion reveals the limited growth scenario with stunted shoot/canopy structure and without leafy canopy for most of the time during a year (Table 11.2). This coupled with the capacity of the plant to grow as trees with a canopy height sufficient to allow sunlight to reach the ground makes this plant very suitable for its integration with other agricultural crops thereby proving its feasibility to be used in agrisilvicultural and agrisilvipastoral systems.

TABLE 11.2 Period During Which Mulberry is Devoid of the Leaf Under Kashmir Conditions

S. No	Duration	Time (months)	Reasons for non-availability of leaf
1.	January to 15th April	3.5	Winter Dormancy
2.	June 2nd week to July last	1.5	Harvesting of the leaf along with shoots for spring rearing (Maincrop)
3.	November–December	02	Leaf fall
Total		7.0	

Under the cold climatic condition in Kashmir mono-cultivation of mulberry is not fruitful for the local farmers in terms of economic gain. Under such circumstances integration of mulberry plantation with other foraging plants along with agricultural crops seems to be the best alternative for the farming community due to the multidimensional benefits of such a system. Benefits include efficient utilization of land area, improvement in the total leaf production and yield, improvement in the nutrient, light and water use efficiency of the mulberry crop and inherent capability to combat the problems of pest, disease, and growth of unwanted plants. Further mulberry plants act as shade plants for other crops of lower strata. Mulberry-based agroforestry system is a very efficient way to make sericulture a sustainable business for the global community. It will help the farmers to get work as well as the cash flow throughout the year besides meeting their daily requirement of vegetables which are becoming very expensive. This can go a long way towards environment amelioration and climate change mitigation.

11.5 FEASIBILITY OF MULBERRY-BASED AGROFORESTRY SYSTEM

In order to understand the feasibility of mulberry-based agroforestry system, a study was conducted at TSRI (Temperate Sericulture Research

Institute), Mirgund and later demonstrated to farmers and other field functionaries using the crop combinations as in Table 11.3.

TABLE 11.3   Crop Combinations in a Mulberry-Based Agroforestry System

Treatment	Tree used	Agricultural crop	
		Rabi season	Kharif season
T-1	Mulberry at 9X9 spacing	Onion	Rajmash
T-2	-----Do-------	Garlic	Rajmash
T-3	-----Do-------	Turnip	Rajmash
T-4	-----Do-------	Peas	Sag
T-5	-----Do-------	Sag	Rajmash
T-6	-----Do-------	Nil	Nil

Observations on various growth and yield parameters were recorded coinciding with the final stage of silkworm rearing both during spring and late summer/early autumn and are presented in Tables 11.4 and 11.5.

TABLE 11.4   Growth and Yield in Mulberry Under Different Crop Combinations During Spring

Treatment	No. of shoot-lets per plant	Total shoot length/ plant	Leaf yield/plant
T-1 (Onion+Rajmash)	59.21	35.56	4.30
T-2 (Garlic+Rajmash)	59.21	34.41	3.81
T-3 (Turnip+Rajmash)	59.98	35.51	3.97
T-4 (Peas+Sag)	60.42	34.66	3.79
T-5 ((Sag+Rajmash)	60.24	35.02	3.76
T-6 (Control)	60.21	34.81	3.79
F.test	NS	NS	NS

While going through the tables, it is clear that there was no adverse effect on the growth and production of the mulberry tree by growing vegetables both during the rabbi and the Kharif seasons (Figure 11.3). This seems that there has not been any competition between the companion crops as far as the nutrient utilization is concerned.

TABLE 11.5 Growth and Yield in Mulberry Under Different Crop Combinations During Autumn

Treatment	No. of branches per plant	Total shoot length/ plant	Leaf yield/ plant
T-1 (Onion+Rajmash)	18.79	26.07	3.92
T-2 (Garlic+Rajmash)	19.25	28.15	4.19
T-3 (Turnip+Rajmash)	18.50	26.82	3.79
T-4 (Peas+Sag)	18.33	26.14	3.60
T-5 ((Sag+Rajmash)	18.29	26.30	4.28
T-6 (Control)	19.01	25.71	3.65
F.test	NS	NS	NS

FIGURE 11.3   (See color insert.) A woman rearer harvesting vegetables are grown under mulberry trees in Kashmir.

The perusal of Table 11.6 indicates that a farmer can earn more than 50,000/ha/annum by following the module. However, the foliage production increases after the trees are fully established and they remain productive even after 50 years. Mulberry leaves can be used to feed the silkworm and produce cocoons twice a year which is very economical for the farming community.

In one hectare of land, approximately 1500 plants are possible at a spacing of 9 x 8 feet spacing. A total of 18,600 kg leaf can be harvested during Spring @ 12 kg per tree and 22,500 kg during late summer/early autumn @15 kg per tree as the leaf production is more during the 2nd crop as compared to the 1st crop.

TABLE 11.6 Income Through Mulberry-Based Agroforestry System

Treatment	Rabi yield (Kg)	Kharif yield (Kg)	Income by sale of		Total Income (Rs)	Production cost (Rs)	Net income (Rs)
			Rabi crop (Rs)	Kharif crop (Rs)			
T-1 (Onion+Rajmash)	20	09	200	90	290	100	190 (58727)
T-2 (Garlic+Rajmash)	18	09	270	90	360	200	160 (49455)
T-3 (Turnip+Rajmash)	25	09	250	90	340	100	240 (74182)
T-4 (Peas+Sag)	15	20	150	200	350	120	230 (71090)
T-5 ((Sag+Rajmash)	20	09	200	90	290	100	190 (58727)
T-6 (Control)	Nil	Nil	Nil	Nil	Nil	Nil	

(Figures in parenthesis represent estimated values for 1 hectare of land).

A farmer can rear 15.5 and 18 ounces of silkworm seed, respectively during the 1st and 2nd crop which will yield 930 kg cocoons during spring @ 60 kg per ounce and 720 kg cocoons during the 2nd crop @ 40 kg/ounce of silkworm seed. This will fetch him an amount of Rs. 1,65,000/ha/year through the sale of cocoons @ Rs. 100 per kg. Further, he can sell the pruned twigs for an amount of Rs. 10,000 as such or thousands after making the charcoal out of it.

11.6 EMPLOYMENT OPPORTUNITY

The various operations scheduled for the successful growth and the consequent employment and revenue generation by virtue of a mulberry-based agroforestry system are furnished hereunder in brief:

- Weeding and digging: 2nd week of March
- Fertilizer application: 1st week of April
- Rearing of worms for spring crop: 10th May to 5th June
- Harvesting of leaf for spring crop: 15th May to 5th June
- Cocoon harvesting, drying, grading, and marketing: 5th June to 20th June
- Preparation of land and sowing of Kharif crop: 5th June to 15th June
- Pruning of mulberry and collection of pruned wood: 5th June to 15th June
- Weeding and hoeing and application: 25th July to 7th August

2nd doze of chemical fertilizers:
- Rearing of worms for 2nd crop: 10th August to 15th September

(Late Summer):
- Harvesting of leaf for 2nd crop: 15th August to 15th September
- Harvesting and processing of Kharif crop: 15th to 25th September
- Application of FYM: 25th to 30th September
- Preparation of land and sowing of Rabi crop: 30th September to 15th October
- Harvesting and processing of Rabi crop: February to March

The cultural activities for this mulberry-based agroforestry systems are such that during the major portion of a year a farmer gets work on his farm.

11.7   POTENTIAL OF MULBERRY-BASED AGROFORESTRY SYSTEM IN MITIGATING CLIMATE CHANGE

Mulberry is a perennial plant which once established remains productive for more than 50 years. It has a great variability and hence a wider adaptability. It can thrive well in different climates and different soils. It can tolerate temperature variations and grows well under temperate, subtropical and tropical conditions and has remarkable features enabling it to thrive well under varied climates. The features include profuse and deep root system, branching pattern, and foliar characteristics.

1. Deep root system helps it to behave well even under rainfed and water scarce conditions.
2. Branching pattern helps it to cope up with other climatic vagaries like heavy snowfall and fast winds.
3. The foliar characteristics give it the capability to adjust to various climatic conditions by adjustments through temperature tolerance (through thick leaf), water retention and reduced transpiration (through stomatal dimensions and frequency) and so on.

Seeing the broad range of features in various foliar characteristics, it seems that mulberry can be a better choice for plantation in the changing climate (Table 11.7).

TABLE 11.7   Foliar Characteristics of Mulberry

S. No.	Feature	Minimum	Maximum
1.	Leaf thickness (μm)	100	341
2.	Upper cuticle Thickness (μm)	3	11
3.	Lower cuticle thickness (μm)	2	11
4	Upper epidermal thickness (μm)	10	46
5.	Lower epidermal thickness (μm)	4	25
6.	Stomatal size (μm)	150	788
7	Stomatal frequency/sq.mm	277	1170

Mulberry can be used for prevention of soil erosion, protective agroforestry systems like windbreaks and shelterbelts, enhancing land use, biomass production for bio-energy, purification of atmosphere and carbon dioxide sequestration and other aesthetic values. Mulberry plants act as a carbon sink by removing the carbon and storing it as cellulose in their trunk, branches, leaves, and roots and releasing Oxygen back to the atmosphere. A healthy tree store about 13 pounds of Carbon annually or 2.6 tons/acre annually and one acre of plantation annually absorbs CO_2 to such an extent that it can supplement the amount of CO_2 produced by a driving car covering 26,000 miles ground distance. Gapuz (2012) provided a comparative account of C sequestration per annum between mulberry plantation of a single year and long duration mulberry stand. The first component sequesters C 44,050 to 172,978 kg/year at the rate 12,666 kg/ha/year in comparison to the second component which sequesters C 53,999 kg/ha/year. This, therefore, indicates the lesser impact of climate change over mulberry plantation in comparison to other plants. On the other hand, mulberry cultivation can help in mitigating the problems arising out of climate change being a hardy, perennial multipurpose plant.

Leaf litter decomposition is a fundamental process of ecosystem functioning and replenishing the soil with organic carbon which is closely linked to the nutrient supply for plant growth. Kaushal et al. (2012) reported that the decomposition rate was highest in *Morus alba* followed by *Grewia optiva, Toona ciliata,* and *Populus deltoids.*

Research conducted by CSRTI (Central Sericulture Research and Training Institute), Mysore, revealed the potentiality of mulberry waste to act as biofuel. Carbon-rich biomass of mulberry throws a challenge of its effective utilization as energy recovery by maintaining carbon stock. Method of carbon estimation in mulberry waste shoots include shredding of plant material through equipment, sun-dried followed by estimating carbon content through waste densification at high temperature in briquette machine through which cylinder biocoal are prepared. Carbon content in the waste shoots has been reported as high as 33–48%. Carbon loss would be greater than 50% when it is left for decomposition. Biofuel in the form of biochar developed through this process reflects good thermal value with uniform burning with the smokeless condition. The replacement value of 3–4 kg biocoal approximates 1 liter of diesel, 8–10 MT biocoal can supplement up to 2,500 liters of diesel giving an economic gain by saving Rs. 16–20 per liter basis along with CO_2 emission reduction (Srikantaswamy et al., 2012). Biochar has been reported as a rich, and more stable soil nutritive amendment as compared to other amendments as the organic carbon content in this amendment has been reported to the extent of 90 percent and thus can be an effective sink for carbon (Ahmad et al., 2017).

Mulberry plantation produces biomass containing a high level of lignocellulose without gestation period capable of capturing atmospheric CO_2 at higher efficiency and therefore acts like CO_2 pump in the process of carbon trapping, carbon sequestration. Such properties make it a suitable alternative as energy crop besides cocoon production which altogether promotes socioeconomic upliftment of the community stakeholders along with income generation and livelihood security.

11.8 CONCLUSION

Mulberry-based agroforestry system seems to be an effective module for livelihood security, effective resource management, and climate change mitigation. The plant having great variability has wider adaptability; annual growth cycle such that there are options for other crops to be grown as its companion crops; high biomass production necessitating utilization of sufficient carbon dioxide from the atmosphere and rapid litter decomposition helps in recycling of materials back to the soil beside providing fruit, fodder, fencing material, and biochar. The plant seems to be a versatile plant having multi-faceted utility, wider adaptability besides rich variability.

KEYWORDS

- agroforestry
- carbon sequestration
- climate change
- cocoons
- employment generation
- livelihood security
- mulberry
- silkworm
- sustainable agriculture

REFERENCES

1. Ahmed, F., Islam, M. S., & Iqbal, M. T. (2017). Biochar amendment improves soil fertility and productivity of the mulberry plant. *Eurasian Journal of Soil Science*, 6(3), 226 – 237.
2. Anonymous (2015). Agroforestry and Climate Change Annual Report 2014–2015: Research, Education, Outreach & Entrepreneurship. Centre for Agroforestry, University of Missouri, 40pp.
3. Anonymous (2015). Economic Survey of J&K, 2014–15. Directorate of Economics and Statistics. Government of Jammu and Kashmir. India. pp. 10–11
4. Anonymous (2015). http://unfccc.int/meetings/paris_nov_2015/meeting/8926.php
5. Dhyani, S. K., Ram, A., & Dev, I. (2016). Potential of agroforestry systems in carbon sequestration in India. *The Indian Journal of Agricultural Sciences, 86*(9), http://epubs.icar.org.in/ejournal/index.php/IJAgS/article/view/61348
6. Gapuz (2012). Silkworm Rearing. *DMMMSU-SRDI Technoguide*. Don Mariano Marcos Memorial State University, Bacnotan, La Union. The Philippines. pp. 1–29
7. Huo, Yongkang (2000). Mulberry cultivation in China. *Proceedings of the Electronic Conference*.
8. Jamison, J. (2014). Mulberry: a plant with exceptional potential. http://heartvillage.org/mulberry-a-plant-with-exceptional-potential/
9. Kaushal, R., Verma, K. S., Chaturvedi, O. P., & Alam, M. N. (2012). Leaf litter decomposition and nutrient dynamics in four multipurpose tree species. *Range Management and Agroforestry*, 33(1), 20–27.
10. Lorenz, K., & Lal, R. (2014). *Agronomy and Sustainable* Development, https://doi.org/10.1007/s13593–014–0212-y.
11. Miyashita, V. (1986). A report on mulberry cultivation and training methods suitable to bivoltine rearing in Karnataka, Central Silk Board, Bangalore, India.
12. Nair, P. K. R. N. (1979). *An Introduction to Agroforestry*. Kluwer Academic Publishers, Norwell, USA. pp.14.
13. Ravindran, S., Tikader, A., Girish Naik, V., Ananda Rao, A., & Mukherjee, P. (1988). Distribution of mulberry species in India and its utilization. Poster paper presented in "National Dialogue."
14. Ravindran, S., Ananda Rao, A., Girish Naik, V., Tikader, A., Mukherjee, P., & Thangavelu, K. (1997). Distribution and variation in mulberry germplasm. *Indian Journal of Plant Genetic Resources, 10*(2), 233–242.
15. Sastry, C. R. (1984). Mulberry varieties, exploitation and pathology. *Sericologia, 24*(3), 333–359.
16. Sinha, R. K. (2001). *Practical Handbook on Characterization, Evaluation and Database Management of Mulberry (Morus sp)*. Central Sericultural Germplasm Resource Centre, Hosur, Tamil Nadu, India.
17. Srikantaswamy, K., Mala V. Rajan and Qadri, S. M. H. (2012). Mulberry Biomass Production in Carbon Removal, Sequestration and its Waste Utilization for Energy–an Option for Sustainable Crop and Environmental Development. http://www.formatex.info/emr2012/abstracts/htm/255.pdf.

18. Tiwari, P., Kumar, R., Thakur, L., & Salve, A. (2017). Agroforestry for Sustainable Rural Livelihood: A Review. *International Journal of Pure and Applied Biosciences,* *5*(1), 299–309.

19. Vijayan, K., Tikader, A., Weiguo, Z., Nair, C. V., Ercisli, S., & Tsou, C. H. (2011). Morus. In: Kole C. (eds.). *Wild Crop Relatives: Genomic and Breeding Resources.* Springer, Berlin, Heidelberg.

Index